Interdisciplinary Evolution Research

Volume 5

About the Series

The time when only biologists studied evolution has long since passed. Accepting evolution requires us to come to terms with the fact that everything that exists must be the outcome of evolutionary processes. Today, a wide variety of academic disciplines are therefore confronted with evolutionary problems, ranging from physics and medicine, to linguistics, anthropology and sociology. Solving evolutionary problems also necessitates an inter- and transdisciplinary approach, which is why the Modern Synthesis is currently extended to include drift theory, symbiogenesis, lateral gene transfer, hybridization, epigenetics and punctuated equilibria theory.

The series Interdisciplinary Evolution Research aims to provide a scholarly platform for the growing demand to examine specific evolutionary problems from the perspectives of multiple disciplines. It does not adhere to one specific academic field, one specific school of thought, or one specific evolutionary theory. Rather, books in the series thematically analyze how a variety of evolutionary fi elds and evolutionary theories provide insights into specific, well-defi ned evolutionary prob-lems of life and the socio-cultural domain.

Editors-in-chief of the series are Nathalie Gontier and Olga Pombo. The Series is edited from within the Applied Evolutionary Epistemology Lab of the Centre for Philosophy of Science, more information on the lab and the Centre is available at http://appeel.fc.ul.pt and http://cfcul.fc.ul.pt

More information about this series at http://www.springer.com/series/13109

Laura Desirèe Di Paolo • Fabio Di Vincenzo •
Francesca De Petrillo
Editors

Evolution of Primate Social Cognition

Springer

Editors
Laura Desirèe Di Paolo
Georg-August-University Goettingen
Lichtenberg-Kolleg
Göttingen, Germany

Fabio Di Vincenzo
Department of Environmental Biology
Sapienzà Università di Roma
Rome, Italy

Francesca De Petrillo
Department of Psychology
University of Michigan
Ann Arbor, MI, Michigan, USA

ISSN 2199-3068 ISSN 2199-3076 (electronic)
Interdisciplinary Evolution Research
ISBN 978-3-030-06721-2 ISBN 978-3-319-93776-2 (eBook)
https://doi.org/10.1007/978-3-319-93776-2

This Springer imprint is published by the registered company Springer Nature Switzerland AG
The registered company address is: Gewerbestrasse 11, 6330 Cham, Switzerland

Preface

Cicero, quoting the Latin poet Ennius, wrote in the *De Natura Deorum*, (I, XXXV): *"Simia quam similis turpissima bestia nobis!"* (The ape, vilest of beasts, how like us!), thereby arguing that we are similar to the apes in the "ugliest" aspects of our conduct. Centuries later, distinguished men and women of the Victorian society were upset by the idea that humans share common ancestry with apes and monkeys, an idea that was implicitly formulated in Charles R. Darwin's theory of evolution by means of natural selection. The long-standing habit to see in monkeys and apes the mirrored and deformed image of our own nature, for many, brought with it a discomfort in accepting evolution. In his works on the descent of man and the expression of emotions in man and animals, Darwin broke this game of mirrors forcing us to see the similarities and the differences between men, apes and monkeys for what they really are, the result of a long historical process of common descent with diversification.

A good 150 years after Charles Darwin's work, studies in primatology, anthropology and comparative psychology have successfully illuminated the differences and the similarities between humans and non-human primates, shedding light on the evolutionary roots of human social behaviour.

Primatological studies are traditionally based on either wild field observations or captive experimental paradigms. The first mainly focuses on the study of primate behaviour and sociality in their natural environments. The latter, instead, tests cognitive skills under experimental settings created specifically to examine aspects of behaviour and social cognition, such as social learning and cooperation. More recently, neuroscientists have joined the investigation of the origin and evolution of higher forms of cognition in primates, including humans. Using digitally aided brain imaging techniques, researchers have begun to shed light on the neurophysiological processes underpinning many socially related cognitive capacities, such as empathy, intentionality, planning, communication and so on.

These joint efforts result in a growing body of evidence that provides important insights into the proximal mechanisms underlying primate social cognition. Moreover, progress has been made in unravelling the ultimate causes. In fact, the proximal data obtained by traditional methods are now being put to use as parameters for experimental modelling.

This peer-reviewed book builds upon these advances, where invited contributors from a wide variety of academic backgrounds examine the intimate relationship between sociality and cognition in primate evolution. Focus lies on the deep interconnections that exist within non-human primate cognition between ecological, social, developmental and learning contexts.

Such a multidisciplinary perspective calls for constant updating in line with advances made, and with this book we therefore hope to provide an extensive review of current research as well as outline open questions in the field that will inspire future research on the topic.

We would like to cordially thank the authors for contributing, as well as the many scholars who were willing to review the chapters. We are also indebted to Nathalie Gontier first for putting her trust in our work and then for assisting and helping us through this adventure. Finally, we thank the entire Springer team for guiding us through the various steps of the editing process.

Göttingen, Germany — Laura Desirè Di Paolo
Rome, Italy — Fabio Di Vincenzo
Ann Arbor, MI — Francesca De Petrillo
2018

Contents

An Evolutionary Perspective on Primate Social Cognition 1
Francesca De Petrillo, Fabio Di Vincenzo, and Laura D. Di Paolo

Part I Aspects of Primate Social Cognition

What Did You Get? What Social Learning, Collaboration, Prosocial Behaviour, and Inequity Aversion Tell Us About Primate Social Cognition 13
Lydia M. Hopper and Katherine A. Cronin

Affective States, Motivation, and Prosocial Behaviour in Primates 27
Hika Kuroshima and Kazuo Fujita

Understanding Empathy from Interactional Synchrony in Humans and Non-human Primates 47
Lira Yu, Yuko Hattori, Shinya Yamamoto, and Masaki Tomonaga

The Cognitive Implications of Intentional Communication: A Multifaceted Mirror 59
David A. Leavens

A Comparison of the Socio-communicative Behavior in Chimpanzees and Bonobos 79
Jared P. Taglialatela, Scott C. Milne, and Robert E. Evans

Part II Studying Primate Social Cognition: Theory, Observation, Experiments, and Modelling

Primate Social Cognition: Evidence from Primate Field Studies 97
Julia Ostner

Contribution of Social Network Analysis and Collective Phenomena to Understanding Social Complexity and Cognition 111
Denis Boyer and Gabriel Ramos-Fernandez

Comparative Economics: Using Experimental Economic Paradigms to Understand Primate Social Decision-Making 129
Julia Watzek, Mackenzie F. Smith, and Sarah F. Brosnan

The Special Case of Non-human Primates in Animal Experimentation . . 143
Augusto Vitale and Marta Borgi

Epigenetics and the Evolution of Human Social Cognition 163
Gillian Ragsdale and Robert A. Foley

Neanderthals and *Homo sapiens*: Cognitively Different Kinds of Human? 181
Eiluned Pearce

Part III Cultural Artefacts and Transmission in Primates

Recognizing Culture in Wild Primate Tool Use 199
Michael Haslam, Tiago Falótico, and Lydia Luncz

Culture and Selective Social Learning in Wild and Captive Primates . . . 211
Stuart K. Watson, Jennifer Botting, Andrew Whiten, and Erica van de Waal

The Zone of Latent Solutions and Its Relation to the Classics: Vygotsky and Köhler 231
Eva Reindl, Elisa Bandini, and Claudio Tennie

Minimal Cognitive Preconditions on the Ratchet 249
Elizabeth Renner and Tadeusz Zawidzki

Emulation, (Over)imitation and Social Creation of Cultural Information 267
Laura Desirèe Di Paolo and Fabio Di Vincenzo

The Acquisition of Biface Knapping Skill in the Acheulean 283
Ceri Shipton and Mark Nielsen

Visuospatial Integration: Paleoanthropological and Archaeological Perspectives 299
Emiliano Bruner, Enza Spinapolice, Ariane Burke, and Karenleigh A. Overmann

About the Editors

Laura Desirèe Di Paolo is a postdoctoral fellow at the Lichtenberg-Kolleg in the Institute for Advanced Study at the Georg-August University of Göttingen and member of the Primate Cognition Research Group at the Leibniz-Science Campus in Göttingen. She is a philosopher of cognitive and life sciences, with a particular interest in primatology and in comparative and developmental psychology. Her work focuses on social cognition and social learning strategies in human and nonhuman primates and on their impact on the evolution of human-like cultural cognition.

Fabio Di Vincenzo is a postdoctoral fellow in the Department of Environmental Biology of the Sapienza University in Rome. He is councillor of the Italian Institute of Human Paleontology (IsIPU) and of the Italian Institute of Anthropology (IsITA). His prevailing scientific interest is related to the evolution of human populations during the Middle Pleistocene in Europe.

Francesca De Petrillo is a postdoctoral researcher in the Department of Psychology at the University of Michigan. Her research examines the evolutionary origins of human cognition by employing a multidisciplinary approach integrating both comparative and developmental research. In doing so, she aims to elucidate which aspects of cognition are unique to humans and how species' differences in life history, ecology and social structure account for differences in their cognitive skills. She received a PhD in environmental and evolutionary biology from Sapienza University of Rome and conducted postdoctoral research at Harvard University and at the University of Michigan.

List of Contributors

Elisa Bandini School of Psychology, University of Birmingham, Birmingham, UK

Marta Borgi Section of Behavioural Neuroscience, Department of Cell Biology & Neuroscience, National Institute of Health, Rome, Italy

Jennifer Botting Centre for Social Learning and Cognitive Evolution, and Scottish Primate Research Group, School of Psychology and Neuroscience, University of St Andrews, St Andrews, UK

Inkawu Vervet Project, Mawana Game Reserve, Swart Mfolozi, KwaZulu Natal, South Africa

Smithsonian National Zoological Park, Washington, DC, USA

Denis Boyer Instituto de Física, Universidad Nacional Autónoma de México, Ciudad de México, México

Centro de Ciencias de la Complejidad, Universidad Nacional Autónoma de México, Ciudad de México, México

Sarah F. Brosnan Department of Psychology, Language Research Center, Georgia State University, Atlanta, GA, USA

Department of Philosophy, Neuroscience Institute, Center for Behavioral Neuroscience, Georgia State University, Atlanta, GA, USA

Emiliano Bruner National Research Centre into Human Evolution, Burgos, Spain

Ariane Burke Department of Anthropology, University of Montreal, Montreal, QC, Canada

Katherine A. Cronin Lester E. Fisher Center for the Study and Conservation of Apes, Lincoln Park Zoo, Chicago, IL, USA

Francesca De Petrillo Department of Psychology University of Michigan, Ann Arbor, MI, USA

Laura Desirèe Di Paolo Lichtenberg-Kolleg Institute for Advanced Study, Georg-August-University, Göttingen, Germany

Fabio Di Vincenzo Department of Environmental Biology, Sapienza University of Rome, Rome, Italy

Italian Institute of Paleoanthropology, Rome, Italy

Robert E. Evans Department of Ecology, Evolution, and Organismal Biology, Kennesaw State University, Kennesaw, GA, USA

Tiago Falótico Research Laboratory for Archaeology and the History of Art, University of Oxford, Oxford, UK

Institute of Psychology, University of São Paulo, São Paulo, Brazil

Robert A. Foley Division of Biological Anthropology and Leverhulme Centre for Human Evolutionary Studies, University of Cambridge, Cambridge, UK

Kazuo Fujita Department of Psychology, Graduate school of Letters, Kyoto University, Kyoto, Japan

Michael Haslam Research Laboratory for Archaeology and the History of Art, University of Oxford, Oxford, UK

Yuko Hattori Primate Research Institute, Kyoto University, Inuyama, Japan

Lydia M. Hopper Lester E. Fisher Center for the Study and Conservation of Apes, Lincoln Park Zoo, Chicago, IL, USA

Hika Kuroshima Department of Psychology, Graduate school of Letters, Kyoto University, Kyoto, Japan

David A. Leavens School of Psychology, University of Sussex, Falmer, UK

Lydia Luncz Research Laboratory for Archaeology and the History of Art, University of Oxford, Oxford, UK

Scott C. Milne Department of Ecology, Evolution, and Organismal Biology, Kennesaw State University, Kennesaw, GA, USA

Mark Nielsen School of Psychology, University of Queensland, Brisbane, QLD, Australia

Faculty of Humanities, University of Johannesburg, Johannesburg, South Africa

Julia Ostner Department of Behavioral Ecology, University of Göttingen, Göttingen, Germany

Primate Social Evolution Research Group, German Primate Center, Leibniz Institute for Primate Research, Leibniz, Germany

Karenleigh A. Overmann School of Archaeology, University of Oxford, Oxford, UK

Center for Cognitive Archaeology, University of Colorado, Colorado Springs, CO, USA

Eiluned Pearce Social & Evolutionary Neuroscience Research Group, Experimental Psychology, University of Oxford, Oxford, UK

Gillian Ragsdale Leverhulme Centre for Human Evolutionary Studies, University of Cambridge, Cambridge, UK

Gabriel Ramos-Fernandez Centro de Ciencias de la Complejidad, Universidad Nacional Autónoma de México, Ciudad de México, México

Centro Interdisciplinario de Investigación para el Desarrollo Integral Regional (CIIDIR) Unidad Oaxaca, Instituto Politécnico Nacional, Oaxaca, México

Eva Reindl School of Psychology, University of Birmingham, Birmingham, UK

Elizabeth Renner Center for the Advanced Study of Human Paleobiology, Department of Anthropology, The George Washington University, Washington, DC, USA

Psychology, University of Stirling, Stirling, UK

Ceri Shipton Centre of Excellence for Australian Biodiversity and Heritage, The Australian National University, Canberra, ACT, Australia

Mackenzie F. Smith Department of Psychology, Language Research Centre, Georgia State University, Atlanta, GA, USA

Enza Spinapolice Department of Science of Antiquities, Sapienza University of Rome, Rome, Italy

Jared P. Taglialatela Department of Ecology, Evolution, and Organismal Biology, Kennesaw State University, Kennesaw, GA, USA

Ape Cognition and Conservation Initiative, Des Moines, IA, USA

Claudio Tennie School of Psychology, University of Birmingham, Birmingham, UK

Masaki Tomonaga Primate Research Institute, Kyoto University, Inuyama, Japan

Erica van de Waal Inkawu Vervet Project, Mawana Game Reserve, Swart Mfolozi, KwaZulu-Natal, South Africa

Anthropological Institute and Museum, University of Zurich, Zurich, Switzerland

Augusto Vitale Section of Behavioural Neuroscience, Department of Cell Biology & Neuroscience, National Institute of Health, Rome, Italy

Stuart K. Watson Centre for Social Learning and Cognitive Evolution, and Scottish Primate Research Group, School of Psychology and Neuroscience, University of St Andrews, Saint Andrews, UK

Julia Watzek Department of Psychology, Language Research Center, Georgia State University, Atlanta, GA, USA

Andrew Whiten Centre for Social Learning and Cognitive Evolution, and Scottish Primate Research Group, School of Psychology and Neuroscience, University of St Andrews, Saint Andrews, UK

Inkawu Vervet Project, Mawana Game Reserve, Swart Mfolozi, KwaZulu-Natal, South Africa

Shinya Yamamoto Institute for Advanced Study, Kyoto University, Kyoto, Japan

Lira Yu Wildlife Research Center, Kyoto University, Kyoto, Japan

Tadeusz Zawidzki Department of Philosophy, The George Washington University, Washington, DC, USA

An Evolutionary Perspective on Primate Social Cognition

Francesca De Petrillo, Fabio Di Vincenzo, and Laura D. Di Paolo

> *Now, if some one man in a tribe, more sagacious than the others, invented a new snare or weapon, or other means of attack or defense, the plainest self-interest, without the assistance of much reasoning power, would prompt the other members to imitate him; and all would thus profit.*
>
> (Charles Darwin, The Descent of Man, and Selection in Relation to Sex, 1871)

Abstract The Machiavellian intelligence hypothesis and the social brain hypothesis have revolutionized traditional views on how primate cognition can be studied. Beyond the study of individual problem-solving capacities of various primates, these hypotheses have demonstrated the close relationship between the complexity of primate social life and the emergence of more sophisticated cognitive skills. The social brain hypothesis demonstrated the existence of a close correlation between the volume of the neocortex and the number of individuals in primate social groups. The amount of studies in this area have increased dramatically and have successfully enhanced our understanding of the evolutionary roots of complex social phenomena, including theory of mind, cultural transmission, social learning, and shared attention. The cognitive capacities present in primates also underlie the evolution of cognitive

F. De Petrillo (✉)
Department of Psychology, University of Michigan, Ann Arbor, MI, USA
e-mail: fdepetri@umich.edu

F. Di Vincenzo (✉)
Department of Environmental Biology, Sapienza University of Rome, Rome, Italy

Italian Institute of Human Paleontology (IsIPU), Rome, Italy
e-mail: fabio.divincenzo@uniroma1.it

L. D. Di Paolo
Lichtenberg-Kolleg Institute of Advanced Study, George-August University of Göttingen, Göttingen, Germany

DPZ, Deutsche Primatenzentrum, Leibniz Institute for Primate Research, Göttingen, Germany

L. D. Di Paolo et al. (eds.), *Evolution of Primate Social Cognition*, Interdisciplinary Evolution Research 5, https://doi.org/10.1007/978-3-319-93776-2_1

capacities in humans. This chapter introduces present avenues taken in research on primate social cognition, and it walks the reader through the chapters of this volume.

Keywords Primates · Cognition · Sociality · Evolution · Behavior · *Homo*

1 From the Social Brains of Primates...

The primate lineage is characterized by an increase in brain size relative to body size. Humans evolved late within that lineage, and their brains are about three times larger than the brains of our closest relatives, the great apes (Jerison 1973). The human brain comes with a unique package of cognitive skills not possessed by other primates, which includes language and mathematics. But our brains take up a lot of energy when performing these cognitive skills. In fact, our brain is the most expensive organ to ever have evolved.

Ever since Darwin, evolutionary biologists and comparative psychologists have therefore asked why we find this trend of increasing brain size in primates and how these brains underlie complex cognitive skills.

Traditional explanations for the evolution of primate cognition focused upon environmental factors that drive cognitive evolution (Clutton-Brock and Harvey 1980; Byrne 1996). Today, however, primate cognition is by and large studied as an outcome of primate sociality. This paradigm shift from ecological to social explanations for the evolution of primate cognition results, on the one hand, from behavioral research on primate behavior as it occurs in natural settings and, on the other, from important research brought forth by the social brain hypothesis.

The *social brain hypothesis* correlates primate cognition to the complexity of social lifestyles (Dunbar 1998; Humphrey 1976) and suggests that an increase in sociality is an important causal factor in the evolution of larger brains in primates (Dunbar and Shultz 2007). Different authors have focused upon different aspects of social life that may have resulted in an increase in brain size. Some state that the ability to predict the behavior of other individuals and to socially outmaneuver other members of the group drove the evolution of social intelligence in primates (Whiten and Byrne 1988; Humphrey 1976). Others have focused on the social pressure associated with maintaining group cohesion and how increasingly large groups bring forth new information-processing demands (Dunbar 1998), and still others have emphasized the important role of social learning (Jolly 1966). That brain size correlates with social behavior today is evidenced by both behavioral and neurobiological data. Neurobiological comparisons between different primates have indicated positive relationships between relative brain and/or neocortex size and mean group size (Dunbar 1992, 1998; Dunbar and Shultz 2007), although some authors have highlighted that this pattern might differ between sexes (e.g., Lindenfors 2005) and across group cohesion metrics (Lehmann and Dunbar 2009). Behavioral data demonstrates that primate species that live complex social lives also have sophisticated cognitive skills, such as social learning (e.g., Bonnie and de Waal 2006), cooperation (e.g., Melis et al. 2006; Boesch 1994; Burkart et al. 2007; Burkart and van Schaik 2010), and possibly a theory of mind (e.g., Call and Tomasello 2008).

2 To the Specificities of Human Social Cognition

Today, the correlation between cognitive evolution and sociality, and sociality and brain size, is supported by a large body of evidence. Humans, in addition, have been defined not only as a social species but as an "ultra-social" species (Boyd and Richerson 1996). According to the *cultural intelligence hypothesis*, the necessity of participating and exchanging knowledge in cultural groups has fostered the emergence of especially sophisticated cognitive skills in humans that allow us to create different cultural groups, each one with a distinctive set of social norms, institutions, languages, and symbols (Herrmann et al. 2007).

Humans possess extremely sophisticated social skills such as theory of mind—the ability to model the subjective, unobservable mental states of other individuals (Wellman 1992; Perner 1991). Evidence exists that demonstrates that infants can understand others as intentional agents (e.g., Woodward 1998), and perhaps they can even understand false beliefs, at least on an "implicit level" (Onishi and Baillargeon 2005). These sophisticated mind-reading abilities may allow humans to engage in complex social interactions including imitation, social learning, and language acquisition (Tomasello 2009; Hare 2017).

Are such complex abilities uniquely human? Over the last few decades, comparative cognitive research has devoted considerable attention to understanding whether or not other species share these abilities. As some work has pointed out, cognitive traits are never all-or-nothing phenomena (De Waal and Ferrari 2010); often they integrate traits found in other species. Several primate species, for example, appear to show some but not all components of humanlike theory of mind (Marticorena et al. 2011). There is evidence that chimpanzees (*Pan troglodytes*)—one of our closest living relatives—appear to attribute desires (Call and Tomasello 2008; Buttelmann et al. 2007), visual perspectives (Okamoto-Barth et al. 2007; Bräuer et al. 2005), intentions (Hare et al. 2000; Kaminski et al. 2008), and possibly false beliefs (Krupenye et al. 2016) to other individuals. Rhesus macaques (*Macaca mulatta*) also exhibit the ability to attribute visual perspective (Flombaum and Santos 2005) and knowledge (Marticorena et al. 2011) to other individuals. Chimpanzees (Whiten et al. 1999), orangutans (van Schaik et al. 2003), Japanese macaques (Huffman 1996), and capuchin monkeys demonstrate forms of cultural transmission (Perry et al. 2003; Fragaszy et al. 2004). Beyond the primate lineage, cultural transmission abilities have also been found in dolphins and whales (Rendell and Whitehead 2001) and New Caledonian crows (Hunt and Gray 2003). Taken together, these works suggest that the complex social abilities that we see in humans might be underpinned by socio-cognitive mechanisms that have their origins in cognitive skills shared with other primates and animals, which were then co-opted for new use (Tomasello and Carpenter 2007).

But our species also possesses social abilities not found in other primates. In a comparison between chimpanzees, orangutans (*Pongo pygmaeus*), and 2.5-year-old children (*Homo sapiens*), children and chimpanzees showed similar cognitive abilities in the physical domain (i.e., reasoning about physical properties, number, and causality), but human children outperformed the ape species in the social domain (i.e., theory of mind, intentional communication, and social learning) (Herrmann et al. 2007). Moreover, humans appear to be the only species that can exhibit flexible cooperation between non-kin and that makes use of punishment (Fehr et al. 2002; Tomasello et al. 2012; Henrich 2015).

These findings have led some scholars to suggest that human and nonhuman primate cognition results from different cognitive mechanisms (e.g., Tennie et al. 2009). When presented with a new problem-solving task in a similar social learning context, for example, human social learners focus on different information than chimpanzees (Tomasello and Call 1997). While chimpanzees focus more on the effect of a certain action on the environment, humans also pay attention to the behavioral sequences of the demonstrator, resulting in a higher copying fidelity. This ability to copy is essential to transmit cultural practice across generations (Richerson and Boyd 2005). Thus, while our species might share some common cognitive mechanisms underlying complex social skills with nonhuman primates, we show unique patterns of cooperative behavior and cultural transmission. We are the only species, for example, that has produced material and symbolic artifacts, such as language or money, and we exhibit sophisticated cooperation between individuals, even between strangers, that has led to the emergence of human societies.

How social skills shared with other primates evolved into new forms of human cognition is the center of one of the most fascinating debates about our evolutionary history. In humans, the relationship between sociality and cognition is well represented in the archaeological record of material culture. Material cultural artifacts exemplify the concepts of "embodied (or situated) cognition" and of the "extended mind." Embodied cognition emphasizes the simultaneity of the processes of acting and of knowing and of the almost complete identity between the action performed in a given context and the perception/knowledge of the objects that are present and that can be used in that context, in a perspective that takes the name of enactivism (Varela et al. 1991). The extended mind (Clark and Chalmers 1998), on the other hand, assumes the equivalence between the "internal" (mental) cognitive processes and the production and management of cultural objects, considering both as deriving from a "projection" or "outsourcing" of the former. From the Paleolithic onward, human beings have used the cultural products of their own cognition to modify, according to their needs, ever larger portions of the natural environment. Humans are "niche constructors" (Odling-Smee et al. 2003). This active "construction" of the external world could only happen within a social-cognitive dimension. Material culture and, more broadly, any active "construction" of the external world exist because of the presence of well-developed forms of social learning (emulation and imitation) and cooperation that are brought forth by a set of socio-cognitive skills.

3 The Volume

Our closest living relatives evolved sophisticated social skills that include deception calling, empathy, grudging, reconciliation, gaze-following, and pointing (e.g., Whiten and Byrne 1988; Tomasello et al. 1999), as well as social traditions (e.g., Goodall 1986; Whiten et al. 1999). Both suggest that chimpanzees are capable of interpreting the social world in mental states. Our own sociocultural specificity skills are homologous to the cognitive skills demonstrated by our cousins, and human cognition thus has deep evolutionary roots. The presence of socio-cognitive capacities in our cousins in turn encourages the study of social cognition in a larger number of primate species, including lemurs, monkeys, and other apes. Investigating the ethological, ecological, and neurophysiological aspects of social cognition is crucial for a better understanding of the complex dynamics that underlie the evolution of the wide range of different adaptations that we observe among nonhuman primates and humans, including the evolution of important aspects of our uniqueness.

A first step toward such research involves finding a good operational definition of primate social cognition. Given that different primate species demonstrate great variability of social skills, often in association with the occupancy of different ecological conditions, a unitary definition of social cognition is hard to find. Social cognition is not reducible to individual cognitive mechanisms. Pertinent to our understanding of social cognition is an understanding of how multiple "nervous systems" interact among each other. The behavioral expression of neural activity in one brain can influence the activity of the neurons in the brains of conspecifics via the making of "social networks."

The way in which multiple individuals are able to interact positively, cooperate, and understand the intention of others enables a sharing of common goals that in turn becomes crucial for transmitting information by means of sophisticated forms of social learning. Cumulative culture depends on social cognition, which in turn influences the fitness of the individuals that participate in the social interaction.

The aim of this volume is to offer a valuable contribution to future research in the field, by bringing together expert researchers coming from primatology, anthropology, ethology, philosophy of cognitive sciences, neurophysiology, mathematics, and psychology to discuss both the foundations of nonhuman primate and human social cognition and the means that currently exist to study its various aspects.

In the first part, Aspects of Primate Social Cognition, authors present a multitude of innovative approaches to the study of several core aspects of social cognition across primates, such as learning, prosocial behavior, and communication. Particular emphasis is given to both the relationship between food and social behavior and the connection with empathy and communication. The first two chapters "What Did You Get? What Social Learning, Collaboration, Prosocial Behaviour, and Inequity Aversion Tell Us About Primate Social Cognition" and "Affective Stages, Motiva tion, and Prosocial Behaviour in Primates" authored, respectively, by Hopper and Cronin and Kuroshima and Fujita are devoted to in-depth argumentations about prosocial behaviors among primates in respect to important aspects of cognition such as social learning, decision-making, and inequity aversion. Kuroshima and Fujita, in particular, analyze the affective states that may underlie prosocial behaviors in capuchin monkeys (*Cebus apella*).

In the chapter "Understanding Empathy from Interactional Synchrony in Humans and Non-human Primates," Yu and colleagues analyze the emergence of the empathic mind from the perspective of movement coordination, with special focus on mimicry and interactional synchrony in humans, chimpanzees, and macaques. The origin and evolutionary development of empathy is one of the most discussed topics in ethology and neurophysiology, and it is the core basis for the establishment of sophisticated forms of communication in primates.

The last two chapters of the first part of the volume further demonstrate this point. In a very provocative manner, Leavens, in the chapter entitled "The Cognitive Implications of Intentional Communication: A Multi-faceted Mirror" critically examines the inconsistency and the "double standard" in interpretation of a number of contemporary scientific practices and protocols used to justify some alleged differences in intentional communication in human and nonhuman primates. Taglialatela, Milne, and Evans in the chapter "A Comparison of Socio-communicative Behavior in Chimpanzees and Bonobos", offer a valuable contribution to our knowledge and understanding of the differences in vocalizations and other aspects of communication between chimpanzees and bonobos. These differences in communicative and social characteristics appear to be related to differences in resource exploitation strategies, and feeding ecology likely played a crucial role in selecting for the advanced cognitive abilities of both species.

The second part of the volume, Studying Primate Social Cognition: Theory, Observation, Experiments, and Modelling, is devoted to the various epistemic and methodological means there exist to study social cognition and in particular focuses on proximate and ultimate mechanisms of social cognition that can be examined through experimental, modeling, and field studies. In the chapter "Primate Social Cognition: Evidence from Primate Field Studies," Ostner emphasizes the importance of emotional bookkeeping as a cognitively less demanding mechanism used by many primates to keep track of their own dominance, as well as of relationships of third parties, indicating triadic awareness.

Social complexity can be also investigated using a modeling approach. Proximate data obtained by traditional methods can be used as parameters in models. According to this point of view, social network analysis can provide a better understanding of the high-level properties (e.g., structure, dynamics) that spontaneously emerge from local interactions. In the chapter on "Contribution of Social Network Analysis and Collective Phenomena to Understanding Social Complexity and Cog nition," Boyer and Ramos-Fernandez review social network literature and discuss the role of social network analysis for deepening our understanding of social complexity and cognition. And Watzek, Smith, and Brosnan, in their chapter on "Comparative Economics: Using Experimental Economics Paradigms to Under stand Primate Social Decision-Making," provide a detailed description of new paradigms used in experimental economics, which allow for direct comparisons about decision-making in cooperative contexts across species.

In the chapter "The Special Case of Non-human Primates in Animal Experimentation," Vitale and Borgi summarize interesting data regarding the use of nonhuman primates in scientific research. They also provide information about the legal status of nonhuman primates in Europe and other countries, by giving accurate information on why these species are considered special to some extent. In the chapter "Epigenetics and the Evolution of Human Cognition," Ragsdale and Foley review the roles that could be played by epigenetic regulations across human diversity and during human evolution. Authors propose that epigenetic variations could have been key players for evolution and adaptability of mankind and be in part responsible for inter-individual diversity.

Pierce, in the chapter entitled "Neanderthals and Homo sapiens: Cognitively Different Kinds of Human?," digs deeper into the field of human evolution. The author emphasizes the key role played by the cognitive capacity to maintain and manage frequent interpersonal interactions and to keep track of, and coordinate with, social partners in the evolutionary success of our own species *Homo sapiens* and in the demise of the Neanderthals.

The third and final part of the book, Cultural Artifacts and Transmission in Primates, focuses on the mechanisms of cultural transmission in primate and human societies. These are investigated in light of the impact social cognition has on the evolution of technological capacities. In the chapter "Visuospatial Integration: Palaeoanthropological and Archaeological Perspectives," Bruner and colleagues stress the importance of visuospatial cognition for social mechanisms during human evolution and in particular in organizing the spatial and temporal interactions between the brain, body, social group, and environment. Using both paleo-neurological and archaeological approaches, the authors show how visuospatial integration is particularly relevant when one considers the evolution of modern human culture in which a tight relationship between inner (e.g., memory, consciousness, simulation capacity) and outer (socially related) cognitive components occurs.

The evolution of culture is the core topic of the other chapters of this section. The chapter on "Recognizing Culture in Primate Tool Use," written by Haslam, Falótico, and Luncz, and the chapter on "Social Learning and Culture: Complementary Studies of Wild and Captive Primates," written by Watson, Botting, Whiten, and de Waal, complement each other. Both offer a broad overview of the cultural phenomena associated with nonhuman primates in conditions of captivity and in the wild and on the main elements of the distinction with respect to human culture.

Cumulative culture in humans is characterized by high-fidelity transmission of information via sophisticated forms of social learning and by the crucial aspect of "the ratchet." As discussed in the chapter written by Renner and Zawidzki and titled "Minimal Cognitive Preconditions on the Ratchet," it is the key difference that distinguishes human from primate culture. The authors present results of empirical research performed from within a background of comparative and developmental psychology, integrating it with a novel theoretical analysis of the bootstrapping problem to defend a hypothesis about the minimal cognitive preconditions on the ratchet. A similar point but more focused on the origin and spread of innovations in human cumulative culture is presented in the chapter "Emulation, (Over)imitation and Social Creation of Cultural Information" of Di Paolo and Di Vincenzo. The early emergence of cultural innovations during human evolution, as witnessed in the Acheulean knapping technology, is analyzed in the chapter on "The Zone of Latent Solutions Concept and its Relationship to the Classics: Vygotsky and Köhler" authored by Reind, Bandini, and Tennie. The role of high-fidelity social transmission in allowing the establishment, spread, and endurance in time of such lithic tradition of the Early Paleolithic is debated by Shipton and Nielsen in their chapter on "The Acquisition of Biface Knapping Skill in the Acheulean."

The present volume results from an effort to join primatologists and human cognitive archaeologists and to integrate experimental data with theory. All the authors provide an accessible state of the art about the evolution of social cognition in primates, making this book a must-read for both young and established researchers.

References

Boesch C (1994) Cooperative hunting in wild chimpanzees. Anim Behav 48(3):653–667

Bonnie KE, de Waal FB (2006) Affiliation promotes the transmission of a social custom: handclasp grooming among captive chimpanzees. Primates 47(1):27–34

Boyd R, Richerson PJ (1996) Why culture is common, but cultural evolution is rare. Proc Br Acad 88:77–94

Bräuer J, Call J, Tomasello M (2005) All great ape species follow gaze to distant locations and around barriers. J Comp Psychol 119(2):145

Burkart JM, van Schaik CP (2010) Cognitive consequences of cooperative breeding in primates? Anim Cogn 13(1):1–19

Burkart JM, Fehr E, Efferson C, van Schaik CP (2007) Other-regarding preferences in a non-human primate: common marmosets provision food altruistically. Proc Natl Acad Sci USA 104(50): 19762–19766

Buttelmann D, Carpenter M, Call J, Tomasello M (2007) Enculturated chimpanzees imitate rationally. Dev Sci 10(4):F31–F38

Byrne RW (1996) Relating brain size to intelligence in primates. In: Mellars PA, Gibson KR (eds) Modelling the early human mind. McDonald Institute for Archaeological Research, Cambridge, pp 49–56

Call J, Tomasello M (2008) Does the chimpanzee have a theory of mind? 30 years later. Trends Cogn Sci 12(5):187–192

Clark A, Chalmers D (1998) The extended mind. Analysis 58(1):7–19

Clutton-Brock TH, Harvey PH (1980) Primates, brains and ecology. J Zool 190(3):309–323

De Waal FB, Ferrari PF (2010) Towards a bottom-up perspective on animal and human cognition. Trends Cogn Sci 14(5):201–207

Dunbar RI (1992) Neocortex size as a constraint on group size in primates. J Hum Evol 22(6): 469–493

Dunbar RI (1998) The social brain hypothesis. Brain 9(10):178–190

Dunbar RI, Shultz S (2007) Evolution in the social brain. Science 317(5843):1344–1347

Fehr E, Fischbacher U, Gächter S (2002) Strong reciprocity, human cooperation, and the enforcement of social norms. Hum Nat 13(1):1–25

Flombaum JI, Santos LR (2005) Rhesus monkeys attribute perceptions to others. Curr Biol 15(5): 447–452

Fragaszy D, Izar P, Visalberghi E, Ottoni EB, de Oliveira MG (2004) Wild capuchin monkeys (*Cebus libidinosus*) use anvils and stone pounding tools. Am J Primatol 64(4):359–366

Goodall J (1986) The chimpanzees of Gombe: patterns of behavior. Belknap Press of Harvard University Press, Cambridge, MA

Hare B (2017) Survival of the friendliest: *Homo sapiens* evolved via selection for prosociality. Annu Rev Psychol 68:155–186

Hare B, Call J, Agnetta B, Tomasello M (2000) Chimpanzees know what conspecifics do and do not see. Anim Behav 59(4):771–785

Henrich J (2015) The secret of our success: how culture is driving human evolution, domesticating our species, and making us smarter. Princeton University Press, Princeton, NJ

Herrmann E, Call J, Hernández-Lloreda MV, Hare B, Tomasello M (2007) Humans have evolved specialized skills of social cognition: the cultural intelligence hypothesis. Science 317(5843): 1360–1366

Huffman MA (1996) Acquisition of innovative cultural behaviors in nonhuman primates: a case study of stone handling, a socially transmitted behavior in Japanese macaques. In: Heyes CM, Galef BG Jr (eds) Social learning in animals: the roots of culture. Academic Press, San Diego, CA, pp 267–289

Humphrey NK (1976) The social function of intellect. In: Bateson PPG, Hinde RA (eds) Growing points in ethology. Cambridge University Press, Cambridge, pp 303–317

Hunt GR, Gray RD (2003) Diversification and cumulative evolution in New Caledonian crow tool manufacture. Proc R Soc Lond B Biol Sci 270(1517):867–874

Jerison HJ (1973) Evolution of the brain and intelligence. Academic, New York

Jolly A (1966) Lemur social behavior and primate intelligence. Science 153(3735):501–506

Kaminski J, Call J, Tomasello M (2008) Chimpanzees know what others know, but not what they believe. Cognition 109(2):224–234

Krupenye C, Kano F, Hirata S, Call J, Tomasello M (2016) Great apes anticipate that other individuals will act according to false beliefs. Science 354(6308):110–114

Lehmann J, Dunbar RIM (2009) Network cohesion, group size and neocortex size in female-bonded Old World primates. Proc R Soc Lond B Biol Sci 276(1677):4417–4422

Lindenfors P (2005) Neocortex evolution in primates: the 'social brain'is for females. Biol Lett 1(4):407–410

Marticorena DC, Ruiz AM, Mukerji C, Goddu A, Santos LR (2011) Monkeys represent others' knowledge but not their beliefs. Dev Sci 14(6):1406–1416

Melis AP, Hare B, Tomasello M (2006) Engineering cooperation in chimpanzees: tolerance constraints on cooperation. Anim Behav 72(2):275–286

Odling-Smee FJ, Laland KN, Feldman MW (2003) Niche construction: the neglected process in evolution. Princeton University Press, Princeton, NJ

Okamoto-Barth S, Call J, Tomasello M (2007) Great apes' understanding of other individuals' line of sight. Psychol Sci 18(5):462–468

Onishi KH, Baillargeon R (2005) Do 15-month-old infants understand false beliefs? Science 308(5719):255–258

Perner J (1991) Understanding the representational mind. MIT Press, Cambridge

Perry S, Panger M, Rose LM, Baker M, Gros-Louis J, Jack K, MacKinnon KC, Manson J, Fedigan L, Pyle K (2003) Traditions in wild white-faced capuchin monkeys. In: Fragaszy D, Perry S (eds) The biology of traditions: models and evidence. Cambridge University Press, Cambridge, pp 391–425

Rendell L, Whitehead H (2001) Culture in whales and dolphins. Behav Brain Sci 24(2):309–324

Richerson PJ, Boyd R (2005) Not by genes alone: how culture transformed human evolution. University of Chicago Press, Chicago, IL

Tennie C, Call J, Tomasello M (2009) Ratcheting up the ratchet: on the evolution of cumulative culture. Philos Trans R Soc B Biol Sci 364(1528):2405–2415

Tomasello M (2009) The cultural origins of human cognition. Harvard University Press, Cambridge, MA

Tomasello M, Call J (1997) Primate cognition. Oxford University Press, New York

Tomasello M, Carpenter M (2007) Shared intentionality. Dev Sci 10(1):121–125

Tomasello M, Hare B, Agnetta B (1999) Chimpanzees, *Pan troglodytes*, follow gaze direction geometrically. Anim Behav 58(4):769–777

Tomasello M, Melis AP, Tennie C, Wyman E, Herrmann E (2012) Two key steps in the evolution of human cooperation: the interdependence hypothesis. Curr Anthropol 53(6):673–692

Van Schaik CP, Ancrenaz M, Borgen G, Galdikas B, Knott CD, Singleton I, Suzuki A, Utami SS, Merrill M (2003) Orangutan cultures and the evolution of material culture. Science 299(5603): 102–105

Varela F, Thompson E, Rosch E (1991) The embodied mind: cognitive science and human experience MIT press. Massachusetts, Cambridge

Wellman HM (1992) The child's theory of mind. MIT Press, Cambridge, MA

Whiten A, Byrne RW (1988) Machiavellian intelligence hypothesis. In: Byrne RW, Whiten A (eds) Machiavellian intelligence. Clarendon, Oxford, pp 1–9

Whiten A, Goodall J, McGrew WC, Nishida T, Reynolds V, Sugiyama Y, Tutin CEG, Wrangham RW, Boesch C (1999) Cultures in chimpanzees. Nature 399(6737):682

Woodward AL (1998) Infants selectively encode the goal object of an actor's reach. Cognition 69(1):1–34

Part I
Aspects of Primate Social Cognition

What Did You Get? What Social Learning, Collaboration, Prosocial Behaviour, and Inequity Aversion Tell Us About Primate Social Cognition

Lydia M. Hopper and Katherine A. Cronin

Abstract Consideration of social cognition—how an individual's decision-making is influenced by her/his social environment—is key to understanding the behaviour of socially living nonhuman primates. In this chapter we discuss primate social cognition by focusing on primates' behavioural responses to the presence and actions of others, how they adjust their behaviour to maximize their own gains, and possibly also the rewards received by a partner. Individuals can observe and replicate the actions of others, or the outcomes of their actions, to accelerate behavioural acquisition of techniques to obtain rewards (social learning). Beyond passively observing others, primates can also work with group mates to obtain rewards more easily or to get rewards that would otherwise be unattainable by a single individual (collaboration). Although not universally seen among primates, one individual may also help another to acquire resources (prosocial behaviour). Prosocial and collaborative interactions may result in an imbalance of benefits received, and certain primate species respond negatively when receiving less than a social partner (inequity aversion), a response which may protect individuals against cheating. Such behaviours demonstrate how the interplay between an individual's desires and those of others can modify behavioural outcomes and the importance of considering cognition from a social perspective in order to understand the decision-making of individuals. However, there is variation both within and between species in their sensitivity to the actions of others and their responses to them. Thus, a comparative framework is needed when studying what is meant by 'primate social cognition'.

Keywords Inequity aversion · Cooperation · Social learning · Prosociality · Collaboration · Fairness

---

L. M. Hopper (✉) · K. A. Cronin
Lester E. Fisher Center for the Study and Conservation of Apes, Lincoln Park Zoo, Chicago, IL, USA
e-mail: lhopper@lpzoo.org

L. D. Di Paolo et al. (eds.), *Evolution of Primate Social Cognition*, Interdisciplinary Evolution Research 5, https://doi.org/10.1007/978-3-319-93776-2_2

1 Introduction

Forty years ago, in his essay *The Social Function of Intellect*, Nicholas Humphrey (1976) described the interaction between the complexity of nonhuman primate society and their cognitive abilities. Specifically, Humphrey described a feedback between primate intelligence and the dynamics of their natural social groups. Humphrey's theory led the way for further discussions of primate cognition within the context of their social world [e.g. the Machiavellian intelligence hypothesis, Byrne and Whiten (1988); and the social brain hypothesis, Dunbar (1998); see also de Waal (1982)]. The idea that cognition evolves in response to social demands continues to gain support from contemporary research (Pasquaretta et al. 2014), although recent analyses have demonstrated that ecological factors, such as home range size and diet, may be better predictors of primate brain size than group size (DeCasien et al. 2017; Powell et al. 2017).

Many primates live in complex social groups, and even those that spend the majority of their time alone or in pairs (e.g. *Pongo* and *Aotus* species) are still influenced by the behaviour of other members of their species. Conspecifics influence many daily activities of primates, including movement patterns (Strandburg-Peshkin et al. 2015), foraging choices (Marshall et al. 2015), and sexual behaviour (Gunst et al. 2015). How social partners influence an individual's behaviour depends on the relationship between the individuals involved—their relative rank, their personalities, and the species in question (Hopper 2017). At the simplest level, the presence of group mates may cause individuals to be less neophobic and more exploratory, but, conversely, the presence of others, especially if they are more dominant, may inhibit an individual's behaviour (Huguet et al. 2014). Thus, there is feedback between an individual and her/his social environment that is likely influenced by both proximate and ultimate mechanisms.

Within this chapter, we focus on four aspects of nonhuman primate social cognition, (1) social learning, (2) prosocial behaviour, (3) collaboration, and (4) inequity aversion, with particular attention given to social learning and collaboration. These four behaviours are connected because they all result from an interaction between two or more primates and the focal animal alters their behaviour in response to the behaviour of the other(s). We focus on research with both captive and wild primate species, but note that such behaviours have also been studied in a number of other non-primate species [for descriptions of this research, see Salwiczek et al. (2012), Vonk and Shackelford (2012), Zentall and Wasserman (2012), and Cronin and Hopper (2017)].

2 Social Learning

Beyond learning new skills via trial and error, primates, and many other species, are also capable of learning from observation of others' actions (i.e. social learning). Social learning, put simply, describes the way in which one individual can learn from the actions of another and is important because it potentially enables individuals to avoid the costs of individual learning. Individual learning can be costly in a number of ways, including the cost of time spent 're-inventing the wheel' or the costs of making a potentially fatal mistake (e.g. eating a poisonous food), which is why social learning is considered advantageous. It should be noted, though, that social learning is only advantageous if an individual can gain relevant information that is not outdated or inaccurate. Thus, primates must also consider when to rely on social learning, and who to copy, in order to maximize their success (Castro and Toro 2004). Whether, and what, primates can socially learn has been studied extensively with both captive and wild populations, and social learning has been demonstrated by all four great ape species, Old World and New World monkeys, and certain prosimians (Hopper 2017 provides a review). Experimental research has shown that primates can socially learn a range of novel behavioural skills, including how to solve novel unfamiliar foraging puzzles, how to order icons presented in a serial learning task, and how to make and use tools to access food.

Highlighting the importance of cognitive ability for social learning to arise, across the order of Primates, social learning skill (and innovative ability) appears to be correlated with relative and absolute brain size (Reader and Laland 2002), although this relationship is not clear-cut (Snowdon and Cronin 2009). At the simplest level, however, the behaviour of primates can be influenced by the presence of others (e.g. by increasing neophilia and exploratory behaviour), which does not require an underlying cognitive understanding of the learning process. Even in cases of more complex social learning, the expert animal does not need to intentionally train a naïve observer and nor does social learning require primates to understand the mental states of one another—the transmission of information can be passive on the part of the expert. This is of note because although primates are capable of learning new skills via observation, they do not appear to regularly engage in teaching. Typically, primates do not appear to slow down or repeat their actions so that others can learn from observing them, even if their behaviour does promote learning by observers (although see Humle and Snowdon 2008 for a convincing case of teaching by tamarins, *Saguinus oedipus*). However, it has been argued that our failure to observe teaching among primates is because the need for targeted instruction is rare (and so teaching is rare) and also because our definitions of teaching are overly constrictive (Byrne and Rapaport 2011). Regardless of primates' ability or need for teaching, their skill at social learning appears universal.

Social learning can arise both from direct observation of another individual ('observational learning') and from changes to the environment caused by their actions ('emulative learning'), and much experimental effort has been expended to tease apart the different mechanisms that describe the process of social learning and the fidelity with which an observer replicates the actions of another (Hopper 2010 provides a review). Because of the potential costs associated with social information, primates should be selective both in when they rely on social learning and who they choose to copy. Therefore, beyond 'what' is copied, primatologists have also studied 'when' and 'who' primates copy, so-called social learning strategies or transmission biases (Kendal et al. 2015). Additionally, the likelihood that a primate will rely on social learning is also influenced by their individual personality and rank. For example, the reliance on social information by foraging chacma baboons (*Papio ursinus*) appears to be influenced by each individual's personality characteristics (Carter et al. 2014), and for chimpanzees, rank appears to influence learning such that subordinate chimpanzees are more likely to copy the actions of others than are dominant individuals (Kendal et al. 2015).

Whether social learning occurs at all is contingent as much on dynamics of an individual's social group as it is on the cognitive prowess of the naïve individual. For example, for rhesus macaques (*Macaca mulatta*), a highly despotic species, an individual's rank within their group can determine both if they can observe social information (i.e. if they can be in close-enough proximity to the 'expert') and also whether they are then able to use any new information that they gain (Drea and Wallen 1999). Thus, even if animals are able to learn from one another, it is still not guaranteed that the new behaviour or innovation will spread widely within a group or be maintained over time as a new tradition (e.g. Cronin et al. 2014; see Brosnan and Hopper 2014 for a review). For example, primates appear less likely to attend to and copy the behaviour of low-ranking or peripheral group members. This means that of the many innovations that arise within primate societies, very few are adopted widely (Nishida et al. 2009). Conversely, if the innovator is very high ranking, low-ranking individuals may not be able to get close enough to learn about or copy their inventions; the dominant individual may monopolize access to an area of food source, or the low rank of a subordinate might inhibit their ability to exploit the invention. In a study with captive squirrel monkeys (*Saimiri sciureus*), Claidière et al. (2013) reported that the monkeys who were more centrally integrated within their group, as identified by social network analysis, adopted the novel foraging technique of a trained group mate more rapidly than peripheral members of the group.

Rather than an academic exercise to ascertain how skilful primates are at copying one another's actions, those studying primate social learning aim to compare our own cultural world (and the mechanisms that support it) with that of our primate cousins. This is because social learning allows for non-genetic inheritance and so is the mechanism that underpins the emergence and maintenance of behavioural traditions and, ultimately, culture (Hopper et al. 2012; Caldwell et al. 2012). Importantly, social learning allows for knowledge and behaviours to be transmitted between individuals, not only down generations (vertical transmission) but also across generations (horizontal transmission), and even among non-kin. More globally, a primate's ability to learn from their group members means that they are able to attend to the actions of others, learn about the outcomes of their actions, and, potentially, assess their value. Such social observations are also integral to the other behaviours that we discuss in this chapter, collaboration, prosocial behaviour, and responses to inequity.

3 Collaboration

While social learning involves observation of group mates to acquire knowledge, collaboration takes place when two or more individuals work together to solve a problem or achieve a common goal (Boesch and Boesch 1989). The cognitive aspect enters into this discussion when researchers attempt to measure whether the primates *understand* the role that the partner plays in the collaborative act (Mendres and de Waal 2000; Noe 2006; Visalberghi et al. 2000). Social influences are again strongly at play, as social relationship impacts the likelihood of collaboration just as it affects social learning (Coussi-Korbel and Fragaszy 1995; Cronin 2016a; Cronin and Sánchez 2012; van Schaik 2003). In fact, a great deal of experimental work has attempted to determine how social dynamics influence collaborative success, considering both within- and between-species variation in relationships.

Although a comparative perspective is embraced in the interpretation of nearly all studies of primate collaboration, it has been rare for a single study to directly compare multiple primate species. One exception is a study by Petit et al. (1992) that compared the collaborative behaviour of two closely related species of macaques: the more tolerant tonkean macaque (*M. tonkeana*) and the more competitive rhesus macaque (*M. mulatta*) (Thierry 1985, 2007). In this study the researchers placed food in a small hole underneath a stone that was too heavy for a single macaque to move. The tolerant tonkean macaques were sometimes able to work together to move the rock and obtain the food, whereas the more competitive rhesus macaques generally avoided conspecifics near the food source. A second comparative study, by Hare et al. (2007), assessed bonobos (*Pan paniscus*) and chimpanzees (*P. troglodytes*). The authors first tested each species on their ability to feed in close proximity, a measure of tolerance, and found that bonobos co-fed more readily than chimpanzees. Both species were then presented with the widely used cooperative barpull task (Crawford 1937; Hirata and Fuwa 2007) that requires two individuals to simultaneously pull opposite ends of a rope to bring a tray with food on it into reach. If the two individuals do not pull the rope simultaneously, the rope unthreads from the tray so that it is no longer possible to pull the tray in. Hare et al. (2007) found that when the tray held a clumped, monopolizable food source, the bonobos were more successful than the chimpanzees at collaborating to obtain the food. One interpretation of this species difference is that, for the chimpanzees, it was less likely that two partners would be willing to work to pull in food given that one chimpanzee was likely to go unrewarded for their efforts. Alternatively (or additionally), unlike the bonobos, the chimpanzees may not have been comfortable in close proximity to a partner when food (which leads to competition) was nearby. The findings from the macaques, bonobos, and chimpanzees demonstrate how low tolerance may impede cooperative opportunities and highlight differences across species. Thus, collaboration may not occur regardless of whether the individuals are cognitively capable of understanding the task.

Given these apparent cross-species differences, there has been growing interest in the social cognition of cooperative breeders, spearheaded in part by the cooperative breeding hypothesis (Burkart et al. 2009; Hrdy 2009). This hypothesis posits that cooperative breeders have evolved psychological adaptations that facilitate prosocial and collaborative behaviours in the context of rearing young that are expressed in other domains as well [but see Thornton and McAuliffe (2015)]. Marmosets and tamarins are cooperative breeders known to be highly tolerant towards group mates and to regularly share food. Cronin and colleagues tested pair-bonded male–female dyads of cottontop tamarins and found that even when there was a single food reward fully monopolizable by one individual, collaboration was frequent and there was no aggression (Cronin et al. 2005; Cronin and Snowdon 2008). These results are similar to those found for apes and macaques described above, in that tolerant group dynamics seem to set the stage for collaboration. Additional studies, which have focused on variation in collaborative problem-solving *within* primate species, also report that more tolerant, egalitarian dyads achieve greater collaborative success in nonhuman primates (de Waal and Davis 2003; de Waal and Suchak 2010; Melis et al. 2006) as well as humans (Cronin et al. 2015a, b).

The examples above provide insight about how social influences impact *opportunities* to collaborate but provide less information about what primates *understand* about collaboration. Of course the latter is the more difficult question to answer. When collaboration is studied from a cognitive perspective, the relevant question is whether actors understand the role of the partner in the collaborative task. That is, do actors act as though their partner is necessary, and do they alter their behaviour flexibly in response to changes in their partners' behaviour? To test this, researchers often provide subjects with a collaborative task both in the presence and absence of their partner, intermittently block the view of the partner, and/or delay the partner's access to the task (reviewed in Cronin 2016a). If subjects are attempting to coordinate actions with their partners, one would expect them not to attempt to solve the task when their partner is absent or delayed and expect their performance to suffer when they cannot see their partner to facilitate coordination. In general this is what has been found. However, it should be noted that the time delays subjects are able to wait are relatively short (i.e. up to 10 s; Melis et al. 2006), and even when the partner is not nearby, subjects still attempt some solutions on their own (e.g. Cronin et al. 2005; Cronin and Snowdon 2008; Melis et al. 2006; Suchak et al. 2014). Whether this is due to a lack of understanding, a lack of inhibition, or the low cost of solo attempts is not clear. However, differences in the subjects' attempts between the test (partner present) and control (partner absent or delayed), while perhaps not extreme, are typically statistically significant and indicate some cognitive understanding of the social aspect of collaboration.

There are two notable limitations of experimental studies of primate collaboration to date. The first is that most experimental studies have focused on dyadic interactions, while many primates live in large social groups, and multiple individuals are involved in cooperative interactions (e.g. to hunt, form alliances, defend territories;

reviewed in Mitani 2009). One notable exception is experimental field work on territorial defence in chimpanzees (Herbinger et al. 2009; Wilson et al. 2002, 2011), which has demonstrated that chimpanzees are capable of monitoring and coordinating with several group mates to achieve a common goal and are capable of evaluating their probability of success (for experimental captive studies, see also Cronin et al. 2015a, b; Suchak et al. 2014). The second limitation is that the majority of research on collaboration has centred on obtaining food (a limitation of many tests of social learning too; Watson and Caldwell 2009). It is not clear whether tolerance (which is often measured as ability to co-feed) would play such a large role in collaborative performance if subjects were not being posed with the challenge to collaborate in order to obtain food rewards. Again, looking to field experiments on collaboration in chimpanzees in the context of territorial behaviour suggests that conclusions drawn from experiments involving food may not extend to non-food contexts.

4 Prosocial Behaviour

Collaboration and prosocial behaviour are commonly considered together as types of cooperative social behaviours. However, while collaboration involves coordinated behaviour of at least two individuals to bring about a benefit, prosocial acts require the behaviour of only a single individual to benefit another. As with collaboration, communication and social relationships again influence the probability of the behaviour being expressed. Specifically, an act is considered prosocial if it improves the welfare of another individual (Cronin 2012; Silk et al. 2005). Prosocial acts can include behaviours such as warning others of danger, sharing food, or facilitating access to a resource for another individual (e.g. Crockford et al. 2012; Feistner and McGrew 1989; Yamamoto et al. 2012). Although we may imagine a wide range of scenarios involving prosocial behaviour in the wild, the experimental approach to studying prosociality has focused largely on one paradigm: the prosocial choice task. This approach has also been used across several nonhuman primate species and with human children (for reviews see Cronin 2016a; House 2013; Padilla-Walker and Carlo 2014).

In the prosocial choice task, subjects are typically presented with a choice between two options: (1) a prosocial option that provides a food reward to herself/himself and to a recipient and (2) another option which provides a reward for the actor only. The same amount of effort is required for both options; the options differ only by whether or not the recipient also receives a benefit. The frequency with which actors choose the prosocial option is compared with a nonsocial control condition in which a recipient cannot benefit. Evidence of prosocial behaviour is assumed if the actor chooses the prosocial option more often when another individual is present to receive the reward than when there is nobody present. Sometimes the two choices are presented on trays, and the subject physically chooses between them; other times the choices are represented symbolically with tokens (e.g. Amici et al. 2014). Another common iteration of this task has removed the reward for the donor completely. Although this makes the prosocial act costly, it has been a method of interest given that some studies report that the donor may either be distracted by food in front of them or have a difficult time processing multiple food locations in their decision (House et al. 2014; Burkart and Rueth 2013; Stevens 2010; see also Duque and Stevens, this volume). Finally, recent work has also questioned whether animals will show costly prosociality, that is, give up a reward to allow another to benefit (e.g. Sterck et al. 2015).

Despite a significant amount of research effort committed to studying prosociality, reports of robust and reliable prosocial behaviours in primates are rare (reviewed in Cronin 2016a). Positive reports of prosociality have been shown to be sensitive to slight methodological differences (Cronin 2012; House et al. 2014), and the magnitude of prosocial effects is typically quite small (reviewed in Stevens 2010; see also Duque and Stevens this volume). As with collaboration and social learning, there has been some evidence that prosocial tendencies may be influenced by social relationships and tolerance (reviewed in Cronin 2016a), but the influence of social relationships is less clear than in the other sociocognitive domains. Thus, testing animals in settings with increased ecological validity will provide us with a deeper understanding of primate prosociality (Cronin 2016b).

5 Inequity Aversion

We note that many experimental designs used to test primates' collaborative and prosocial tendencies result in the test subjects receiving rewards of different quality or quantity. Whether the subjects recognize these differences, and attempt to provide their partner with better rewards, is the core of all tests of prosocial behaviour. A parallel area of research that has also tested whether primates recognize and respond to differences in reward value includes studies designed to investigate responses to inequity. Indeed, this connection is not incidental, and it has been proposed that certain primate species developed a response to inequity (recognizing unequal reward distributions) in order to recognize who are reliable social partners with whom to collaborate (Brosnan 2011).

Even from a young age, humans respond negatively to receiving a less-preferred outcome than that received by a social partner, a response shared by a number of different primate species (Cronin and Hopper 2017). In typical tests of inequity, two primates are rewarded differentially by an experimenter for completing the same task. Thus, although the effort each primate exerts is equal, the rewards they receive for doing so are not. Negative responses to inequity were first tested with nonhuman primates 15 years ago, although there are much earlier reports that primates respond negatively to receiving a reward of a lower value than what they had anticipated ('contrast effects', Tinklepaugh 1928). Brosnan and de Waal (2003) reported that capuchins (*Cebus apella*) were less likely to accept the food rewards they were given or to complete the task (in this case, exchange a token with an experimenter) when they were given a less-preferred reward than a social partner who also had to exchange tokens with an experimenter. In the years since this first study, a number of different primate species have been tested in studies of inequity, which have revealed interesting differences in primates' responses across species and studies (reviewed in Brosnan 2013).

Previous research suggests that primates are more likely to show a negative reaction to inequity if they and their partner have to work to receive rewards than if they are simply given rewards of different values for 'free'. However, the kind of task that they have to complete to earn their rewards does not appear to be as important (e.g. exchange task or target task, Freeman et al. 2013; but see McAuliffe et al. 2015). Furthermore, the responses of nonhuman primates tested under the same experimental conditions suggest that species that cooperate with non-kin under natural conditions (e.g. chimpanzees) appear more likely to react to inequity than species that typically cooperate with kin (e.g. owl monkeys, Brosnan 2013) again highlighting the interaction between these two aspects of primate social cognition.

Intriguingly, primates' responses to inequity do not appear to be universal. For example, extensive studies with chimpanzees and capuchins, the species tested most frequently in tests of inequity aversion, have revealed variance in their responses to inequity both across and within species (Brosnan et al. 2015). For example, while one study reported that male chimpanzees responded more strongly than females (Brosnan et al. 2010), another found that only females responded to receiving a less-preferred reward than their test partner (Hopper et al. 2014). Furthermore, other research has reported no response at all to inequity by chimpanzees (Bräuer et al. 2009). Addressing this variance, Brosnan et al. (2015) analysed the relative impact of chimpanzees' age, sex, rank, social relationships, and personality on their responses to inequity. While demographic factors did not correlate with the chimpanzees' behavioural responses to inequity, their personality characteristics did; however, there is clearly much we have yet to learn about both these proximate and ultimate predictors of primates' responses to inequity.

6 Conclusions

Much of the recent research on primate social cognition has focused heavily on identifying what species are capable of and under what circumstances certain behaviours are elicited. For example, research has questioned how faithfully primates copy the actions of others (Hopper 2010), who they are likely to copy (Horner et al. 2010), and which individuals are more likely to rely on social information (Kendal et al. 2015), yet far less effort has been given to understanding how social learning occurs and what primates understand about the actions of others (but see Hecht et al. 2012). Similarly, it has been shown that more tolerant or closely bonded primates are also more likely to socially learn from one another (Claidière et al. 2013) and to collaborate (Hare et al. 2007). The most parsimonious explanation for this is that both behaviours are more likely to occur in animals that are in closer proximity to one another, but this tells us very little about primates' understanding of these processes. Thus, to date, more focus has been put on the 'social' aspects of sociocognitive behaviours than on the underlying 'cognition'.

Much research supports the hypothesis that primates can learn from one another and work together to solve problems. Importantly, however, it seems that the design of the experiment matters quite a bit; tasks that mirror challenges primates face in the wild and are minimally mechanistic seem to elicit the strongest collaborative and prosocial skills in primates (Chalmeau et al. 1997; Mendres and de Waal 2000; Visalberghi et al. 2000; Burkart et al. 2014). Testing in a social setting also allows for partner choice, which may impact success (Noe 2006; Suchak et al. 2014) and increase ecological validity (Cronin et al. 2017). Studies of social learning run in a naturalistic group setting not only stimulate social learning but are able to provide richer data sets as they allow the primates to choose who to copy and when to use social information (Drea 2006; Kendal et al. 2015). Thus, we call for more researchers to employ experimental designs that have increased ecological validity, present tasks that are not cognitively opaque, and test aspects of social cognition that extend beyond typical tasks that involve learning how to obtain a food source (whether via social learning or collaboration). Of course, any research that is explicitly comparative in nature also provides us with a clearer understanding of the ultimate, as well as the proximate, mechanisms for such behaviours.

Research to date may also have been hampered by the separation of studies of cooperation and competition, when in fact these sociocognitive domains probably overlap and research would benefit from considering them in concert (Schmelz and Call 2016). Furthermore, and as mentioned above, many studies of prosocial behaviour and collaboration create uneven food distributions for the primates being tested (e.g. Claidière et al. 2015), tapping into the question of how primates respond to inequity. Perhaps only by considering these influences collectively and simultaneously investigating the various factors that may influence primates' behaviour and decision-making can we gain a full understanding of their sociocognitive capacity (Brosnan et al. 2015). Taking a more holistic approach may help to explain the variance across studies that has been reported to date, both across and within species, for each of the four topics we discussed in this chapter.

References

Amici F, Visalberghi E, Call J (2014) Lack of prosociality in great apes, capuchin monkeys and spider monkeys: convergent evidence from two different food distribution tasks. Proc R Soc B 281(1793). https://doi.org/10.1098/rspb.2014.1699

Boesch C, Boesch H (1989) Hunting behavior of wild chimpanzees in the Tai National Park. Am J Phys Anthropol 78:547–573

Bräuer J, Call J, Tomasello M (2009) Are apes inequity averse? New data on the token-exchange paradigm. Am J Primatol 71(2):175–181

Brosnan SF (2011) A hypothesis of the co-evolution of cooperation and responses to inequity. Front Neurosci 5. https://doi.org/10.3389/fnins.2011.00043

Brosnan SF (2013) Justice and fairness related behaviors in non-human primates. Proc Natl Acad Sci USA 110:10416–10423

Brosnan SF, de Waal FBM (2003) Monkeys reject unequal pay. Nature 425(6955):297–299

Brosnan SF, Hopper LM (2014) Psychological limits on animal innovation. Anim Behav 92: 325–332

Brosnan SF, Talbot C, Ahlgren M, Lambeth SP, Schapiro SJ (2010) Mechanisms underlying responses to inequitable outcomes in chimpanzees, *Pan troglodytes*. Anim Behav 79(6): 1229–1237

Brosnan SF, Hopper LM, Richey S, Freeman HD, Talbot CF, Gosling S, Lambeth SP, Schapiro SJ (2015) Personality influences responses to inequity and contrast in chimpanzees. Anim Behav 101:75–87

Burkart JM, Rueth K (2013) Preschool children fail primate prosocial game because of attentional task demands. PLoS One 8(7):e68440

Burkart JM, Hrdy SB, van Schaik CP (2009) Cooperative breeding and human cognitive evolution. Evol Anthropol 18:175–186

Burkart JM, Allon O, Amici F, Fichtel C, Finkenwirth C, Heschl A, Huber J, Isler K, Kosonen ZK, Martins E, Meulman EJ, Richiger R, Rueth K, Spillmann B, Wiesendanger S, van Schaik CP (2014) The evolutionary origin of human hyper-cooperation. Nat Commun 5:4747

Byrne RW, Rapaport LG (2011) What are we learning from teaching? Anim Behav 82(5): 1207–1211

Byrne RW, Whiten A (1988) Machiavellian intelligence. Oxford University Press, Oxford

Caldwell CA, Schillinger K, Evans CL, Hopper LM (2012) End state copying by humans (Homo sapiens): implications for a comparative perspective on cumulative culture. J Comp Psychol 126:161–169

Carter AJ, Marhsall HH, Heinsohn R, Cowlishaw G (2014) Personality predicts the propensity for social learning in a wild primate. Peer J 2:e283

Castro L, Toro MA (2004) The evolution of culture: from primate social learning to human culture. Proc Natl Acad Sci USA 101(27):10235–10240

Chalmeau R, Visalberghi E, Gallo A (1997) Capuchin monkeys, *Cebus apella*, fail to understand a cooperative task. Anim Behav 54(5):1215–1225

Claidière N, Messer EJE, Hoppitt W, Whiten A (2013) Diffusion dynamics of socially learned foraging techniques in squirrel monkeys. Curr Biol 23:1251–1255

Claidière N, Whiten A, Mareno MC, Messer EJE, Brosnan SF, Hopper LM, Lambeth SP, Schapiro SJ, McGuigan N (2015) Selective and contagious prosocial resource donation in capuchin monkeys, chimpanzees and humans. Sci Rep 5:7631

Coussi-Korbel S, Fragaszy DM (1995) On the relation between social dynamics and social learning. Anim Behav 50(6):1441–1453

Crawford MP (1937) The cooperative solving of problems by young chimpanzees. Comp Psychol Mono 14:1–88

Crockford C, Wittig R, Mundry R, Zuberbuhler K (2012) Wild chimpanzees inform ignorant group members of danger. Curr Biol 22(2):142–146

Cronin KA (2012) Cognitive aspects of prosocial behavior in nonhuman primates. In: Seel NM (ed) Encyclopedia of the sciences of learning, vol 3. Springer, Berlin, pp 581–583

Cronin KA (2016a) Comparative studies of cooperation: collaboration and prosocial behavior in animals. In: Call J, Burghardt GB, Pepperberg I, Snowdon CT, Zental T (eds) APA handbook of comparative psychology. American Psychological Association, Washington, DC

Cronin KA (2016b) Unnatural paradigm calls into question whether macaques' social decisions represent empathy. Proc Natl Acad Sci USA 113(10):E1331

Cronin KA, Hopper LM (2017) A Comparative perspective on helping and fairness. In: Sommerville J, Decety J (eds) Social cognition, frontiers in developmental science series. Routledge Taylor and Francis Group, New York, NY, pp 26–45

Cronin KA, Sánchez A (2012) Social dynamics and cooperation: the case of nonhuman primates and its implications for human behavior. Adv Complex Sys 15:1250066

Cronin KA, Snowdon CT (2008) The effects of unequal reward distributions on cooperative performance by cottontop tamarins, *Saguinus oedipus*. Anim Behav 75(1):245–257
Cronin KA, Kurian AV, Snowdon CT (2005) Cooperative problem solving in a cooperatively breeding primate (*Saguinus oedipus*). Anim Behav 69(1):133–142
Cronin KA, Pieper BA, van Leeuwen EJC, Mundry R, Haun DBM (2014) Problem solving in the presence of others: how rank and relationship quality impact resource acquisition in chimpanzees (*Pan troglodytes*). PLoS One 9(4):e93204
Cronin KA, Acheson DJ, Hernandez P, Sanchez A (2015a) Hierarchy is detrimental for human cooperation. Sci Rep 5:18634
Cronin KA, Pieper B, van Leeuwen EJC, Crockford C, Haun DBM (2015b) Cooperating to compete: evaluating behavioral coordination in response to simulated territorial intrusion in chimpanzees (*Pan troglodytes*). Am J Primatol 77(S1):77
Cronin KA, Jacobson SL, Bonnie KE, Hopper LM (2017) Studying primate cognition in a social setting to improve validity and welfare: a literature review highlighting successful approaches. PeerJ 5:e3649
de Waal FBM (1982) Chimpanzee politics: power and sex among apes. Jonathon Cape, London
de Waal FBM, Davis JM (2003) Capuchin cognitive ecology: cooperation based on projected returns. Neuropsychol 41(2):221–228
de Waal FBM, Suchak M (2010) Prosocial primates: selfish and unselfish motivations. Phil Trans R Soc B 365(1553):2711–2722
DeCasien AR, Williams SA, Higham JP (2017) Primate brain size is predicted by diet but not sociality. Nat Ecol Evol 1:112
Drea CM (2006) Studying primate learning in group contexts: tests of social foraging, response to novelty and cooperative problem solving. Methods 38:162–177
Drea CM, Wallen K (1999) Low-status monkeys "play dumb" when learning in mixed social groups. Proc Natl Acad Sci USA 96(22):12965–12960
Dunbar RIM (1998) The social brain hypothesis. Evol Anthropol 6:178–190
Feistner ATC, McGrew WC (1989) Food-sharing in primates: a critical review. In: Seth PK, Seth S (eds) Perspectives in primate biology, vol 3. Today & Tomorrow's Printers and Publishers, New Delhi, pp 21–36
Freeman HD, Sullivan J, Hopper LM, Talbot CF, Holmes AN, Schultz-Darken N, Williams LE, Brosnan SF (2013) Different responses to reward comparisons by three primate species. PLoS One 8(10):e76297
Gunst N, Leca JB, Vasey PL (2015) Influence of sexual competition and social context on homosexual behavior in adolescent female Japanese macaques. Am J Primatol 77(5):502–515
Hare B, Melis AP, Woods V, Hastings S, Wrangham R (2007) Tolerance allows bonobos to outperform chimpanzees on a cooperative task. Curr Biol 17:619–623
Hecht EE, Gutman DA, Preuss TM, Sanchez MM, Parr LA, Riling JK (2012) Process versus product in social learning: comparative diffusion tensor imaging of neural systems for action execution–observation matching in macaques, chimpanzees, and humans. Cereb Cortex 23(5): 1014–1024
Herbinger I, Papworth S, Boesch C, Zuberbühler K (2009) Vocal, gestural and locomotor responses of wild chimpanzees to familiar and unfamiliar intruders: a playback study. Anim Behav 78(6): 1389–1396
Hirata S, Fuwa K (2007) Chimpanzees (*Pan troglodytes*) learn to act with other individuals in a cooperative task. Primates 48:13–21
Hopper LM (2010) 'Ghost' experiments and the dissection of social learning in humans and animals. Biol Rev 85:685–701
Hopper LM (2017) Social learning and decision making. In: Schapiro SJ (ed) Handbook of primate behavior management. CRC Press, Taylor & Francis Group, Boca Raton, FL, pp 225–242
Hopper LM, Marshall-Pescini S, Whiten A (2012) Social learning and culture in child and chimpanzee. In: de Waal FBM, Ferrari PF (eds) The primate mind: built to connect with other minds. Harvard University Press, Cambridge, pp 99–118

Hopper LM, Lambeth SP, Schapiro SJ, Brosnan SF (2014) Social comparison mediates chimpanzees' responses to loss, not frustration. Anim Cogn 17(6):1303–1311
Horner V, Proctor D, Bonnie KE, Whiten A, de Waal FBM (2010) Prestige affects cultural learning in chimpanzees. PLoS One 5(5):e10625
House BR (2013) Ontogenetic and phylogenetic variation in prosocial behavior: differences in prosociality across human development and primate species. Retrieved from https://escholarship.org/uc/item/7hz650k1.pdf
House BR, Silk JB, Lambeth SP, Schapiro SJ (2014) Task design influences prosociality in captive chimpanzees (*Pan troglodytes*). PLoS One 9(9):e103422
Hrdy SB (2009) Mothers and others. Belknap Press, Cambridge, MA
Huguet P, Barbet I, Belletier C, Monteil JM, Fagot J (2014) Cognitive control under social influence in baboons. J Exp Psychol Gen 143(6):2067–2073
Humle T, Snowdon CT (2008) Socially biased learning in the acquisition of a complex foraging task in juvenile cottontop tamarins, *Saguinus oedipus*. Anim Behav 75:67–277
Humphrey NK (1976) The social function of intellect. In: Bateson PPG, Hinde RA (eds) Growing Points in Ethology. Cambridge University Press, Cambridge, pp 303–317
Kendal RL, Hopper LM, Whiten A, Brosnan SF, Lambeth SP, Schapiro SJ, Hoppitt W (2015) Chimpanzees copy dominant and knowledgeable individuals: implications for cultural diversity. Evol Hum Behav 36(1):65–72
Marshall HH, Carter AJ, Ashford A, Rowcliffe JM, Cowlishaw G (2015) Social effects on foraging behavior and success depend on local environmental conditions. Ecol Evol 5(2):475–492
McAuliffe K, Chang LW, Leimgruber KL, Spaulding R, Blake PR, Santos LR (2015) Capuchin monkeys, *Cebus apella*, show no evidence for inequity aversion in a costly choice task. Anim Behav 103:65–74
Melis AP, Hare B, Tomasello M (2006) Engineering cooperation in chimpanzees: tolerance constraints on cooperation. Anim Behav 72(2):275–286
Mendres KA, de Waal FBM (2000) Capuchins do cooperate: the advantage of an intuitive task. Anim Behav 60(4):523–529
Mitani JC (2009) Cooperation and competition in chimpanzees: current understanding and future challenges. Evol Anthropol 18:215–227
Nishida T, Matsusaka T, McGrew CW (2009) Emergence, propagation or disappearance of novel behavioral patterns in the habituated chimpanzees of Mahale: a review. Primates 50(1):23–36
Noe R (2006) Cooperation experiments: coordination through communication versus acting apart together. Anim Behav 71(1):1–18
Padilla-Walker LM, Carlo G (2014) The study of prosocial behavior: past, present and future. In: Padilla-Walker LM, Carlo G (eds) Prosocial development: a multidimensional approach. Oxford University Press, Oxford, pp 3–16
Pasquaretta C, Levé M, Claidière N, van de Waal E, Whiten A, Macintosh A, Pelé M, Bergstrom M, Borgeaud C, Brosnan F, Crofoot M, Fedigan L, Fichtel C, Hopper LM, Mareno M, Petit O, Schnoell AV, Polizzi di Sorrentino E, Thierry B, Tiddi B, Sueur C (2014) Social networks in primates: smart and tolerant species have more efficient networks. Sci Rep 4:7600
Petit O, Desportes C, Thierry B (1992) Differential probability of "coproduction" in two species of macaque (*Macaca tonkeana, M. mulatta*). Ethology 90(2):107–120
Powell LE, Isler K, Barton RA (2017) Re-evaluating the link between brain size and behavioural ecology in primates. Proc R Soc B 284:20171765
Reader SM, Laland KN (2002) Social intelligence, innovation, and enhanced brain size in primates. Proc Natl Acad Sci USA 99(7):4436–4441
Salwiczek LH, Prétôt L, Demarta L, Proctor D, Essler J, Pinto AI, Wismer S, Stoinski T, Brosnan SF, Bshary R (2012) Adult cleaner wrasse outperform capuchin monkeys, chimpanzees and orang-utans in a complex foraging task derived from cleaner-client reef fish cooperation. PLoS One 7(11):e49068

Schmelz M, Call J (2016) The psychology of primate cooperation and competition: a call for realigning research agendas. Phil Trans R Soc B 371(1686):20150067. https://doi.org/10.1098/rstb.2015.0067

Silk JB, Brosnan SF, Vonk J, Henrich J, Povinelli DJ, Richardson AS, Lambeth SP, Mascaro J, Schapiro SJ (2005) Chimpanzees are indifferent to the welfare of unrelated group members. Nature 437:1357–1359

Snowdon CT, Cronin KA (2009) Comparative cognition and neuroscience. In: Berntson G, Cacioppo J (eds) Handbook of neuroscience for the behavioral sciences. Wiley, Hoboken, NJ, pp 32–55

Sterck EH, Olesen CU, Massen JJ (2015) No costly prosociality among related long-tailed macaques (*Macaca fascicularis*). J Comp Psychol 129(3):275–282

Stevens JR (2010) Donor payoffs and other-regarding preferences in cotton-top tamarins (*Saguinus oedipus*). Anim Cogn 13:663–670

Strandburg-Peshkin A, Farine DR, Couzin ID, Crofoot MC (2015) Shared decision-making drives collective movement in wild baboons. Science 348(6241):1358–1361

Suchak M, Eppley TM, Campbell MW, de Waal FBM (2014) Ape duos and trios: spontaneous cooperation with free partner choice in chimpanzees. PeerJ 2:e417

Thierry B (1985) Patterns of agonistic interactions in three species of macaque (*Macaca mulatta, M fascicularis, M tonkeana*). Aggress Behav 11:223–233

Thierry B (2007) Unity in diversity: lessons from macaque societies. Evol Anthropol 16(6): 224–238

Thornton A, McAuliffe K (2015) Cognitive consequences of cooperative breeding? A critical appraisal. J Zool 295(1):12–22

Tinklepaugh OL (1928) An experimental study of representative factors in monkeys. Comp Psychol 8:197–236

van Schaik CP (2003) Local traditions in orangutans and chimpanzees: social learning and social tolerance. In: Fragaszy D, Perry SE (eds) The biology of animal traditions. Cambridge University Press, Cambridge, pp 297–328

Visalberghi E, Pellegrini Quarantotti B, Tranchida F (2000) Solving a cooperation task without taking into account the partner's behavior: the case of capuchin monkeys (*Cebus apella*). J Comp Psychol 114(3):297–301

Vonk J, Shackelford TK (2012) The Oxford handbook of comparative evolutionary psychology. Oxford University Press, New York, NY

Watson CFI, Caldwell CA (2009) Understanding behavioral traditions in primates: are current experimental approaches too focused on food? Int J Primatol 30:143–167

Wilson ML, Britton NF, Franks NR (2002) Chimpanzees and the mathematics of battle. Proc R Soc Lond B 269(1496):1107–1112

Wilson ML, Kahlenberg SM, Wells M, Wrangham RW (2011) Ecological and social factors affect the occurrence and outcomes of intergroup encounters in chimpanzees. Anim Behav 83(1):277–291

Yamamoto S, Humle T, Tanaka M (2012) Chimpanzees' flexible targeted helping based on an understanding of conspecifics' goals. Proc Natl Acad Sci USA 109(9):3588–3592

Zentall TR, Wasserman EA (2012) The Oxford handbook of comparative cognition. Oxford University Press, New York, NY

Affective States, Motivation, and Prosocial Behaviour in Primates

Hika Kuroshima and Kazuo Fujita

Abstract Humans regularly help others upon seeing them in trouble, often almost impulsively. Help may be given even to strangers or to different species, with no return expected. Studies have reported that not only humans but also non-human primates show prosocial behaviours, such as helping and cooperation, in both natural and captive settings. Some researchers suggest that these behaviours relate to a sense of fairness and an ability to empathize. However, it is still not clear what kinds of affective states and motivation non-human species have and how they are involved in prosocial behaviours. In this chapter, we introduce our studies on cooperative behaviour, food sharing, and third-party evaluation in captive tufted capuchin monkeys (*Sapajus apella*), a New World monkey species. These studies demonstrated that the dominant monkeys behave prosocially towards subordinate monkeys, that capuchins are sensitive to partners' efforts and rewards, and that they refuse food from humans who show anti-social behaviour towards others. From these results, we propose that advantageous inequality aversion may promote prosocial behaviour only towards subordinates and that empathy may not be important in food contexts.

Keywords Prosocial behaviour · Cooperative behaviour · Food sharing · Third-party evaluation · Capuchin monkey

1 Introduction

When we find others in trouble, we often offer a helping hand, for example, by sharing our possessions, working together, giving information, or giving encouragement. Sometimes we do this for nothing in return. Moreover, this helping behaviour extends to unrelated and unfamiliar others and even other species. Why do we feel such a strong desire to help others? Many studies have reported that not only humans but also non-human primates show prosocial behaviours, such as

H. Kuroshima (✉) · K. Fujita
Department of Psychology, Graduate School of Letters, Kyoto University, Kyoto, Japan

L. D. Di Paolo et al. (eds.), *Evolution of Primate Social Cognition*, Interdisciplinary Evolution Research 5, https://doi.org/10.1007/978-3-319-93776-2_3

cooperation, helping, and sharing. For example, wild chimpanzees (*Pan troglodytes*) hunt together and share meat with group members (Boesch and Boesch 1989; Boesch 1994; Mitani and Watts 2001). Also, captive chimpanzees fetch out-of-reach objects for humans and conspecifics (Warneken et al. 2006, 2007) and, when requested, give appropriate tools that others need to procure a reward (Yamamoto et al. 2009; Yamamoto and Tanaka 2009). Prosocial behaviour has been defined as "voluntary actions that are intended to help or benefit another individual or group of individuals" (Eisenberg and Mussen 1989, p. 3). What kinds of motivation or affect drive prosocial behaviour in non-human primates? This question lies at the core of understanding evolution of prosocial behaviour and its underlying mechanisms.

Some researchers suggest that prosocial behaviour is closely related to empathy and a sense of fairness. de Waal (2008) defined empathy as the capacity to (1) be affected by and share the emotional state of another (e.g. emotional contagion), (2) assess the reasons for the other's state, and (3) identify with the other, adopting his or her perspective. He proposed a Russian doll model of multilayered empathy that has a perception-action mechanism (PAM) as its inner core structure (de Waal 2012). It starts with state-matching based on motor mimicry. Shared neural representations (Preston and de Waal 2002) automatically elicit emotional contagion, and sympathetic concern arising from perceiving others' mental states gives rise to consolation behaviour. Finally, development of perspective-taking abilities makes it possible to perform targeted helping (de Waal et al. 2008, 2012). Recently, from a neurobehavioural view, Decety et al. (2016) have proposed how empathy motivates prosocial behaviour. According to this theory, witnessing another's distress (step 1) can lead to an aversive affective arousal combined with a physiological stress response (step 2). When appropriate, a prosocial drive is triggered (step 3) which, depending on the context, can lead to prosocial behaviour. By opposing the classic tension-reduction model (Hull 1943) in which helping others in distress serves to regain homeostasis of the helper, they argued that helping and caring per se are inherently rewarding.

Fehr and Schmidt (1999) distinguished two aspects of the sense of fairness: advantageous and disadvantageous inequity aversions (IA). Both are negative reactions, but the former occurs when an individual earns a better reward than a partner for the same effort, whereas the latter occurs when the individual earns a poorer reward than the partner. Over the last decade, empirical studies have demonstrated disadvantageous IA in several non-human primate species including chimpanzees (Brosnan et al. 2005; Hopper et al. 2014), brown (tufted) capuchin monkeys (*Cebus apella*) (Brosnan and de Waal 2003; van Wolkenten et al. 2007), cotton-top tamarins (*Saguinus oedipus*) (Cronin et al. 2010), rhesus macaques (*Macaca mulatta*) (Hopper et al. 2013), and long-tailed macaques (*Macaca fascicularis*) (Massen et al. 2012). Typically, individuals in these species refuse to work or reject a reward when the reward is poorer than the one that their conspecific partner receives for doing the same task. Humans show both types of IA, but in non-human primates empirical evidence for advantageous IA is scarce. Brosnan and de Waal (2014), in their review paper, suggested that disadvantageous IA (first-order IA) plays a role in maintaining a successful cooperative relationship with partners, while advantageous IA (second-order IA) plays a role in partner choice to forestall inequitable reward distribution. Thus, both types of inequity aversion function to balance rewards among participants in cooperative situations. Also, Yamamoto and Takimoto (2012) have hypothesized that empathy may promote prosocial behaviour, while a sense of fairness may stabilize cooperation. Although there may be strong links among prosocial behaviour, empathy, and a sense of fairness, it is still unclear how affective states operate in prosocial behaviour.

In this chapter, we introduce our empirical studies on food sharing, cooperation and division of labour, and third-party social evaluation in tufted capuchin monkeys (*Sapajus* (formerly *Cebus*) *apella*), a species of New World monkeys. Tufted capuchin monkeys are well known to be socially tolerant and to learn various skills including tool use by observing others (Coelho et al. 2015; Custance et al. 1999; Fragaszy and Visalberghi 2004; Kuroshima et al. 2008; Ottoni et al. 2005). They are also sensitive to fairness (Brosnan and de Waal 2003; van Wolkenten et al. 2007). Studying their prosocial behaviour in food sharing and cooperative situations and their third-party social evaluation can help to throw light on the kinds of affect that support capuchins' prosocial behaviour, and how empathy and a sense of fairness affect their prosocial tendencies, from an evolutionary perspective.

2 Prosocial Tendencies in Feeding Contexts

Two types of experimental paradigms have been used to determine primates' prosocial tendencies in feeding contexts: a token-exchange task and a platforms task.

2.1 Token-Exchange Task

In the token-exchange procedure, subjects first learn that different tokens can be exchanged for different outcomes. Then a partner is introduced into an adjacent compartment. In the test session, if a subject chooses a "selfish token", only the subject receives a reward for exchanging the token with an experimenter, but if the subject chooses a "prosocial token", both the subject and the partner are rewarded. Thus, in this paradigm the subject controls who benefits.

de Waal et al. (2008) trained eight adult female brown capuchin monkeys to discriminate between "selfish" and "prosocial" tokens. Then the experimenter offered subjects a choice of the two tokens on 30 trials. The mean frequency of choosing the prosocial token increased precipitously in the final ten trials. The subjects chose the prosocial token significantly more often than the selfish one when the partner was a group mate, visible, and likely to get the same reward. In this experiment, all subjects were more dominant or older than the partners. However, Amici et al. (2014) pointed out that the rate of choosing prosocial tokens should be compared with that in a partner-absent condition. They (Amici et al. 2014) tested six primate species, namely, chimpanzees, bonobos (*Pan paniscus*), gorillas (*Gorilla gorilla*), orangutans (*Pongo abelii*), tufted capuchin monkeys, and spider monkeys (*Ateles geoffroyi*), in a similar token choice task and found that none of the species choose the prosocial token significantly more often when there was a partner in the adjacent cage than when there was none. Thus, the subjects did not use significantly more often the prosocial token for their partner when compared with the partner-absent condition.

Suchak and de Waal (2012) found that reciprocity enhanced prosocial choices in capuchins. In their test, both participants in alternation could choose between the prosocial and the selfish tokens. Monkeys chose the prosocial token significantly more often in this reciprocal condition than when only one monkey continued choosing. We also tested capuchin monkeys using an alternating token-exchange paradigm (Kuroshima et al. 2012). In our procedure, the subjects (two adult males and four adult females) always had access to one selfish and one prosocial token, whereas their partner (an adult male) had access to different types of tokens according to the test conditions. The experimenter requested each monkey in turn to exchange a token for a reward (Fig. 1). Five test conditions were run as follows:

Selfish-token condition: The partner had two selfish tokens and no prosocial one. The partner had no choice but to give a selfish token upon the experimenter's request.

Fig. 1 Test setting for token-exchange task. One monkey responded to the experimenter's exchange request. The other monkey was rewarded according to the type of the token chosen by the first monkey

Prosocial-token condition: The partner had two prosocial tokens and no selfish one. The partner had no choice but to give a prosocial token upon the experimenter's request.

Both tokens condition: The partner had both selfish and prosocial tokens. The partner could choose to give either one upon the experimenter's request.

No token condition: The partner had no tokens. The partner could give no token upon the experimenter's request unless the subject transferred a token through the separating grid.

Partner-absent condition: No partner was in the adjacent cage.

Contrary to our expectation, no subjects ever exchanged the prosocial token significantly more often than they did in the *partner-absent condition*. Thus, reciprocity had no effect in our study. Additionally, in the *no token condition*, we examined whether subjects transferred tokens when the experimenter requested a token from their partner, who had no token. A few monkeys actively tossed their token into the partner's cage or appeared to intentionally bring it within the partner's reach. However, this kind of giving behaviour was infrequent (19 of 118 trials in 3 of 6 subjects), and the type of token transferred was random. The subjects did not appear to actively help other individuals or exploit another's opportunity to increase their own gain. Individual relationships may be another important factor that supports prosocial behaviour. Results from the "platforms task" described below suggest that capuchin monkeys indeed change their prosocial behaviour depending on the social relationship with their partners.

2.2 Platforms Task

In the platforms task, subjects were presented with two options on platforms or containers. The subjects always obtained the same reward regardless of their choice, prosocial or selfish. Choosing the former delivered rewards both to the subject and the partner, but choosing the latter rewarded only the subject.

Takimoto et al. (2010) tested whether tufted capuchin monkeys would choose the prosocial option. Two monkeys faced each other. The subjects were four adult monkeys (two males), and the partners were either a dominant male (the dominant condition) or a subordinate female (the subordinate condition). With no effort required, the partners received either high- or low-value food depending upon the subjects' choice, whereas the subjects obtained the same food irrespective of their choice (Fig. 2). Results showed that subjects chose the prosocial option more often when the partner was subordinate compared to when there was no partner, but they chose randomly in the presence of a dominant partner (Exp. 1). Interestingly, when an opaque screen blocked visual access between the monkeys, the subjects' preference for the prosocial option in the presence of the subordinate disappeared (Exp. 3). The authors concluded that "…capuchin monkeys are sensitive to others' food rewards and that they change their food choice strategies depending on the social rank of the recipient" (p. 258).

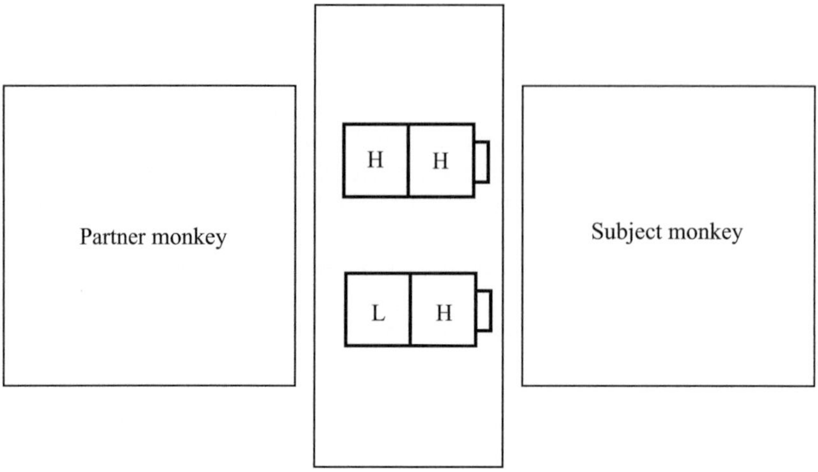

Fig. 2 Test setting for the platforms task. Two monkeys (subject and partner) were placed face to face. There were two food containers on the table. Both food containers could be opened only from the subject's side. When the subject pulled the handle, both monkeys could collect a reward. "H" indicates a high-value reward; "L" indicates a low-value reward. In this study, the subject always obtained high-value food regardless of his/her choice

It is noteworthy that in Takimoto et al. (2010) subjects showed more prosocial choices towards subordinate than dominant partners. However, because only one dominant monkey was used as the partner, it cannot yet be concluded that social rank modulates monkeys' prosocial behaviour. However, Lakshminarayanan and Santos (2008) demonstrated prosocial responding in tufted capuchin monkeys when a low-ranking monkey was the partner. In long-tail macaques (*Macaca fascicularis*), Massen et al. (2010, 2011) also demonstrated that it was not relationship quality, i.e. "friends" vs. "nonfriends", but relative difference in dominance between subjects and partners that influenced subjects' prosocial choice. The directional prosocial tendency from dominants towards subordinates implies that the threat of punishment by dominants is unlikely to explain prosocial choice in feeding contexts.

Dominant monkeys are not always prosocial to subordinates; however, they appear to be sensitive to the value of their own food. Takimoto (2012) tested whether capuchin monkeys would provide a high-value food to a partner by diminishing the value of their own food. In this study, choosing the prosocial option resulted in both subject and partner receiving a middle-value food, whereas choosing the selfish option led to a high-value food for the subject but a low-value food for the partner. Results showed that subjects chose the selfish option in over 90% of trials, regardless of the presence of a partner; there was no prosocial tendency even towards the subordinate partner. Also, in Sterck et al. (2015), when long-tailed macaques had to choose a low-value food to give a food towards a partner who was kin, subjects failed to choose this option. These results suggest that monkeys prioritize the quality of their own reward over others' welfare.

Research has also shown that capuchin monkeys take their partner's labour into account when deciding whether to be prosocial or not. In Takimoto and Fujita (2011), subjects sometimes required help from their partner. In the equal labour condition, the partner first had to correctly manipulate an apparatus before the subject was able to open a drawer. In the unequal labour condition, the subject did everything. Results showed that subjects chose the prosocial option to deliver high-value food to the partner only in the former condition. By modifying this procedure, Takimoto (2012) demonstrated that capuchins would give a reward to the partner even at the expense of their own reward, when they needed the partner's help. That is, in the equal labour condition, subjects occasionally delivered high-value food to the partner, but they did not do so in the unequal labour condition. This suggests that capuchin monkeys may even accept some costs to acknowledge help by a partner. Taken together, these results suggest that relative social ranks, value of own rewards, and partners' labour affect tufted capuchin monkeys' prosocial tendencies.

2.3 Summary of Prosocial Tendencies in Feeding Contexts

Our results showed that dominant subjects quite often chose a prosocial option or deliver food to subordinate partners. For tufted capuchin monkeys, social relationships may strongly influence the likelihood of prosocial choices. Moreover, they were sensitive to a partner's effort in the prosocial choice task. In feeding contexts, capuchin monkeys may take into account who the partner is and how much effort the partner contributes. However, in both our test conditions, the subjects could almost always keep what they owned (i.e. their reward). Would they be willing to share their possession (i.e. their food) with the partner? Capuchin monkeys may spontaneously transfer some food to a partner who has none (de Waal, 1997), but it is unknown whether capuchin monkeys would offer their own food if resulted in them remaining with none. At the same time, questions arise about the extent to which they pay attention to their own effort and about the importance of the food rewards being visible or not. This species may not be particularly sensitive to the amount of effort required of them; they may not care about it if it is trivial. In a study by Paukner et al. (2006), for example, capuchin monkeys always looked into multiple tubes before choosing the baited one, regardless of their knowledge of the baited locations, presumably because they discounted this additional labour. Weak evidence for metamemory in this species (Basile et al. 2009) may be due to their strong impulse to perform target tasks with high inherent uncertainty (Beran and Smith 2011). Future studies should more closely examine whether capuchins' prosocial responses in fact simply reflect impulsivity as opposed to flexible adjustment to the partner's identity and situation. Below, we describe how monkeys spontaneously divide efforts with a partner to get food.

3 Prosocial Tendencies in Work Sharing: Cooperation and Division of Labour

To investigate whether they can collaborate with other individuals to get food rewards, Mendres and de Waal (2000) trained tufted capuchin monkeys to pull together to bring within reach a counterweighted tray from which one or both could collect a reward. Five same-sex pairs of adult subjects learned to synchronize their pulling behaviour to succeed on the task. Even if the reward was placed on only one side of the tray, monkeys maintained their collaboration by sharing the food through a mesh partition. The monkeys often coordinated their own pulling behaviour the partner's by glancing at the partner when they required his/her help. The authors argued that capuchin monkeys understand that a partner is required and can work together. In this type of task, the partners perform the same actions in synchrony, even if only one obtains the reward. An interesting question is whether they would collaborate by performing different actions, i.e. whether they would show division of labour.

3.1 Spontaneous Division of Labour

Hattori et al. (2005) tested whether capuchin monkeys spontaneously divide labour with partners to get rewards. Six capuchin monkeys (three males and three females) were first trained individually to collect food by performing a two-action sequence: pulling a tab in one compartment then sliding a block in the other (Fig. 3). The block could not move without the tab being pulled first. After sliding the block, the monkey could collect two rewards, one from under the block and the other from a food cup attached to the apparatus. After all subjects learned this sequential task, a transparent panel was inserted between the two compartments. Then, two monkeys were introduced, one in each compartment, so that the monkeys had to act in the appropriate sequence to get the rewards: the first monkey pulling out the tab and the second pushing the block. All three pairs spontaneously solved this cooperative task (Exp. 1).

We further found that the monkeys continued to cooperate even when only one of them obtained a reward on each trial if on every trial their roles switched (Exp. 3). In test trials a single reward was placed under the block, which meant that the tab operator monkey had to work for nothing, as the transparent panel prevented any sharing of the reward. In sum, the monkeys continued working for the partner without immediate reward.

However, in Hattori et al. (2005), it was the experimenter who assigned the monkeys' respective roles; the monkeys' division of labour was not a consequence of any negotiation. To explore whether and how capuchins might spontaneously assign roles to each other, Arahori et al. (2013) used a multiple-task apparatus. This task required six actions. To collect rewards, monkeys had to manually knock down

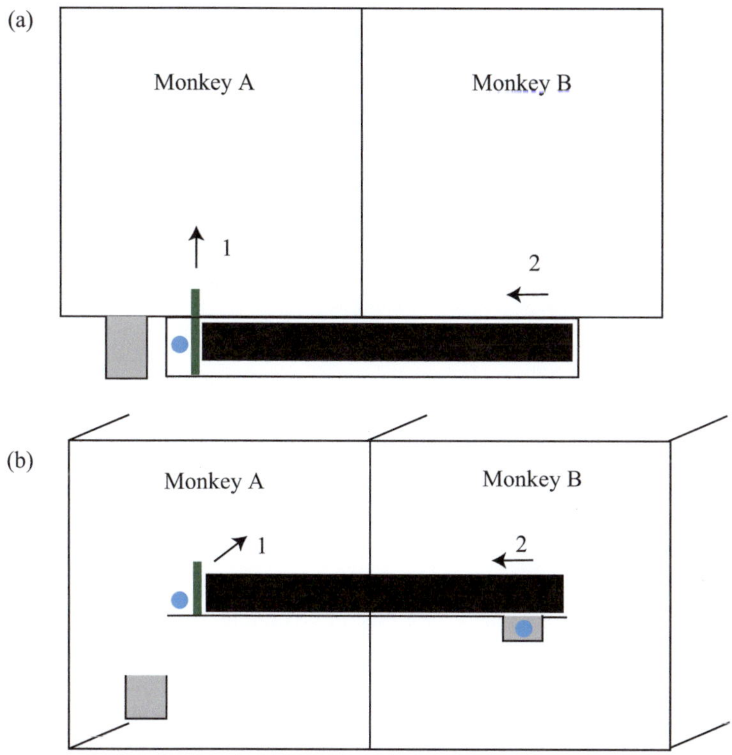

Fig. 3 A top view (**a**) and front view (**b**) of the apparatus for division of labour task. In the test trial, two monkeys (*A* and *B*) were in adjacent compartments. When monkey *A* pulled out the tab (shown as a green rod), monkey *B* could slide the block (a black long rectangle). After these two actions were done in sequence, both monkeys were able to collect reward (blue circles)

five rods and then pull a baited tray towards their cage (Fig. 4). The rods and the tray could be operated from each of two adjacent compartments. Thus, to collect the rewards, the monkeys had to make the six actions in total, with any specific allocation of actions between them.

After all the nine monkeys individually learned to operate the apparatus, paired monkeys were placed one in each compartment, so that they could see each other through the transparent wall. The subjects were seven middle-ranked monkeys, and the partners were either the most dominant (dominant condition) or the most subordinate monkey (subordinate condition) in their social group. Rewards were always the same for each monkey. The results showed that subjects knocked down more rods in the dominant condition than in the subordinate condition. But both monkeys pulled the tray equally often. The monkeys spontaneously allocated the six actions between them, apparently by separating the two components of knocking down the rods and pulling the tray. Interestingly, the monkey who pulled in the tray often received a prior glance from the other monkey. It is possible that this glance was a request or signal for the final action, but this requires further study.

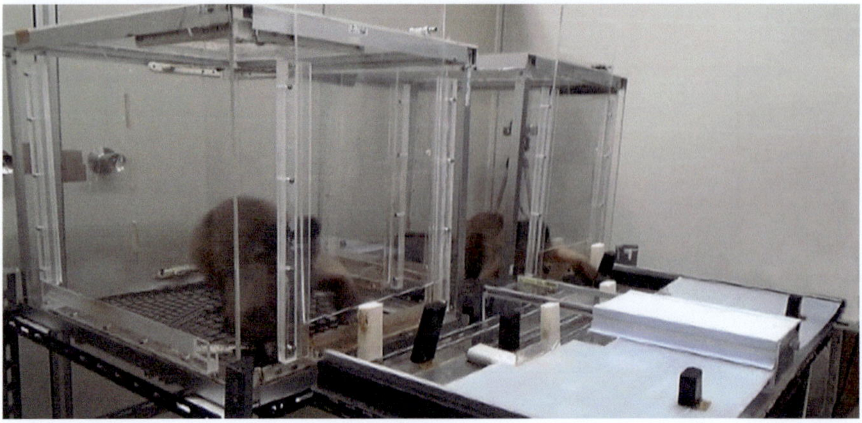

Fig. 4 Test situation for division of labour task. The apparatus had five rods and a baited tray. The monkeys had to manually knock down all five rods then pull in the tray. After completing all six actions, each monkey could collect its own reward on the tray

Perhaps the most important finding in this study was that no monkey was a free rider. There may be several reasons for this. For example, free-riding might not be in capuchin monkeys' repertoire, or the lack of anonymity might have kept them from adopting such a strategy, or perhaps the monkeys simply enjoyed working, and so on. Further research is needed to clarify how division of labour comes about, by carefully checking who works with whom and how rewards are shared and influences of individual personalities.

3.2 Summary of Prosocial Tendencies in Work Sharing

In this section, we described tufted capuchin monkeys' spontaneous sharing of the work required to obtain food alongside partners. Both subjects and partners engaged in the task, without free-riding. This suggests that the monkeys spontaneously share their efforts among partners in not only a collaborative pulling task (Mendres and de Waal 2000) but also a division of labour task (Hattori et al. 2005; Arahori et al. 2013). However, it is not clear how they divided work and why they did not free-ride on the partner's effort. It can be asked what kinds of situation might encourage free-riding. Further studies are required to address the "rules" the monkeys use for division of labour.

4 Evaluation of Third-Party Prosocial Tendencies

We have already seen that capuchin monkeys appear to have some sense of fairness regarding rewards and labour between partners. This can be seen as a rudimentary form of morality. In the case of humans, morality applies to third-party interactions that are irrelevant to the observers' direct benefit. Indirect reciprocity (Nowak and Sigmund, 1998) is sometimes thought to be a uniquely human style of helping that is based on the third-party evaluation and reputation. We examined capuchins sensitivity to third parties' prosocial tendencies and their fairness in the form of reciprocity in exchanges.

4.1 Third-Party Evaluation Based on Cooperative Attitude and Fairness

Anderson et al. (2013a) tested tufted capuchin monkeys' ability to evaluate prosocial tendencies of human actors in a context in which one actor asked for help. In the test situation, two actors, the requester (*A*) and the responder (*B*), sat side by side in front of a monkey across a table. *A* unsuccessfully tried to remove the lid of a transparent container to take out an object (Fig. 5). After several attempts, *A* looked at and showed the container to *B* to request help. In the helper condition, *B* supported thc container with both hands, thus enabling *A* to open it and take out the object, which *A* then showed to the monkey. In the non-helper condition, in response to *A*'s request for help, *B* briefly turned away and refused to help. *A* then continued trying to open the container, in vain. After these demonstrations, an opaque screen was placed between actors and monkey, and the container was removed from the table. Each actor placed a small piece of identical food on the palm of one hand, and then the screen was removed to allow the monkey to take the reward from one actor. The monkeys showed no preference for either actor in the helper condition, but they were significantly less likely to take the food offered by the unhelpful actor *B* than the requester in the non-helper condition (Exp. 1). Three additional experiments confirmed that the monkeys negatively evaluated an actor who explicitly refused to help. These results suggest that capuchin monkeys evaluate others by observing their interactions even if these are irrelevant to the monkeys' own benefit and that this evaluation translates into behaviour, namely, avoiding anti-social individuals.

Anderson et al. (2013b) also asked whether capuchin monkeys are sensitive to the reciprocity of exchanges between human actors. Again, two actors, *A* and *B*, sat side by side in front of a monkey across a table. In front of each actor was a pair of containers, one empty and the other containing three identical small balls. To start a trial, *A* held out her empty container towards *B* to request her balls. *B* picked up and put all three of her balls one by one into *A*'s container (Fig. 6). *B* then held out her empty container to *A* to request the latter's balls in return.

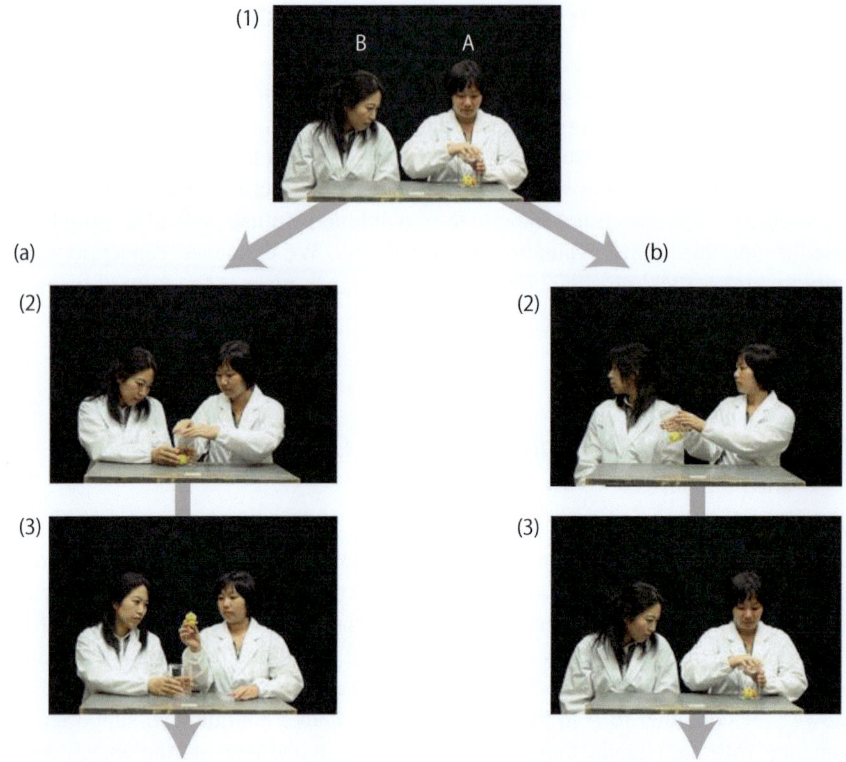

Fig. 5 The left panel (**a**) shows helper condition: (1) *A* tried but failed to open the container; (2) *A* asked *B* for help, *B* helped; (3) *A* succeeded and took the toy out of the container. The right panel (**b**) shows non-helper condition: (1) *A* tried but failed to open the container, (2) *A* asked *B* for help but *B* turned her face away to reject *A*'s request, (3) *A* tried but again failed to open the container

In the reciprocity condition, *A* transferred all three of her balls into *B*'s container; thus, in this condition, the actors simply exchanged their balls equally. In the non-reciprocity condition, *A* turned her head away then simply picked up each of her balls and returned them to the original container one by one; thus *A* ended up with six balls and *B* with none. In this condition, the final distribution of the possessions was not equal. Then, as in Anderson et al. (2013a), the monkeys were offered a choice between the two actors.

In the reciprocity condition, the monkeys showed no preference between the actors, but in the non-reciprocity condition, they were clearly less willing to take food from the non-reciprocating actor *A* (Exp. 1). These results suggest that capuchin monkeys have the ability to judge inter-individual reciprocity when they observe object exchanges between third parties and more specifically that they negatively evaluate non-reciprocators. Further comparative studies are required to clarify the evolution, mechanisms, and functions of such third-party evaluation.

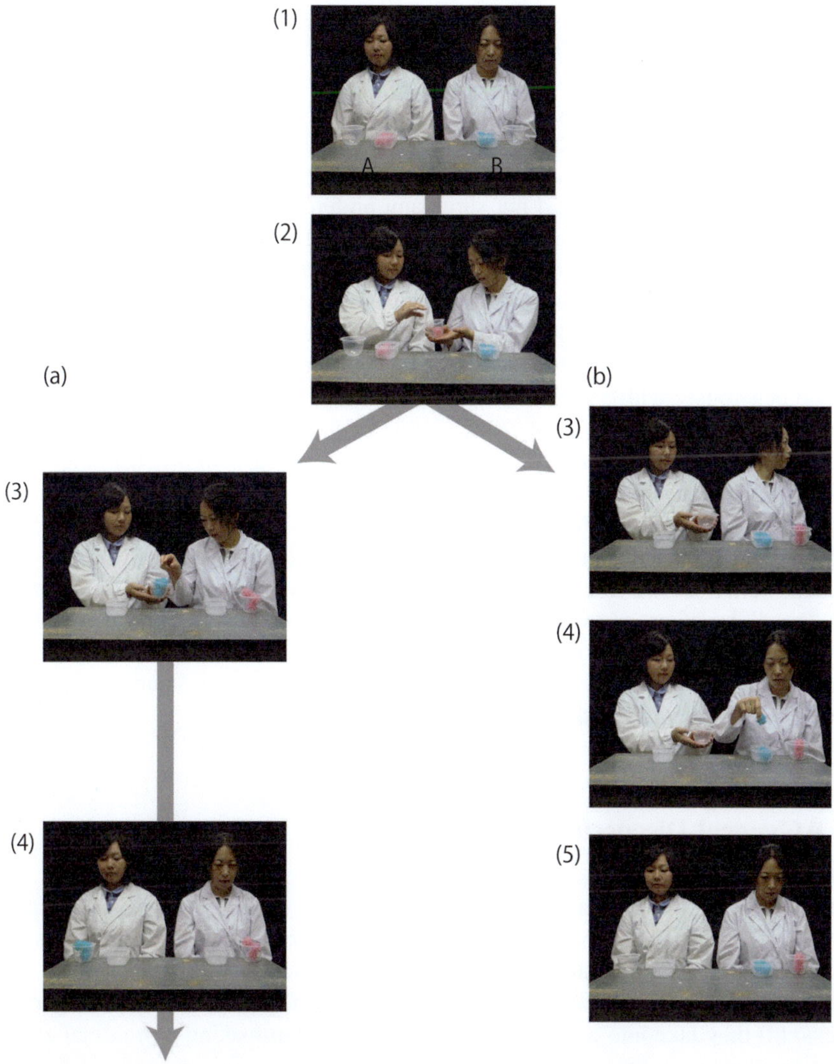

Fig. 6 The left panel (**a**) shows reciprocal condition: (1) each actor had three balls; (2) *B* gave all three balls to *A* upon request by *A*; (3) *B* then requested *A*'s balls, and *A* transferred her three balls to *B*; (4) *A* and *B* each ended up with three balls. The right panel (**b**) shows non-reciprocal condition: (1) each actor had three balls, (2) *B* gave all three balls to *A* upon request by *A*, (3) *B* then requested *A*'s balls but *A* turned away from *B* to reject the request, (4) *A* simply picked up each of her balls, (5) *A* ended up with all six balls, *B* none

4.2 Summary of Third-Party Evaluations

In this section, we presented studies showing that tufted capuchin monkeys avoided engaging with non-reciprocating and uncooperative human actors. This kind of negativity bias has also been reported in human infants (Baumeister et al. 2001;

Vaish et al. 2010; Hamlin et al. 2007, 2010). For example, Hamlin et al. (2010) showed that when a geometric figure (a sphere) appeared to try to climb a steep hill and another figure (a triangle) appeared to assist (the helping event) or a third figure (a square) appeared to hinder the sphere (the hindering event), infants as young as 3 months of age avoided looking at the hinderer. This kind of negativity bias might be shared between humans and some non-human primate species.

By contrast, capuchins did not show any preference for an actor who behaved helpfully or reciprocally to another. They may detect and judge only another's anti-social behaviour. Human children by preschool age have developed a sense of self (Aquino and Reed 2002; Boseovski, 2010) that includes the desire to be evaluated as "good" (Burhans and Dweck, 1995), and they behave in ways to maintain a positive reputation (Bryan et al. 2014; Fu et al. 2016). Moreover, 5-year-olds share goods more generously with a recipient when they are observed by the latter than when they are not (Engelmann et al. 2012, 2013; Leimgruber et al. 2012). In contrast, chimpanzees failed to modify their behaviour irrespective of whether they were observed by a groupmate or not (Engelmann et al. 2012). Although tufted capuchin monkeys are sensitive to being looked at by others (Hattori et al. 2007) and flexibly adjust their requesting behaviours depending on the observer's attentional state (Hattori et al. 2010), it is unknown whether they have any desire to be perceived as "good" by others or even whether they behave more prosocially when they are watched. If non-human primates lack a desire to be evaluated as "good" or "nice" by third parties, they may have no reason to evaluate prosocial third parties positively. Future studies could test this hypothesis.

5 General Discussion

In this chapter we summarized our empirical studies on food sharing, cooperation and division of labour, and third-party-based social evaluation of helpfulness and reciprocity in tufted capuchin monkeys. These studies have shown that in food contexts, (1) dominant monkeys may behave prosocially towards subordinate monkeys, (2) they take into account a partner's effort when making a prosocial choice and spontaneously divide effort with partners without free-riding and (3) they are less willing to accept food from humans they judge to be anti-social on the basis of the humans' third-party interactions. Below we discuss how a sense of fairness, empathy, and other affective processes may promote capuchin monkeys' prosocial behaviours.

5.1 Role of Sense of Fairness

Our results demonstrated that tufted capuchin monkeys are sensitive to differences in rewards between partners (Takimoto et al. 2010, 2012; Takimoto and Fujita 2011), which is consistent with previous studies (Brosnan and de Waal 2003; van

Wolkenten et al. 2007). The monkeys also took into account partners' help in their decision to behave prosocially (Takimoto and Fujita 2011; Takimoto 2012) and spontaneously divided the amount of labour required to obtain a reward with the partner, without free-riding (Arahori et al. 2013). Moreover, the monkeys' sensitivity extends to detection of anti-social third parties (Anderson et al. 2013a, b).

How does their sense of fairness promote capuchins' prosocial behaviour? In the platforms task (Takimoto et al. 2010), their prosocial behaviours were directed by dominants towards subordinates. This appears consistent with concepts of concern for "others' welfare" or "other-regarding preferences" (de Waal and Suchak 2010; Lakshminarayanan and Santos 2008). However, an alternative interpretation is possible. Instead of being tolerant of subordinates, monkeys might simply be more competitive towards partners of closer social rank to themselves. The monkeys' basic sense of equality concerning rewards and effort between partners might promote delivering same-quality reward to themselves and subordinates. But in the presence of a dominant partner, increased stress or increased psychological value of food for oneself might block any concern for equality. So far there is no demonstration of advantageous inequity aversion (IA) in capuchins. Advantageous IA arises from a negative emotion when an individual receives better rewards than a partner; it contributes to equitable outcomes among members of a group (Fehr and Schmidt 1999; Brosnan and de Waal 2014). This advantageous IA may be expressed in subjects' prosocial choice towards only a subordinate, but not a dominant partner. We should note that, in the platforms task, monkeys had only two options: not choosing the anti-social option of delivering low-value food to the partner meant choosing the only alternative of the same food for both monkeys. Thus, the apparent equity in this situation may be an artefact.

In future work, subjects should have three options that lead to the partner receiving a high-, middle-, or low-value food. Choosing the high-value option for the partner (and for themselves) would be strong evidence for advantageous IA. Such a sense of fairness, if it exists, ought to be effective in eliminating anti-social behaviour and maintaining fair distributions of food and effort within a group.

5.2 *Role of Empathy*

How does empathy relate to prosocial behaviour? Yamamoto et al. (2009) demonstrated that chimpanzees understand conspecific partners' goals and need for help, but they seldom help without first receiving a request from the partner. Liebal et al. (2014) found that with the exception of orangutans, prosocial behaviour by great apes (gorillas, chimpanzees, and bonobos) towards a human was not differentially affected by having observed that human stealing food from a conspecific (harm condition) or witnessing no such theft (no harm condition). The authors concluded that the apes' prosocial behaviour was not motivated by concern for others. Evidence suggests that young humans are generally more prosocial than apes. For example, Vaish et al. (2009) found that 18-month-old infants behaved prosocially towards a person victim by destroying or taking away her possessions. Kanakogi et al. (2013) reported that 10-month-olds showed rudimentary sympathy towards an animation character after observing it being attacked by another character. Does empathy facilitate prosocial behaviour in non-human animals as well as humans? To date the best evidence for this comes from rodents (Atsak et al. 2011; Ben-Ami Bartal et al. 2011, 2014, Sato et al. 2015). For instance, Ben-Ami Bartal et al. (2011) reported that rats opened a door to liberate a cagemate from an enclosed acrylic tube. The rats did not open the door when the tube was empty or when it contained an inanimate object. Perhaps in non-human animals empathy may promote prosocial behaviour in contexts where they can rescue others from pain or distress. We have focused on tufted capuchin monkeys' prosocial behaviour in contexts involving food. It remains unclear precisely how empathy relates to prosocial behaviour more generally. Further research is required to clarify similarities and differences in the functions of empathy between human and non-human primates.

5.3 *Other Affective States That May Underlie Prosocial Behaviour and Future Directions*

Affective states that promote prosocial behaviour in humans include trust, gratitude, guilt, and so on. In a recent study, Tasimi and Young (2016) demonstrated that 6- to 8-year-old children who recounted their own previous nice behaviour distributed their stickers more generously than children who recounted nice behaviour of another person. Thus, self-evaluation affects generosity in humans. We do not know to what extent non-human primates might engage in similar self-evaluation, trust others, feel gratitude towards others, expect to be appreciated by others, or feel guilty about not helping others. It will be important and revealing to examine how these affective states (or some precursors) promote prosocial behaviour in non-human primates.

Acknowledgements We are grateful for constructive comments and collaboration during these projects to Dr. James R. Anderson and Dr. Ayaka Takimoto. We also thank Minori Arahori, Benoit Bucher, and Risa Nagasawa for collaborating on projects. All our studies were supported by the Grants-in-Aid for Scientific Research Nos. 20220004, 25240020, and 15K12047 to KF and Nos. 21700289 to HK.

References

Amici F, Visalberghi E, Call J (2014) Lack of prosociality in great apes, capuchin monkeys and spider monkeys: convergent evidence from two different food distribution tasks. Proc R Soc B 281:20141699. https://doi.org/10.1098/rspb.2014.1699

Anderson JR, Kuroshima H, Takimoto A, Fujita K (2013a) Third-party social evaluation of humans by monkeys. Nat Commun 4:1561. https://doi.org/10.1038/ncomms2495

Anderson JR, Takimoto A, Kuroshima H, Fujita K (2013b) Capuchin monkeys judge third-party reciprocity. Cognition 127:140–146
Aquino K, Reed A 2nd (2002) The self-importance of moral identity. J Pers Soc Psychol 83: 1423–1440
Arahori M, Kuroshima H, Fujita K (2013) Capuchin monkeys (*Cebus apella*) spontaneously divide their efforts with a partner. Presented at 126th Congress of Kansai Psychological Association, Osaka, Japan, Nov. 2013.
Atsak P, Orre M, Bakker P, Cerliani L, Roozendaal B, Gazzola V et al (2011) Experience modulates vicarious freezing in rats: a model for empathy. PLoS One 6:e21855
Baumeister RF, Bratslavsky E, Finkenauer C, Vohs KD (2001) Bad is stronger than Good. Rev Gen Psychol 5:323–370
Basile BM, Hampton RR, Suomi SJ, Murray EA (2009) An assessment of memory awareness in tufted capuchin monkeys (*Cebus apella*). Anim Cogn 12:169–180
Beran MJ, Smith JD (2011) Information seeking by rhesus monkeys (*Macaca mulatta*) and capuchin monkeys (*Cebus apella*). Cognition 120:90–105
Ben-Ami Bartal I, Decety J, Mason P (2011) Empathy and pro-social behavior in rats. Science 334:1427–1430
Ben-Ami Bartal I, Rodgers DA, Bernardez Sarria MS, Decety J, Mason P (2014) Pro-social behavior in rats is modulated by social experience. Elife 3:e01385
Boesch C (1994) Cooperative hunting in wild chimpanzees. Anim Behav 48:653–667
Boesch C, Boesch H (1989) Hunting behavior of wild chimpanzees in the Tai National Park. Am J Phys Anthropol 78:547–573
Boseovski JJ (2010) Evidence for “rose-colored glasses”: an examination of the positivity bias in young children’s personality judgments. Child Dev Perspect 4:212–218
Brosnan SF, de Waal FB (2003) Monkeys reject unequal pay. Nature 425:297–299
Brosnan SF, de Waal FB (2014) Evolution of responses to (un)fairness. Science 346:1251776
Brosnan SF, Schiff HC, de Waal FB (2005) Tolerance for inequity may increase with social closeness in chimpanzees. Proc R Soc B 272:253–258
Bryan CJ, Master A, Walton GM (2014) “Helping” versus “being a helper”: invoking the self to increase helping in young children. Child Dev 85:1836–1842
Burhans KK, Dweck CS (1995) Helplessness in early childhood: the role of contingent worth. Child Dev 66:1719–1738
Coelho CG, Falotico T, Izar P, Mannu M, Resende BD, Siqueira JO et al (2015) Social learning strategies for nut-cracking by tufted capuchin monkeys (*Sapajus spp.*). Anim Cogn 18:911–919
Cronin KA, Schroeder KK, Snowdon CT (2010) Prosocial behaviour emerges independent of reciprocity in cottontop tamarins. Proc R Soc B 277:3845–3851
Custance D, Whiten A, Fredman T (1999) Social learning of an artificial fruit task in capuchin monkeys (*Cebus apella*). J Comp Psychol 113:13–23
de Waal FB (1997) Food transfers through mesh in brown capuchins. J Camp Psychol 111:370–378
de Waal FBM (2008) Putting the altruism back into altruism: the evolution of empathy. Annu Rev Psychol 59(1):279–300
de Waal FB (2012) The antiquity of empathy. Science 336:874–876
de Waal FB, Leimgruber K, Greenberg AR (2008) Giving is self-rewarding for monkeys. Proc Natl Acad Sci USA 105:13685–13689
De Waal FB, Suchak M (2010) Prosocial primates: selfish and unselfish motivations. Phil Trans R Soc B 365:2711–2722
Decety J, Bartal IB, Uzefovsky F, Knafo-Noam A (2016) Empathy as a driver of prosocial behaviour: highly conserved neurobehavioural mechanisms across species. Phil Trans R Soc B 371: 20150077. https://doi.org/10.1098/rstb.2015.0077
Eisenberg N, Mussen PH (1989) The roots of prosocial behavior in children. Cambridge University Press, Cambridge
Engelmann JM, Herrmann E, Tomasello M (2012) Five-year-olds, but not chimpanzees, attempt to manage their reputations. PLoS One 7:e48433

Engelmann JM, Over H, Herrmann E, Tomasello M (2013) Young children care more about their reputation with ingroup members and potential reciprocators. Dev Sci 16:952–958

Fehr E, Schmidt KM (1999) A theory of fairness, competition, and cooperation. Q J Econ 114:817–868

Fragaszy D, Visalberghi E (2004) Socially biased learning in monkeys. Learn Behav 32:24–35

Fu G, Heyman GD, Qian M, Guo T, Lee K (2016) Young children with a positive reputation to maintain are less likely to cheat. Dev Sci 19:275–283

Hamlin JK, Wynn K, Bloom P (2007) Social evaluation by preverbal infants. Nature 450:557–559

Hamlin JK, Wynn K, Bloom P (2010) Three-month-olds show a negativity bias in their social evaluations. Dev Sci 13:923–929

Hattori Y, Kuroshima H, Fujita K (2005) Cooperative problem solving by tufted capuchin monkeys (*Cebus apella*): spontaneous division of labor, communication, and reciprocal altruism. J Comp Psychol 119:335–342

Hattori Y, Kuroshima H, Fujita K (2007) I know you are not looking at me: capuchin monkeys (*Cebus apella*) sensitivity to human attentional states. Anim Cogn 10:141–148

Hattori Y, Kuroshima H, Fujita K (2010) Tufted capuchin monkeys (*Cebus apella*) show understanding of human attentional states when requesting food held by a human. Anim Cogn 13: 87–92

Hopper L, Lambeth S, Schapiro S (2013) The ontogeny of social comparisons in rhesus macaques (*Macaca mulatta*). J Primatol 2:109. https://doi.org/10.4172/2167-6801.1000109

Hopper LM, Lambeth SP, Schapiro SJ, Brosnan SF (2014) Social comparison mediates chimpanzees responses to loss, not frustration. Anim Cogn 17:1303–1311

Hull CL (1943) Principles of behavior: an introduction to behavior theory. Appleton-Century-Crofts, New York

Kanakogi Y, Okumura Y, Inoue Y, Kitazaki M, Itakura S (2013) Rudimentary sympathy in preverbal infants: preference for others in distress. PLoS One 8(6):e65292

Kuroshima H, Kuwahata H, Fujita K (2008) Learning from others mistakes in capuchin monkeys (*Cebus apella*). Anim Cogn 11:599–609

Kuroshima H, Nagatomo R, Fujita K (2012) Occurrence factors of prosocial behavior in capuchin monkeys (Cebus apella): by using token-exchange paradigm. 72th Annual meeting of the Japanese Society for Animal Psychology, Hogo, Japan, May 2012

Lakshminarayanan VR, Santos LR (2008) Capuchin monkeys are sensitive to others welfare. Curr Biol 18:R999–R1000

Leimgruber KL, Shaw A, Santos LR, Olson KR (2012) Young children are more generous when others are aware of their actions. PLoS One 7:e48292

Liebal K, Vaish A, Haun D, Tomasello M (2014) Does sympathy motivate prosocial behaviour in great apes? PLoS One 9:e84299

Massen JJ, Luyten IJ, Spruijt BM, Sterck EH (2011) Benefiting friends or dominants: prosocial choices mainly depend on rank position in long-tailed macaques (*Macaca fascicularis*). Primates 52:237–247

Massen JJ, van den Berg LM, Spruijt BM, Sterck EH (2010) Generous leaders and selfish underdogs: pro-sociality in despotic macaques. PLoS One 5:e9734

Massen JJ, van Den Berg LM, Spruijt BM, Sterck EH (2012) Inequity aversion in relation to effort and relationship quality in long-tailed macaques (*Macaca fascicularis*). Am J Primatol 74: 145–156

Mendres KA, de Waal FB (2000) Capuchins do cooperate: the advantage of an intuitive task. Anim Behav 60:523–529

Mitani JC, Watts DP (2001) Why do chimpanzees hunt and share meat? Anim Behav 61:915–924

Nowak MA, Sigmund K (1998) Evolution of indirect reciprocity by image scoring. Nature 393(6685):573–577

Ottoni EB, de Resende BD, Izar P (2005) Watching the best nutcrackers: what capuchin monkeys (*Cebus apella*) know about others tool-using skills. Anim Cogn 8:215–219

Paukner A, Anderson JR, Fujita K (2006) Redundant food searches by capuchin monkeys (*Cebus apella*): a failure of metacognition? Anim Cogn 9:110–117

Preston SD, de Waal FB (2002) Empathy: its ultimate and proximate bases. Behav Brain Sci 25: 1–72

Sato N, Tan L, Tate K, Okada M (2015) Rats demonstrate helping behavior toward a soaked conspecific. Anim Cogn 18:1039–1047

Sterck EH, Olesen CU, Massen JJ (2015) No costly prosociality among related long-tailed macaques (*Macaca fascicularis*). J Comp Psychol 129:275–282

Suchak M, de Waal FBM (2012) Monkeys benefit from reciprocity without the cognitive burden. Proc Natl Acad Sci USA 109:15191–15196

Takimoto A (2012) Origin of moral: experimental analysis of higher order emotion in primates. (Unpublished doctoral dissertation), Kyoto university, Kyoto, Japan

Takimoto A, Kuroshima H, Fujita K (2010) Capuchin monkeys (*Cebus apella*) are sensitive to others reward: an experimental analysis of food-choice for conspecifics. Anim Cogn 13: 249–261

Takimoto A, Fujita K (2011) I acknowledge your help: capuchin monkeys sensitivity to others labor. Anim Cogn 14:715–725

Tasimi A, Young L (2016) Memories of good deeds past: the reinforcing power of prosocial behavior in children. J Exp Child Psychol 147:159–166

Vaish A, Carpenter M, Tomasello M (2009) Sympathy through affective perspective taking and its relation to prosocial behavior in toddlers. Dev Psychol 45:534–543

Vaish A, Carpenter M, Tomasello M (2010) Young children selectively avoid helping people with harmful intentions. Child Dev 81:1661–1669

van Wolkenten M, Brosnan SF, de Waal FB (2007) Inequity responses of monkeys modified by effort. Proc Natl Acad Sci USA 104:18854–18859

Warneken F, Chen F, Tomasello M (2006) Cooperative activities in young children and chimpanzees. Child Dev 77:640–663

Warneken F, Hare B, Melis AP, Hanus D, Tomasello M (2007) Spontaneous altruism by chimpanzees and young children. PLoS Biol 5:e184

Yamamoto S, Humle T, Tanaka M (2009) Chimpanzees help each other upon request. PLoS One 4: e7416

Yamamoto S, Tanaka M (2009) Do chimpanzees (*Pan troglodytes*) spontaneously take turns in a reciprocal cooperation task? J Comp Psychol 123:242–249

Yamamoto S, Takimoto A (2012) Empathy and fairness: psychological mechanisms for eliciting and maintaining prosociality and cooperation in primates. Soc Justice Res 25:233–255

Understanding Empathy from Interactional Synchrony in Humans and Non-human Primates

Lira Yu, Yuko Hattori, Shinya Yamamoto, and Masaki Tomonaga

Abstract Humans are highly responsive to social partners. Either consciously or unconsciously, humans adjust one's own behavior to match with those of others. Moreover, humans can predict others' intention or desire and often behave in a prosocial way to help themselves. A comparative approach is one of the powerful tools to understand the evolutionary origins of these social behaviors in humans, including emotional contagion, sympathetic concern, perspective-taking and targeted helping behaviors, and its relationship to each other. Among these social behaviors, an ability for the coordinated movements with others (i.e., interpersonal coordination) has been postulated as the most essential to the other social behaviors. Interpersonal coordination includes two types of behavior: mimicry and interactional synchrony. Mimicry is matching a type of the behavior, such as body postures or facial expressions. In contrast, interactional synchrony is matching the timing of behavior. Relative to the studies on mimicry, there are few studies on interactional synchrony in non-human primate species possibly due to the difficulties in establishing a methodology. In this chapter, we present gradually the increasing number of studies on interactional synchrony in humans and non-human primates. The findings demonstrate that the ability for interactional synchrony is shared across humans, chimpanzees, bonobos, and macaques. The latest findings from a direct comparison between humans and chimpanzees further demonstrate that there are both similarities and significant differences on the ability between two species. At the end of this chapter, we suggest a future direction of the comparative studies on interactional synchrony.

L. Yu (✉)
Wildlife Research Center, Kyoto University, Kyoto, Japan
e-mail: yu.lira.2u@kyoto-u.ac.jp

Y. Hattori · M. Tomonaga
Primate Research Institute, Kyoto University, Inuyama, Japan
e-mail: tomonaga.masaki.4m@kyoto-u.ac.jp

S. Yamamoto
Institute for Advanced Study, Kyoto University, Kyoto, Japan

L. D. Di Paolo et al. (eds.), *Evolution of Primate Social Cognition*, Interdisciplinary Evolution Research 5, https://doi.org/10.1007/978-3-319-93776-2_4

Keywords Social interaction · Rhythmic coordination · Comparative cognition · Empathy · Synchrony

1 Introduction

Our body and mind begin to act differently in the existence of other people nearby. Either consciously or unconsciously, humans adjust one's own body postures, body movements, or facial expressions to match with those of others. Humans can even predict others' intention or desire. Moreover, they often behave in a prosocial way to help themselves. These social behaviors suggest that humans are highly responsive to social partners.

1.1 Social Behaviors in Theoretical Studies

Both action and cognition involved in social interaction have been discussed in psychology. The term *empathy* was often used as an umbrella term that included multiple levels of social behaviors such as emotional contagion, sympathetic concern, perspective-taking, and targeted helping behavior.

In theoretical studies, primatologists and ethologists have proposed that such social behaviors have a certain relationship to each other. For example, the Russian doll model (de Waal 2008, 2009), which is among the best-known frameworks for understanding the evolution of empathy, structures the social behaviors in a layered nature. The three layers are core layer, which consists of emotional contagion induced by automatic perception and action mechanism; middle layer, which covers sympathetic concern and consolation; and the most outer layer, which comprises perspective taking and targeted helping. This model suggests a stepwise development of empathy in that the outer layers evolve later and require more complex cognitive abilities than its core.

Recently, studies with chimpanzees have showed that the ability for perspective taking does not necessarily lead to proactive helping behavior (Yamamoto et al. 2012). Thus, another model has been proposed, called three-factor model (Yamamoto 2017). Unlike the Russian doll model, this model does not assume any simple linear progression in social behaviors. Instead, the model suggests that the three factors, namely, matching with others, understanding of others, and prosociality, are independent of each other and thus must be separated to understand empathy-related phenomena.

Despite the difference in structure of social behaviors, the two models share the idea that matching one's behavior with others (herein after, interpersonal coordination) is widely found in various animals (i.e., mammals, birds, and fishes), and it is essential for the other social abilities or behaviors in humans and non-human animals.

1.2 Relationship Between Interpersonal Coordination and Other Social Behaviors

Several experimental studies have reported the relationship between interpersonal coordination and other social behaviors in humans. When one's movements are mimicked or synchronized by others, one tends to demonstrate increased affiliation or rapport toward them (Chartrand and Bargh 1999; Hove and Risen 2009) or show more prosocial behavior (Van Baaren et al. 2004; Wiltermuth and Heath 2009). These positive social consequences of interpersonal coordination are also reported in human children aged 4 years old (Kirschner and Tomasello 2010), and they seem to occur beginning from infancy at 14 months old (Cirelli et al. 2014). In sum, the findings suggest that interpersonal coordination plays a positive social role in maintaining harmonious social relationship in humans from an early development. Here, we hypothesized that if this behavior is a socially adaptive behavior, it may be shared with other highly social animal species. Moreover, we assumed that investigating similarity and difference in interpersonal coordination between humans and non-human primates may help us to understand more in depth about our ability to adapt in social environment.

1.3 Types of Interpersonal Coordination

Interpersonal coordination includes two types of behavior: mimicry and interactional synchrony (Bernieri and Rosenthal 1991). In mimicry, which is a matching type of behavior, there are several studies from both humans and non-human primates: neonatal facial mimicry in humans (Meltzoff and Moore 1977), chimpanzees (Myowa-Yamakoshi et al. 2004), and macaques (Ferrari et al. 2006); automatic facial mimicry in humans (Dimberg 1982; Dimberg et al. 2000; Sato and Yoshikawa 2007); and contagious yawning in chimpanzees (Anderson et al. 2004; Campbell et al. 2009) and macaques (Paukner and Anderson 2006). Comparative studies from these non-human primates suggest that an ability for matching facial expressions is shared between humans and non-human primates (chimpanzees and macaques). In contrast, there is limited knowledge on interactional synchrony, which is matching the timing of a movement. This might be attributed to the difficulties in establishing a methodology that quantifies the behavior, especially in non-human primate species.

In this chapter, we will introduce gradually the increasing number of studies on interactional synchrony in humans and non-human primates. First, interactional synchrony in human literatures will be introduced. And then, the studies on non-human primates will be presented. Lastly, we will summarize the findings on both humans and non-human primates from a comparative approach.

2 Interactional Synchrony in Humans

Humans have a remarkable ability to synchronize their movement with that of others (Sebanz et al. 2006). For example, in one's daily life, a person tends to match his/her stride and speed with those of others when he/she walk with them at a close distance (Zivotofsky and Hausdorff 2007; van Ulzen et al. 2008). This synchronized walking occurs even though people have distinctive spontaneous walking speed from each other. How does synchronized walking occur then, and what is a typical form of this coordination? In human literatures, two main experimental methods were used to examine the automatic nature of interactional synchrony in humans. In one method, the participants were instructed to maintain their own movements, such as speed, trajectory, or position of the movements, while facing an interacting partner, who is also producing the same or similar movements (Schmidt and O'Brien 1997; Kilner et al. 2003; Issartel et al. 2007; Richardson et al. 2007; Yun et al. 2012). In another method, the participants received no specific verbal instruction, to examine their spontaneous and dynamic changes in movements (Néda et al. 2000; Zivotofsky and Hausdorff 2007; van Ulzen et al. 2008; Oullier et al. 2008). Here, we present a study from each experimental method in more detail.

Schmidt and O'Brien (1997) first demonstrated automatic interactional synchrony in humans under a laboratory setup. In the experiment, adult subjects were instructed to swing a handheld pendulum with their own comfortable frequency as if they could continue doing so for the entire day. An interesting part of this study is that the authors prepared different lengths of the pendulums so that natural frequencies of the pendulums between two participants in a pair become different. In the test, the two participants sat side-by-side and were asked to maintain the comfortable frequency while looking at a partner's moving pendulum. Results showed that the participants could not keep their own pendulum movement when watching a partner's pendulum movement. Moreover, the findings demonstrated that the two pendulums tended to fall into either in- or antiphase synchrony, which were reported to be stable forms of interpersonal coordination (Schmidt et al. 1990; Schmidt and Turvey 1994). The findings suggest that the changes toward synchronized movements are inevitable in humans.

In Oullier et al.'s (2008) investigation of the dynamic nature of interactional synchrony in humans, participants were asked to move their right index finger up and down with their own preferred frequency. After measuring each participant's intrinsic frequency independently, two participants who showed different frequencies of movements were paired. The two participants were then asked to perform the finger movements concurrently, without any further specific instruction, while facing each other. This was a different procedure from the study mentioned above (c.f., Schmidt and O'Brien 1997). However, results were similar with the previous studies, showing that frequency change and phase matching (mostly, inphase synchrony) emerge spontaneously between the interacting participants. The authors also demonstrated that the coordinated movement remains even when visual information of a partner's movement is no longer available, suggesting a function of "social memory" on interactional synchrony in humans.

3 Interactional Synchrony in Non-human Primates

Field observations have reported that wild chimpanzees show spatiotemporal coordination during territorial boundary patrol (Goodall 1986; Boesch and Boesch 2000; Watts and Mitani 2001) or produce vocal chorusing (Arcadi 1996; Mitani and Gros-Louis 1998; Fedurek et al. 2013). How is it differ from or similar to the movement coordination observed in humans? Here, we present three studies examining an ability for interactional synchrony in non-human primate species from a comparative approach.

Nagasaka et al. (2013) first demonstrated spontaneous interactional synchrony in Japanese macaques (*Macaca fuscata*) in a laboratory setup. The monkeys were trained independently to produce repetitive left-to-right button-pressing movements with a preferred hand. Auditory feedbacks were presented for each button-pressing. After the monkeys were able to produce the repetitive movements with their own preferred frequency, two monkeys were set in a pair, faced each other, and performed the movement concurrently. Results showed that the monkeys in a pair change their frequencies of the movements and show inphase synchrony spontaneously. The results also demonstrated that visual cues of a conspecific partner's movement were required for the synchronized movement in Japanese macaques.

Yu and Tomonaga (2015) demonstrated that one chimpanzee (*Pan troglodytes*) gradually and spontaneously augment her rhythmic tapping movement to a conspecific partner's tapping movement. In this experiment, a pair of the chimpanzees sat side-by-side, and auditory feedbacks, which correspond to each participant's tapping movement, were presented. In Yu and Tomonaga (2016), a similar method from the study performed on Japanese macaques was conducted to further investigate whether spontaneous interactional synchrony can occur in chimpanzees. Results demonstrated that chimpanzees show spontaneous tapping tempo change in response to a conspecific partner's tapping movement, when the chimpanzees in a dyad face each other and produce rhythmic tapping movements concurrently.

Large and Gray (2015) showed spontaneous synchronized drumming in a bonobo (*Pan paniscus*). Unlike the studies mentioned above, their study examined cross-species interaction between a human experimenter and one bonobo individual. The human experimenter drummed with a certain speed and examined the bonobo's spontaneous drumming response. Results demonstrated that the bonobo can flexibly change the drumming speed and show inphase synchrony, especially when the experimenter's drumming was near to her spontaneous drumming speed.

4 Comparison of Species

The broadest summary so far is that the ability for interactional synchrony is shared across humans, chimpanzees, bonobos, and macaques. All those primate species showed spontaneous tempo change and/or phase matching of the rhythmic movements when they interact with a social partner. These findings demonstrate that humans share the basic ability with other primate species. However, the differences in quality between species are not yet clear. This may be because few studies have directly compared humans and non-human primates under the same experimental conditions.

Here, we present our previous study of a direct comparison between humans and chimpanzees (Yu and Tomonaga, unpublished). Four pairs of human adults were examined under the same experimental setup, as that used for the chimpanzees (see Yu and Tomonaga 2016). Dyads consisted of the participants who were familiar with each other. This was consistent with the chimpanzees, who were also familiar with each other as they were living in a group in an enriched in- and outdoor compounds (Matsuzawa et al. 2006). To make the procedure consistent with the chimpanzees, no specific verbal instruction was given to the human participants. Results from the current direct comparison showed that both humans and chimpanzees are similar in that they spontaneously change the tempo of the rhythmic movements in response to a partner's movement. As a result, tempo convergence between the participants was observed in each dyad. Interestingly, there was a general tendency in this tempo convergence. In both species, one individual in each pair significantly adjusted tempo of the tapping movement toward a partner's tapping movement. Besides these similarities, the current study also showed the difference between humans and chimpanzees. They were differed in their strength of the tempo convergence (see Fig. 1). This finding suggests that humans are more precise and quick in their tempo adaptation than the chimpanzees, regardless of initial differences in the rhythmic movements between the participants in a pair.

In a following study, we examined the effect of visual cues of an interacting partner's movement on the spontaneous tempo adjustment (Yu and Tomonaga 2018). Two pairs were tested for humans and chimpanzees, respectively. Two test conditions, visible and invisible, were prepared (see Fig. 2). In the visible condition, two participants in a pair produced the tapping movement while they were able to see each other's movement. In contrast, in the invisible condition, the participants were not able to see each other's movement because a visual barrier was placed between them. Auditory feedbacks, corresponding to each participant's tapping movements, were presented in both conditions. Results showed that both humans and chimpanzees significantly change the tapping tempo even in the invisible condition. Moreover, the tempo convergence was observed in both invisible and visible condition. This finding suggests that auditory cues were sufficient for the tempo adaptation, whereas additive visual cues of an interacting partner's movement have little effect on both species. This finding was consistent with the previous studies, reporting that humans are more strongly attracted to auditory cues than visual cues when producing

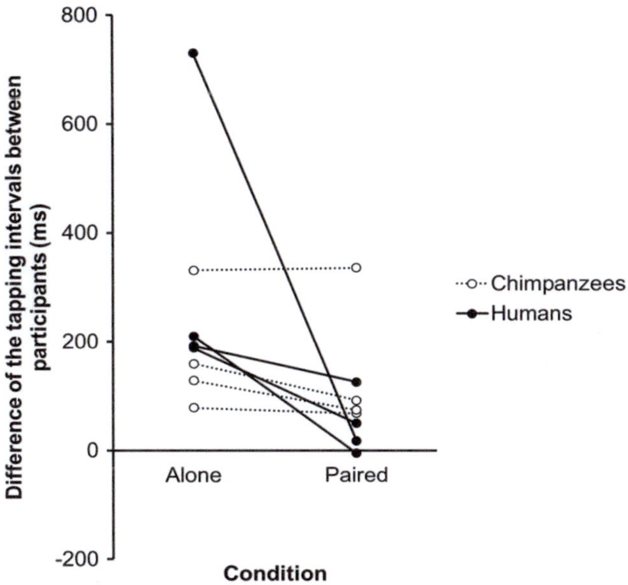

Fig. 1 A direct comparison between humans and chimpanzees showed a difference on strength of the tempo convergence. Each dot represents a difference of the mean tapping intervals between two participants in each pair for alone (baseline) and paired (test) conditions. The dependent variable for both conditions was calculated by subtracting the mean tapping intervals of a relatively fast-tapping individual in the alone condition from those of his/her partner. The dots connected by lines represent data from the same pairs. Chimpanzee's data are from Yu and Tomonaga (2016)

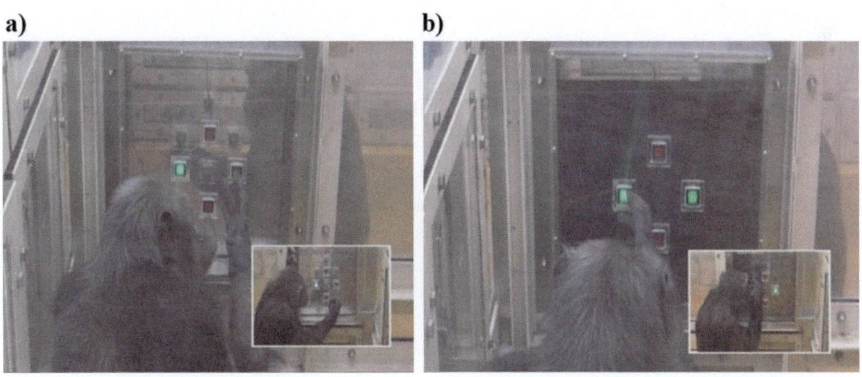

Fig. 2 Two chimpanzees facing each other are concurrently producing the rhythmic tapping movements under two conditions: (**a**) visible and (**b**) invisible conditions. In the invisible condition, a black sheet blocked visual cues of a partner's movement. The small photos in the right-down corner show the other tested chimpanzee of the dyad

rhythmic movement (Repp and Penel 2002, 2004) and chimpanzees can synchronize their rhythmic movements to a nonsocial auditory rhythm (Hattori et al. 2013, 2015).

4.1 Synchrony in a Nonsocial Context

We have presented the studies on *interactional synchrony* in humans and non-human primates. Meaning, we focused on the synchronized movement that often occurs when individuals *interact* with another individual. However, the synchronized movement also occurs notably in a nonsocial context.

Many of the studies have examined the synchronized movement in a nonsocial context and focused on the effect of auditory cues. The underlying hypothesis of these studies is that a neural mechanism for the ability to synchronize movements with external cues is shared with vocal learning ability (Patel 2006; Patel et al. 2009). Recently, however, there is growing evidence that chimpanzees and macaques can also coordinate their movement to auditory rhythms, even though they are not active vocal learners. Here, we introduce the findings from humans, chimpanzees, and macaques.

When hearing musical beat, humans tend to produce head bobbing, body rocking, or finger- or foot-tapping to the beat (Wallin et al. 2000). This spontaneous musical response suggests that humans have a strong sensitivity in perceiving auditory rhythms. In chimpanzees, Hattori et al. (2013, 2015) demonstrated that chimpanzees can also spontaneously synchronize their finger-tapping movements to auditory rhythms, especially the rhythms close to their own spontaneous tapping tempo. In macaques, a review article by Merchant and Honing (2014) well described the major findings from the macaques in comparison with humans and chimpanzees. The following characteristics were observed in macaques: (1) macaques show a large bias toward visual cues rather than auditory cues, to drive their rhythmic tapping movements (Zarco et al. 2009), and (2) macaques rarely show inphase synchrony, although they can adjust their tempo of movements in response to nonsocial rhythmic cues (Zarco et al. 2009; Konoike et al. 2012). For further discussion, please refer to Merchant and Honing (2014).

4.2 Social Partner Versus Nonsocial Partner

Humans respond differently toward movements from a social partner and a nonsocial partner. For example, Kilner et al. (2003) reported that the participants' movement was more affected by a human partner's movement than a robot arm's. Moreover, Kirschner and Tomasello (2009) reported that preschool children produce more precise joint drumming when they drum with a human partner than with a drumming machine or a drum sound coming from a speaker. Relative to the human literatures, few behavioral studies in non-human primates exist. However, there is evidence in macaques that their brain processes biological and non-biological movements differently (Oram and Perrett 1994), like in humans (Frith and Frith 1999; Allison et al. 2000; Grossman et al. 2000; Grèzes et al. 2001).

5 Conclusion

In this chapter, we reviewed empirical evidences showing interactional synchrony in humans and non-human primates (chimpanzees, bonobos, and macaques). Although there are a limited number of studies on non-human primates, they provided valuable evidences to understand the nature of interactional synchrony in primate species. Here, we note that most studies mentioned here required upper-limb movements (i.e., a finger-tapping, button-pressing, or drumming). This may be because of the dominance on using hands in humans when we interact with others. In further comparative studies, we expect to consider more on non-human primates' natural behavior repertoire used in their social context when establishing a methodology. This may provide an opportunity to understand another aspect of social interactions in primate species.

Both *interactional synchrony* and *mimicry* have adaptive function in humans. However, there is one notable difference between these two behaviors. Mimicry shows positive social consequences, such as increased affiliation or rapport, only when it occurs in an unnoticed way (c.f., Chartrand and Bargh 1999). In contrast, interactional synchrony shows the increased rapport or cooperative behavior even when it is produced actively or noticed explicitly (c.f., Hove and Risen 2009; Wiltermuth and Heath 2009; Kirschner and Tomasello 2010). These findings suggest that interactional synchrony can produce more continuous positive social relationship with an interacting partner than mimicry. Therefore, like rhythmic play or dance in human cultures, the ability for interactional synchrony can be used actively for maintaining harmonious social relationships in primate species. We believe that this is among the important reasons for examining interactional synchrony in humans and non-human primate species. Both observational and experimental studies from a comparative approach may help us to further understand how and why humans have acquired the ability and developed it in their evolutionary history.

In the beginning of this chapter, two theoretical models, namely, the Russian doll model (de Waal 2008, 2009) and three-factor model (Yamamoto 2017), were discussed, to emphasize that interpersonal coordination, including interactional synchrony and mimicry, is considered essential for other social behaviors in both models. We assume that sharing time and space with others through interpersonal coordination may facilitate the ability for sharing emotions with others. However, we believe that comparative studies and discussions are needed further to understand how actions involved in social interaction relate to other social behaviors and social cognition in primate species.

Acknowledgments We thank T. Matsuzawa and other staff members at the Language and Intelligence Section and Center for Human Evolution Modeling Research of Kyoto University Primate Research Institute for their support and daily care of the chimpanzees. We would like to acknowledge financial supports by Japan Society for the Promotion of Science Grants-in-Aid for Scientific Research (244525 and 16F16001 to L.Y.; 16H01487 to Y.H.; 26118509, 15H05309, 15H01619, and 17H05862 to S.Y.; and 15H05709, 16H06283, 20002001, 23220006, and 24000001 to M.T.), JSPS-CCSN, Global COE programs (A06, D07), and the JSPS Leading Graduate Program in Primatology and Wildlife Science (U04) at Kyoto University.

References

Allison T, Puce A, McCarthy G (2000) Social perception from visual cues: role of the STS region. Trends Cogn Sci 4(7):267–278

Anderson JR, Myowa-Yamakoshi M, Matsuzawa T (2004) Contagious yawning in chimpanzees. Proc R Soc Lond B Biol Sci 271(Suppl 6):S468–S470

Arcadi AC (1996) Phrase structure of wild chimpanzee pant hoots: patterns of production and interpopulation variability. Am J Primatol 39(3):159–178

Bernieri FJ, Rosenthal R (1991) Interpersonal coordination: behavior matching and interactional synchrony. In: Feldman RS, Rimé B (eds) Fundamentals of nonverbal behavior. Cambridge University Press, Cambridge, pp 401–432

Boesch C, Boesch H (2000) The chimpanzees of the Tai forest. Oxford University Press, Oxford

Campbell MW, Carter JD, Proctor D, Eisenberg ML, de Waal FB (2009) Computer animations stimulate contagious yawning in chimpanzees. Proc R Soc Lond B Biol Sci. https://doi.org/10.1098/rspb.2009.1087

Chartrand TL, Bargh JA (1999) The chameleon effect: the perception-behavior link and social interaction. J Pers Soc Psychol 76(6):893–910

Cirelli LK, Einarson KM, Trainor LJ (2014) Interpersonal synchrony increases prosocial behavior in infants. Dev Sci 17(6):1003–1011

de Waal FBM (2008) Putting the altruism back into altruism: the evolution of empathy. Annu Rev Psychol 59:279–300

de Waal FBM (2009) The age of empathy: nature's lessons for a kinder society. Harmony Books, New York, NY

Dimberg U (1982) Facial reactions to facial expressions. Psychophysiology 19(6):643–647

Dimberg U, Thunberg M, Elmehed K (2000) Unconscious facial reactions to emotional facial expressions. Psychol Sci 11(1):86–89

Fedurek P, Machanda ZP, Schel AM, Slocombe KE (2013) Pant hoot chorusing and social bonds in male chimpanzees. Anim Behav 86(1):189–196

Ferrari PF, Visalberghi E, Paukner A, Fogassi L, Ruggiero A, Suomi SJ (2006) Neonatal imitation in rhesus macaques. PLoS Biol 4(9):e302

Frith CD, Frith U (1999) Interacting minds-a biological basis. Science 286(5445):1692–1695

Goodall J (1986) The chimpanzees of Gombe. Harvard University Press, Cambridge, MA

Grèzes J, Fonlupt P, Bertenthal B, Delon-Martin C, Segebarth C, Decety J (2001) Does perception of biological motion rely on specific brain regions? NeuroImage 13(5):775–785

Grossman ED, Donnelly M, Price R, Pickens D, Morgan V, Neighbor G, Blake R (2000) Brain areas involved in perception of biological motion. J Cogn Neurosci 12(5):711–720

Hattori Y, Tomonaga M, Matsuzawa T (2013) Spontaneous synchronized tapping to an auditory rhythm in a chimpanzee. Sci Rep 3:1566

Hattori Y, Tomonaga M, Matsuzawa T (2015) Distractor effect of auditory rhythms on self-paced tapping in chimpanzees and humans. PLoS One 10(7):e0130682

Hove MJ, Risen JL (2009) It's all in the timing: interpersonal synchrony increases affiliation. Soc Cogn 27(6):949–960
Issartel J, Marin L, Cadopi M (2007) Unintended interpersonal co-ordination: "can we march to the beat of our own drum?". Neurosci Lett 411(3):174–179
Kilner JM, Paulignan Y, Blakemore SJ (2003) An interference effect of observed biological movement on action. Curr Biol 13(6):522–525
Kirschner S, Tomasello M (2009) Joint drumming: social context facilitates synchronization in preschool children. J Exp Child Psychol 102(3):299–314
Kirschner S, Tomasello M (2010) Joint music making promotes prosocial behavior in 4-year-old children. Evol Human Behav 31(5):354–364
Konoike N, Mikami A, Miyachi S (2012) The influence of tempo upon the rhythmic motor control in macaque monkeys. Neurosci Res 74(1):64–67
Large EW, Gray PM (2015) Spontaneous tempo and rhythmic entrainment in a bonobo (Pan paniscus). J Comp Psychol 129(4):317
Matsuzawa T, Tomonaga M, Tanaka M (2006) Cognitive development in chimpanzees. Springer, Tokyo
Meltzoff AN, Moore MK (1977) Imitation of facial and manual gestures by human neonates. Science 198(4312):75–78
Merchant H, Honing H (2014) Are non-human primates capable of rhythmic entrainment? Evidence for the gradual audiomotor evolution hypothesis. Front Neurosci 7:274
Mitani JC, Gros-Louis J (1998) Chorusing and call convergence in chimpanzees: tests of three hypotheses. Behaviour 135:1041–1064
Myowa-Yamakoshi M, Tomonaga M, Tanaka M, Matsuzawa T (2004) Imitation in neonatal chimpanzees (Pan troglodytes). Dev Sci 7(4):437–442
Nagasaka Y, Chao ZC, Hasegawa N, Notoya T, Fujii N (2013) Spontaneous synchronization of arm motion between Japanese macaques. Sci Rep 3:1151
Néda Z, Ravasz E, Brechet Y, Vicsek T, Barabási A-L (2000) Self-organizing processes: the sound of many hands clapping. Nature 403(6772):849–850
Oram MW, Perrett DI (1994) Responses of anterior superior temporal polysensory (STPa) neurons to biological motion stimuli. J Cogn Neurosci 6(2):99–116
Oullier O, de Guzman GC, Jantzen KJ, Lagarde J, Kelso JAS (2008) Social coordination dynamics: measuring human bonding. Soc Neurosci 3(2):178–192
Patel AD (2006) Musical rhythm, linguistic rhythm, and human evolution. Music Percept Interdisc J 24(1):99–104
Patel AD, Iversen JR, Bregman MR, Schulz I (2009) Experimental evidence for synchronization to a musical beat in a nonhuman animal. Curr Biol 19(10):827–830
Paukner A, Anderson JR (2006) Video-induced yawning in stumptail macaques (Macaca arctoides). Biol Lett 2(1):36–38
Repp BH, Penel A (2002) Auditory dominance in temporal processing: new evidence from synchronization with simultaneous visual and auditory sequences. J Exp Psychol Hum Percept Perform 28(5):1085
Repp BH, Penel A (2004) Rhythmic movement is attracted more strongly to auditory than to visual rhythms. Psychol Res 68(4):252–270
Richardson MJ, Marsh KL, Isenhower RW, Goodman JRL, Schmidt RC (2007) Rocking together: dynamics of intentional and unintentional interpersonal coordination. Hum Mov Sci 26(6): 867–891
Sato W, Yoshikawa S (2007) Spontaneous facial mimicry in response to dynamic facial expressions. Cognition 104(1):1–18
Schmidt RC, O'Brien B (1997) Evaluating the dynamics of unintended interpersonal coordination. Ecol Psychol 9(3):189–206
Schmidt RC, Turvey MT (1994) Phase-entrainment dynamics of visually coupled rhythmic movements. Biol Cybern 70(4):369–376

Schmidt RC, Carello C, Turvey MT (1990) Phase transitions and critical fluctuations in the visual coordination of rhythmic movements between people. J Exp Psychol Hum Percept Perform 16(2):227

Sebanz N, Bekkering H, Knoblich G (2006) Joint action: bodies and minds moving together. Trends Cogn Sci 10(2):70–76

Van Baaren RB, Holland RW, Kawakami K, Van Knippenberg A (2004) Mimicry and prosocial behavior. Psychol Sci 15(1):71–74

van Ulzen NR, Lamoth CJ, Daffertshofer A, Semin GR, Beek PJ (2008) Characteristics of instructed and uninstructed interpersonal coordination while walking side-by-side. Neurosci Lett 432(2):88–93

Wallin NL, Merker B, Brown S (2000) The origins of music. MIT Press, Cambridge

Watts DP, Mitani JC (2001) Boundary patrols and intergroup encounters in wild chimpanzees. Behaviour 138(3):299–327

Wiltermuth SS, Heath C (2009) Synchrony and cooperation. Psychol Sci 20(1):1–5

Yamamoto S (2017) Primate empathy: three factors and their combinations for empathy-related phenomena. Wiley Interdiscip Rev Cogn Sci 8(3). https://doi.org/10.1002/wcs.1431

Yamamoto S, Humle T, Tanaka M (2012) Chimpanzees' flexible targeted helping based on an understanding of conspecifics' goals. Proc Natl Acad Sci USA 109(9):3588–3592

Yu L, Tomonaga M (2015) Interactional synchrony in chimpanzees: examination through a finger-tapping experiment. Sci Rep 5:10218

Yu L, Tomonaga M (2016) Unidirectional adaptation in tempo in pairs of chimpanzees during simultaneous tapping movement: an examination under face-to-face setup. Primates 57: 181–185

Yu L, Tomonaga M (2018) Effect of visual cues in addition to moderate auditory cues on temporal coordination: a comparative study in humans and chimpanzees. bioRxiv:290379. https://doi.org/10.1101/290379

Yun K, Watanabe K, Shimojo S (2012) Interpersonal body and neural synchronization as a marker of implicit social interaction. Sci Rep 2:959

Zarco W, Merchant H, Prado L, Mendez JC (2009) Subsecond timing in primates: comparison of interval production between human subjects and rhesus monkeys. J Neurophysiol 102(6): 3191–3202

Zivotofsky AZ, Hausdorff JM (2007) The sensory feedback mechanisms enabling couples to walk synchronously: an initial investigation. J Neuroeng Rehabil 4(1):1

The Cognitive Implications of Intentional Communication: A Multifaceted Mirror

David A. Leavens

A circle looks at a square and sees a badly made circle.
Jeff Vandermeer, *Authority*

Abstract There is a central concern in contemporary cognitive science with the validity of the use of epistemic and intentional terms to interpret the communication patterns of non-human animals. Here I argue (a) that the human developmental transition to intentional communication is a well-described phenomenon, from an empirical standpoint; (b) that the behavioural patterns that characterise intentional communication in our own species are also well-described in the communication of our nearest living relatives, the great apes; (c) that the presence of the behavioural markers for intentional communication in non-human primates does not unambiguously implicate any particular one of a large number of often mutually contradictory hypothetical psychological process models; and (d) that intentional communication by young humans is also consistent with hypothetical process models that are, themselves, mutually contradictory. Intentional communication is a class of behaviour that is open to public, objective measurement. In contrast, the hypothetical cognitive processes supporting intentional communication in both human and non-human animals are not specified by the fact that intentional communication has occurred—they could not be, except when there is an unambiguous behavioural index of invisible psychological processes, which is a contradiction in terms. In this chapter, I will examine a number of contemporary scientific practices that purportedly reveal aspects of psychological processes underlying intentional communication and demonstrate the deficiencies of these protocols. In general, these methodological infelicities support a systematic, discipline-wide double standard of interpretation of the communication of animals and humans. I will conclude that there is no convincing evidence extant of different psychological processes in the intentional communication of apes and preverbal humans.

D. A. Leavens (✉)
School of Psychology, University of Sussex, Falmer, East Sussex, UK
e-mail: davidl@sussex.ac.uk

L. D. Di Paolo et al. (eds.), *Evolution of Primate Social Cognition*, Interdisciplinary Evolution Research 5, https://doi.org/10.1007/978-3-319-93776-2_5

Keywords Social cognition · Hominoidea · Species comparisons · Idols of the Theatre · Great apes · Intentional communication · Experimental methods

Cheetah, a matronly chimpanzee, pointed to a banana, looked at me, and then pointed into her open mouth. With Bill Hopkins, I reported this as an example of an intentionally communicative gesture (Leavens and Hopkins 1998). What else could we do? If you came across a human being in a cage who pointed to a banana somewhere in your vicinity, looked directly into your eyes, and then pointed to her mouth, would you be cast into an epistemological tempest of doubt over the question of whether that human being was intentionally communicating with you? Most of us would accept without further evidence that we had been the recipient of a communicative signal that was directed at us and that was manifestly about a banana that existed in our shared perceptual space. But claim that an ape pointed intentionally and, all of a sudden, the claimant is faced with highly educated, sober researchers who grimly shake their heads as though it is some kind of deep mystery whether the animal is communicating intentionally, in the way that humans allegedly do. My central claim in this chapter is that cognitive scientists simply find it easier to attribute higher-order reasoning skills to humans than to animals, notwithstanding that the psychological processes supporting pointing gestures have never been directly measured in any organism of any species, ever. There is, thus, a double standard of proof applied to animals in comparison with humans, notwithstanding the fact that neither animals nor preverbal humans can assert the psychological processes purportedly underlying their overt behaviour.

Bacon (1620) noted that scientists were vulnerable to a number of baleful influences that warped their perceptions and their argumentation. One of these classes of error he termed, "Idols of the Theatre", which he described as "manifestly instilled and cherished by the fictions of theories and depraved rules of demonstration" (Aphorism LXI). He wrote that, as science advanced, even more numerous such errant dogmas would bedevil those who sought to understand the natural world. This is because scientists, according to Bacon,

> take for the groundwork of their philosophy either too much from a few topics, or too little from many; in either case their philosophy is founded on too narrow a basis of experiment and natural history, and decides on too scanty grounds (Aphorism LXII).

One of the current dogmas that, in my opinion, is significantly impeding our understanding of animal cognition is the widespread assumption that human children evince advanced awarenesses of others' mental processes, even before they can speak (e.g. Tomasello et al. 2007). As we will see, many contemporary researchers employ a "bag of tricks"—what Bacon referred to as "perverted rules of demonstration" (1620, Aphorism XLIV)—to make human children look smart and apes look less clever. These tricks include (a) administering more difficult tasks to the apes than to the humans and interpreting the better human performance as evidence of an evolutionarily based cognitive advantage in humans (fallacious equivalence of protocol); (b) handicapping the apes by denying them adequate exposure to crucial task-relevant experiences and comparing them to human children who have not been isolated from those formative experiences (fallacious comparisons); (c) interpreting superior performance by apes as evidence of cognitive inferiority (fallacious interpretations); (d) repeatedly subjecting high-performing apes to remedial testing until their performance breaks down (fallacious remediation); and (e) mixing nonhandicapped apes with handicapped apes, taking a group average, and then comparing these dubiously aggregated averages to the average performances of nonhandicapped human children (fallacious aggregation). Results from these kinds of fallacious sampling protocols, research designs, and arguments are very popular in the contemporary literature—it is as though scientists are driven by a deep insecurity about humanity's intellectual superiority over the other animals, and therefore conspire to produce findings that make humans look good in comparison. To be fair, humans are extraordinary animals capable, in groups, of stunning technological and other collaborative feats. The paradox is that when you compare an ape infant with a human infant, the ape is a far more physically competent, more socially aware creature, until about the end of the first year of life, when speech begins to become a unique mode of social expression and learning in our species (e.g. Bard et al. 2014). Yet, the consensus among contemporary scientists and philosophers is that humans have unique competencies in social cognition that are manifest as early as the first year of life. In my opinion, nowhere is the chasm between theories of early human socio-cognitive superiority and empirical evidence so wide as contemporary research into the production and comprehension of pointing.

1 Intentional Communication

Pointing is the quintessential intentionally communicative gesture, and it is worth reviewing why that is. Historically, human children's development of the ability to integrate their object-directed behaviour with communication towards a social partner was heralded as a cognitive milestone variously labelled as *coordinated secondary circular reactions* (Bates et al. 1975), *intentional communication* (Bard 1992; Leavens and Hopkins 1998), or *dual-directional signalling* (Masur 1983). Children in Western cultures, near the end of their first years or in the beginning of their second years of life, begin to visually fixate social partners in smoothly orchestrated triadic communicative episodes in which they more or less simultaneously point to entities in the world. For the first time, near the end of the first year of life, children, thus, use pointing to integrate communication about something (the gesture as a directed ray) with some particular other (the visual fixation of the social partner).

In the foundational theoretical work of Elizabeth Bates and her colleagues (e.g. Bates et al. 1975, 1977), pointing was characterised as an imperative, a request, or demand that the interlocutor act in some way. *Protoimperatives* were defined as requests for an interlocutor to act on the world, usually to retrieve and deliver objects to the signaller, and *protodeclaratives* were defined as requests for interlocutors to engage with the child—here, the child uses the object to obtain adult attention and emotional engagement ("laughter, comment, smiles and eye contact—which we have termed 'attention'" Bates et al. 1975, p. 216). An example of pointing to elicit engagement would be a child pointing to a number of distant trucks and turning to his interlocutor with a smile and the verbalisation, "look at all the trucks!" (This example originates in my childhood: evidently, as a toddler I could not yet pronounce "tr-" and replaced it with "f-", and, according to family lore, while I did elicit an affective response, it might not have been the affective response I was expecting.) Thus, according to this view, pointing to request delivery of an object and pointing to request engagement are both instrumental acts to manipulate the behaviour of an interlocutor; hence, I will refer to this, here, as the Instrumental Hypothesis, the idea that communicative acts are intentional by virtue of their goal-directed function (Bard 1992; Moore and Corkum 1994; Leavens 2012a; Racine 2012; Racine and Carpendale 2007). Note that this definition does not appeal to sub-organismal processes—irrespective of its neural correlates, in this point of view, intentional communication is identified by communicative acts displayed at the level of the whole organism and its praxic context, including manual gestures, vocalisations and other auditory signals, and facial expressions (e.g. Hopkins et al. 2011; Leavens et al. 2005a). Organisms either do or do not work to achieve objectively measurable changes in the state of the world. If we accept that the present author and the reader hereof are capable of displaying intentional communication, then we acquired that ability in the complete absence of any direct access to the concomitant neurological, endocrinological, or psychological processes of our conspecifics. There is, thus, no reasonable sense in which such privileged access to unobservables can have influenced our own skill acquisition in this domain, and therefore no necessity to include such unobservables in any scientific operationalisation of intentional communication. We all manage to learn how to do intentional communication in the absence of brain scans, blood plasma assays, and any direct measure of psychological processes, and these early scientific approaches to the study of the development of these skills were tightly grounded in the perceptual input to both developing signallers and their social partners.

At about the same time, however, that Bates and her colleagues were developing the interpretation of pointing as a kind of social tool use, other researchers were characterising protodeclaratives as psychologically distinct from mere requests for infant-directed caregiver responses. Rheingold et al. (1976) argued that the claim that children's objective with protodeclarative pointing is "[to] seek attention for themselves is refuted by the observations: they direct the attention of people to objects external to themselves" (p. 1157). Later researchers developed this position to what is now the canon in developmental psychology, that (proto)declarative signalling indexes an early human perception of states of mind in others, and children use these signals to attempt to manipulate the states of mind of their interlocutors. This is the Representational Hypothesis (e.g. Mundy et al. 1986; Baron-Cohen 1989; Tomasello 1995; Tomasello et al. 2007), and this stance includes a variety of different theoretical perspectives on intentionality which all nevertheless agree that internal, mental representations of the world (a) exist and (b) cause communicative behaviour. Thus, according to representational perspectives, intentional communication is defined by invisible, unverifiable features of the signaller's motivation and conception of mentality in others. I'll return to this difference in interpretation of protodeclarative gestures, below, but the essential point, here, is that whenever intentionality is defined in imaginary, hypothetical terms, such as attempting to manipulate the contents of others' minds, as Baron-Cohen (1989) and many others have construed it, then it follows necessarily that empirical study cannot either confirm or disconfirm the hypothetical elements of the definition (e.g. Leavens et al. 2008, 2014, 2017). While this representational interpretive stance might be useful to philosophers, as Leavens et al. (2008) pointed out, there is an immensely large set of mutually inconsistent hypothetical psychological process models that, nevertheless, do not make different predictions at the level of whole organisms and therefore cannot be subject to falsification by empirical study; the Representational Hypothesis in its many guises is, inherently, non-scientific, by virtue of the unfalsifiability of its claims about psychological process. The question under consideration, here, is not whether one view of intentional communication is ontologically more appropriate than other views; rather my concerns in this chapter are with the scientific utility of certain contemporary practices in comparative psychology: a practice is scientifically useful if generates falsifiable claims. A corollary of this utility criterion is that measurements ought to be applied using the same measuring scales across different groups—measurements across species must be fairly applied, not because of any ethical principle of fair treatment, but because only fair applications of a measuring instrument will yield accurate results. If the measurements are unfair, then the resulting findings are scientifically unreliable.

Whether one views pointing as an intentionally communicative gesture depends critically on what one means by "intentional". In an elegant analysis, Gómez (2009) contrasted Cartesian, "theory of mind" approaches with his interpretation of Brentano's formulation of intentionality as the quality of a mind "being about or pointing to things other than itself" (p. 191). As Gómez described the Cartesian view:

> it . . . keeps a sharp conceptual distinction between behaviour and mental phenomena, with the latter being private, unobservable, detachable from their physical manifestations, but ultimately the cause and explanation of observable behaviour, there is a fundamental dualism between minds and behaviour (2009, p. 191).

In terms of intentional communication, the Cartesian perspective, then, forces mental beings to hypothesise and reason about unobservable, causal mental states, and for some proponents of this Cartesian view, evidence for intentional communication does constitute evidence for this kind of reasoning about invisible, mental processes. In contrast, Brentano's formulation, according to Gómez, emphasises the directedness of minds in relation to things in the world and other mental phenomena. Thus, in the Brentanian perspective, intentionality is a kind of bridge between mental activity and either events in the world or other mental states. Gómez noted that the Brentanian view is compatible with Cartesian approaches—invisible psychological entities form a foundation or platform for meaning and mentality, but he proposed that "Brentano's intuition of intentionality as the mark of the mental can be developed in a different direction—one in which dualism becomes irrelevant and intentionality is conceptualized as an 'embodied' relation with 'potentially real' objects" (p. 191). Gómez goes on to describe how gaze direction embodies attentional focus and provides direct, perceptual access to an observer of the focus of attention of an animal or human, where attention is not conceived of as a purely disembodied, mental phenomenon, in a Cartesian sense, but as an act of casting visual attention by embodied creatures comporting themselves in an objective, potentially shared reality with an observer. Gómez concludes that both this Brentanian sense of intentionality a lá Gómez and the representational, Cartesian sense of intentionality are possessed by humans but that nonlinguistic creatures experience intentionality only in the Brentanian sense (see also a related argument by Moore 2015, 2017). My own view is that this is a theoretical distinction without an empirical assay (see, esp., Leavens et al. 2017; Sehon 2000). Goal-directed *behaviour*, on the other hand, is easily measured in objective terms (Bard 1992), and the remarks that follow are based on the Popperian idea that scientifically useful hypotheses must be potentially falsifiable. To meet the criterion of falsifiability, scientific claims must be grounded in variables that physically exist. Otherwise, as I've argued elsewhere, the presence or influence of hypothetical entities cannot be empirically demonstrated (Leavens et al. 2017), and, of course, there is a long tradition in psychology that calls for a strict conceptual distinction between hypotheses defined entirely by empirical variables (intervening variables) and those that appeal to hypothetical entities, which cannot be directly confirmed (hypothetical constructs, MacCorquodale and Meehl 1948). This is not to advocate for a reactionary return to Behaviourism, but simply to observe that the behavioural (and cognitive) sciences have very substantially constrained rules of inference in comparison with more humanistically oriented scholarly activities, and these constraints are attributable to the scientific requirement to manipulate and measure physical entities.

In the contemporary scientific literature on human psychological development, which is grounded in Western cultures, there is a long-standing set of objective criteria by which most researchers agree that humans have attained the capacity for intentional communication (see Bard 1992; Bates et al. 1975; Leavens et al. 2005a, for elaboration). In these Western cultures, typical development is characterised, late in the first year of life, by pointing that seems to be used to manipulate others to act on the world. For example, Lock (2001) described from film footage a 12-month-old girl in a feeding chair. At first, she reached ineffectively for a distant apple, but then she turned to her mother (offscreen), who was not attending to the child, and the girl vocalised, displaying a putative attention-getting signal. The girl monitored her mother, and when the mother turned to look at the child, the girl pointed to the distant apple, in a clear request for the mother to retrieve the apple for her. Like many children of her age, this child acted *as if* she had a goal in advance of her actions, she tactically chose auditory (vocalisation) and visual (pointing) signals to match the attentional availability of her mother, and in the process, she alternated her gaze from the referent (apple) to the mother (recipient). Babies of this age will also persist in or repeat their signals if the recipient is unresponsive and even elaborate their signals in the face of recipient inaction (Golinkoff 1986). In Western cultures, then, there is a large body of empirical evidence that describes human children's developmental transition in terms of the following behavioural innovations: (a) they begin to discern when they have the visual attention of an interlocutor, (b) they tailor their own signalling to the attentional status of an observer (using auditory or tactile signals to attract attention before displaying visual signals), and (c) they develop tactics of repair of communication, when initial signalling attempts fail—they persist and/or elaborate in their signalling behaviour. Thus, human children at about a year of age act *as if* they have goals in advance of signalling, they clearly discriminate different attentional states in other agents, and they have tactics for signal manipulation in the face of recipient incomprehension. All of these patterns in signalling are publicly available and amenable, therefore, to objective measurement. Intentional communication can be, thus, completely defined in objective terms.

Bates and her colleagues (1975) were among the first to apply Austin's Speech Act Theory (1962) to the communication of preverbal children, arguing that that these kinds of phenomena (gaze alternation, signal choice) were consistent with the idea of babies displaying illocutionary force in their preverbal communication. They defined an illocution as requiring "the intentional use of a conventional signal to carry out some socially recognized function" (Bates et al. 1975, p. 206). In their view, intentionality was a kind of "social tool use": "intentional use of the listener as an agent or tool in achieving some end" (p. 208). Thus, the scientific basis on which humans are said to attain intentional communication is comprised of evidence that children use pointing to particular ends (behaviour ceases on attainment of goal—e.g. a point to an apple results in delivery of the apple; Lock 2001), to influence particular people (behaviour ceases on response of particular people: Povinelli and Eddy 1996; Woodruff and Premack 1979), and that their pointing is about (under the stimulus control of) specific entities in the world (behaviour ceases on attainment of a specific goal, not just any response: Cartmill and Byrne 2007; Leavens et al. 2005a). All of these objective characteristics of intentional communication have been well-described in the communication of young humans (Bates et al. 1975) and great apes (Leavens et al. 2005a).

Although for some decades, pointing was portrayed as a human-unique, species-specific gesture (e.g. Butterworth and Grover 1988), in a programme of research spanning a number of decades, it has been amply demonstrated that chimpanzees and other great apes use pointing to particular ends (Leavens et al. 2004a; Menzel 1999), to influence particular people (Woodruff and Premack 1979), with respect to specific entities in the world (Call and Tomasello 1994; Cartmill and Byrne 2007; Leavens et al. 2005a; Menzel 1999; Roberts et al. 2014). It is in this sense—the use of these gestures to manipulate others to act on the world, or social tool use—that researchers claim to have demonstrated intentional communication in great apes (e.g. Bard 1992; Leavens and Hopkins 1998): great apes display the same objectively measurable patterns in their social tool use that are displayed by human children in Western cultures, near the end of the first years of their lives; these are goal-directed sequences of behaviour in which the animals use their signals to manipulate people to act on the world, and the animals' signal choices display the same kind of sensitivity to the attentional state of the recipient (e.g. Cartmill and

Byrne 2007; Hopkins et al. 2007; Leavens et al. 2004b, 2010) that is displayed by human children. Today, it is widely accepted that chimpanzees request objects through protoimperative pointing, and that this is an intentionally communicative signal for these animals, when applying the same objective behavioural criteria to humans and apes (Leavens 2012a, b; Tomasello et al. 2007). It is in this sense of these physically grounded patterns of behaviour that animals and preverbal humans communicate intentionally—obviously, if the reader is wedded to Cartesian conceptions of behaviour as consequences of mental activities, then the nature of the evidence might seem a bit thin, but it is no more inadequate for non-human animals than it is for preverbal humans (Leavens et al. 2017). Thus, while the Instrumental Hypothesis allows one to interpret communicative behaviour in terms of measurable, physical events, the Representational Hypothesis requires interpretation in terms that cannot be verified: human infants' discernment of epistemic states in others, for example, is not something that can be subjected to direct, objective measurement.

It could be, of course, that neither the Instrumental nor the Representational perspectives are ontologically appropriate for any species, and there are other theoretical perspectives on pointing beyond these two. Two examples, which I cannot summarise adequately, here, due to space constraints, are the accounts of Carpenter and Liebal (2012) and Gómez (2007). Carpenter and Liebal (2012) argued that what makes pointing a psychologically unique phenomenon in humans is the state of joint attention that ensues between people, even babies, upon producing or perceiving a point; their claim is that it is not that the two individuals are simultaneously looking at a common focus, nor is it that a signaller exerts control over the visual orienting behaviour of another, but that this state of affairs constitutes a kind of communion or psychological interdigitation of attention between the two interlocutors. Although I think this idea suffers from some of the same issues of verifiability to which the Representational Hypothesis is vulnerable, it is a potentially non-representational perspective on intentional communication. Developing this idea of "knowing together", Carpenter and Call (2013) argued that this state of we-ness is unique to humans; with respect to great apes, as they put it, "there is no unequivocal evidence that they do so with the sole (and spontaneous) goal of sharing attention and interest with others about something" (p. 57). Of course, as noted by Leavens et al. (2017) in their discussion of this paper, (a) there is no unambiguous evidence that humans do this (see e.g. Liszkowski et al. 2004, for evidence that human babies did not do this in a joint attention context) and (b) any time a scientist claims that an act was *spontaneous*, what this means, in reality, is that the scientist is ignorant of the developmental pathway to that behaviour—it signifies nothing more than that. (A reviewer thought this summation was too strong, asking whether all "spontaneous" behaviour must have a developmental history; the short answer is "yes"; all living creatures begin life as single cells that are manifestly incapable of the kinds of communicative behaviour under discussion, here—this does not necessarily imply that all communicative signals are socially learned, per se, but they do have developmental histories.)

In another example of an alternative view to the instrumental vs. representational contrast under discussion, here, Gómez (2007) argued that the development and comprehension of pointing in human children could be manifested on a psychological platform that is implicitly, but not explicitly Gricean (i.e. Cartesian or representational): "that behaviours are directly perceived as intentional, that is, as being directed to things other than themselves in the essential sense of Brentano's notion of intentionality ..., but without necessarily representing that they are driven by unobservable mental states" (Gómez 2007, p. 730; and see Moore 2015, 2017). Only later in development, according to Gómez, are human children and adults able to engage in a more complex psychological process that would constitute a full-fledged system of reasoning about inner psychological states in others.

The psychological processes that allegedly support pointing *are nothing more than* interpretations by some researchers committed to a particular model of psychological functioning; these interpretations do not necessarily exist in the brains or the minds of the pointing individuals—human or otherwise—they exist in the belief systems of the observers, and are not subject to interrogation by introspection or objective measurement (see, esp., Nisbett and Wilson 1977). One might imagine two cognitive scientists pointing for each other, and it might be the case that each of them has a belief about their own and their interlocutor's psychological functioning, and they might use similar language to describe these imaginary psychological processes. Empirically, however, neither scientist can discover the psychological causes of their own pointing through introspection (if they could, there would be no scientific debate on the psychological causes of pointing—we would have an easy unanimity on a taxonomy of causes of pointing, which we lack, Leavens 2012a), nor can either scientist directly discern these psychological causes in the other; there is no such thing as an Intentionality-O-Meter that a person can go down to the hardware store and buy. Thus, the psychological mechanisms that support pointing gestures are, in all cases, *imaginary* or hypothetical, and they come into play only in contexts in which one or both interlocutors are engaging in the practice of causal ascription; this is axiomatic. Pointing does not reveal its causes (Racine 2012; Racine and Carpendale 2007).

If Cartesian, Gricean, or Brentanian-a lá-Gómezian causes of pointing cannot be revealed through scientific inquiry, through the manipulation and measurement of variables that have physical existence, then this has significant implications for how intentional communication is studied in animals. One implication is that all contemporary claims for insight into the psychological processes of animals or young humans who point are unsupported by direct, empirical data. Now, of course, it is entirely possible that the present author's stance on what constitutes the legitimate and necessary constraints on scientific theory development is unduly and perhaps unrealistically pessimistic. In the following sections, I will briefly outline how even if this was true, even if we reject the epistemological concerns briefly sketched above, there remains a persistent and widespread methodological bias against animals' performances in tests of their social-cognitive abilities. The effect of these biased assays is that everybody, of every theoretical stripe or school or stance, is hampered in their theory development by a grave misrepresentation of animals' abilities; in short, the empirical record pertaining to the cognitive abilities of animals is wildly inaccurate due to unfair comparisons between–and double standards of interpretation of–the communication and cognition of human children and great apes.

2 Fallacious Equivalence of Methods

Notwithstanding the fact that pointing does not unambiguously specify its own psychological causes, many contemporary researchers attempt to explain pointing in psychological process terms. A set of rules of demonstration of these psychological processes has emerged in the scientific literature, and here I wish to examine these rules critically. For example, even while conceding that apes do display protoimperative gestures, some have argued that human protoimperative pointing by 12-month-old babies is cognitively more sophisticated than the protoimperative pointing of much older, even adult great apes (Tomasello et al. 2007; van der Goot et al. 2014). van der Goot et al. argued that because human babies sometimes signalled to a desired toy from distances of approximately 1–2 m, this constituted evidence that these children had a psychological appreciation of a state of common ground with their social partners. In their study, none of the ten great apes in their sample pointed or otherwise gestured towards food from a distance—they always approached as closely as the cage mesh would allow to the desired food, and they signalled from there. van der Goot et al. interpreted this as evidence that great apes do not understand common ground, but it is worth pointing out that the apes in their study were situated significantly further from the desired food—about 6 m—than were the human babies from their desired toys. Moreover, the human and ape testing protocols differed in numerous additional features, and there was no attempt to match the two groups on either life history stage or task-relevant pre-experimental experience. Leavens and his colleagues (2015) placed an experimenter and desirable food approximately 1.5 m apart (thus, approximating the distances presented to the human babies in van der Goot et al. 2014), outside the enclosures of 166 chimpanzees, and they found that, on any given trial, about half of the apes pointed or otherwise gestured from a distance, which is the same finding van der Goot and her colleagues reported for human babies. Thus, simply matching the distance between the subjects and an unreachable, but desirable object seemed to eradicate the "species difference" reported by van der Goot et al. Because a difference in approach behaviour was the evidence for the alleged psychological differences van der Goot and her colleagues claimed to exist between human babies and much older great apes, therefore, based on Leavens et al.'s findings, either great apes also appreciate common ground or—and I think this is far more likely—differences in propensity to approach distant entities during requests do not diagnose the psychological mechanisms supporting this approach behaviour. This curious design feature, making the distance of the target and the subject three to six times further way than the distances used with the human infants, is an example of administering a more difficult task to apes than to humans, finding a difference, and then attributing this difference to the cognitive superiority of humans, with their different evolutionary histories (fallacious equivalence of methods). Close consideration of van der Goot et al. reveals numerous, systematic differences in sampling protocols, test procedures, pre-experimental task-relevant preparation, and age (the humans were 12 months old, and the apes were, on average, 18 years old) that are inextricably intertwined with their preferred explanation for the group differences they reported—evolutionary history; in fact, it is logically impossible to isolate species classification (evolutionary history) as the relevant factor in these apparent group differences, and this constraint on interpretation exists for a large proportion of the scientific literature in this area (see Leavens et al. 2015, 2017, for extended discussion). But the current scientific literature on intentional communication, particularly the development of intentional communication in humans, largely ignores this truth.

3 Fallacious Comparisons

Another perverted rule of demonstration emerges from the direct comparison of humans and apes: differences between representatives of humans and apes are implicitly interpreted as though the responses elicited are completely uninfluenced by the individual learning histories (i.e. life experiences) of both the humans and the apes, so that any differences that may become apparent are attributable to evolutionary, rather than developmental processes (see Bard and Leavens 2014; Leavens and Bard 2011, for discussion). For example, it is widely yet incorrectly held that great apes—humans' nearest living relatives—have difficulty comprehending the intentional signals of humans (e.g. Burkart et al. 2009; Tomasello et al. 2007). Burkart et al. (2009) attribute this alleged superiority in understanding deictic cues to the cooperative breeding of humans. Tomasello and colleagues (2007) attribute this putative superiority to what they believe is a human-unique capacity for shared intentionality, by which they seem to mean that human babies perceive both (a) a joint attentional context and (b) a cooperative motivation, and they bring these perceptions to the interpretation of pointing and other deictic cues. I (Leavens 2012b, 2014; Leavens et al. 2010, 2017) and others (e.g. Lyn et al. 2010) have noted that when some slight effort is made to match apes with humans' early rearing histories, so that apes gain something approximating the same level of exposure to human communication practices that our children experience, then they have no difficulty passing tests of comprehension of human pointing cues. This widespread belief that great apes have some kind of difficulty understanding simple deictic cues (e.g. Miklósi and Soproni 2006; Povinelli et al. 1997; Tomasello et al. 2007) is predicated on poorly designed experimental comparisons in which the ape and human subjects are not matched on task-relevant pre-experimental experiences (e.g. Leavens 2014, Leavens et al. 2017; Lyn et al. 2010). This collapse in methodological rigour is not limited to studies of apes' comprehension of pointing or other deictic cues; in fact, virtually every contemporary claim for the superiority of human children over apes in socio-cognitive reasoning is based on comparisons between typically reared human children in postindustrial, Western cultures, and great apes that have been, to varying degrees, isolated from these kinds of rich enculturation experiences, either because the animals have been raised in zoos or biomedical research centres (e.g. Povinelli et al. 1997; van der Goot et al. 2014) or raised in their natural habitats for their early years and then subsequently taken to sanctuaries after having witnessed their entire social groups exterminated in their presence (e.g. Herrmann et al. 2007; Kirchhofer et al. 2012). Thus, the majority of the evidence brought in to support claims of human uniqueness in social cognition involve comparisons of psychologically handicapped apes with nonhandicapped humans (fallacious comparisons).

4 Fallacious Interpretations

Failure to match apes and human on early rearing experiences is not, by any means, the only infelicity in the majority of ape-human comparisons. Another common tactic used by researchers who seem to have a vested interest in demonstrating human superiority at all costs is to invoke double standards of interpretation. For example, even when apes display superior performance, compared with human children, researchers will attribute their superior performance to their cognitive inferiority. A classic example is a study by Povinelli et al. (1999). These researchers compared 3-year-old human children with chimpanzee adolescents on their ability to use human gaze cues to identify which of two containers was baited (a standard object-choice task). In the Head + Eyes condition, the experimenter looked directly at the baited container whilst simultaneously turning her head in that direction; the apes and the children had no apparent difficulty following this cue, as evidenced by their above-chance performance. In a second cue presentation, the Eyes Only condition, the experimenter kept her head oriented straight ahead but fixated her gaze on the baited container. In this condition, neither the apes nor the human children performed above chance, and hence, again, there was no difference in performance between the humans and the apes. In the third, Above Target condition, the experimenter looked to the same hemispace in which the baited container was located but oriented her head and eyes in such a way as to fixate at a point that was substantially above the baited container. In this condition, the apes were significantly above chance in choosing the baited container, whereas the human children performed at random, chance levels; thus, the apes, but not the humans, were able to use this cue. Povinelli and his colleagues interpreted this superior performance by the apes as evidence for their "low-level" understanding of visual attention, arguing that the humans had a sophisticated, mentalistic comprehension of gaze as a referential cue that apes lack (i.e. that the human children's poor performance was attributable to the children's superior understanding of the psychological bases of visual attention, which they termed a "high-level model"). Some years after this study was published, we reasoned that, under this theoretical perspective, human adults must have the most sophisticated extant reasoning skills to bring to bear on these kinds of tasks, so we administered this battery of cues to adult, human university students; we found that, like the adolescent chimpanzees in Povinelli et al. (1999), human university students had no difficulty using the critical Above Target cue to find hidden treats (Thomas et al. 2008). Thus, great apes are damned if they do succeed and damned if they don't succeed. Povinelli and his colleagues (1999) interpreted any systematic difference between their human subjects and their ape subjects as evidence *for* greater human cognitive sophistication; in this case, they interpreted *superior* performance by chimpanzees as evidence for *inferior* cognitive processing capabilities (fallacious interpretations). Had they simply validated their protocol by including some human adults, then they would have realised that it was, in fact, the chimpanzees and the human adults who were displaying the advanced response profiles, not the much younger human children.

5 Fallacious Remediation

Another tactic in the comparative study of cognition is that of testing until failure. Povinelli et al. (1997), again, provide an example of this. They submitted human children and adolescent chimpanzees to a series of object-choice tasks, with a variety of different cue types. In the first experiment, two of the chimpanzees (Kara & Apollo) demonstrated good comprehension of the deictic cues. When these two apes performed more poorly in a second series of different cues, the authors used this finding to argue that they did not truly understand deictic cues in the first experiment! They argued that because the index finger was close to the baited container in the first series of tests, the apes were using them as associative cues in the absence of an understanding of these cues as intentional signals. Yet, it is well-demonstrated that, developmentally, human children proceed from early stages of understanding pointing as a cue to objects that are close to the pointing hand to much later generalisation of this understanding to more distant entities (e.g. Butterworth 2003). Here, the assumption is that these institutionalised apes housed in a biomedical research centre are not disadvantaged in any way, compared with the preschool human children, with respect to having had pre-experimental opportunities to learn to generalise from proximal to more distal pointing cues, despite their obvious isolation from these European conventions of nonverbal signalling (e.g. Leavens et al. 2017).

In the second experiment, one of the chimpanzees, Candy, performed very well, and rather than conclude that therefore Candy understood the second experiment, she (alone) was remedially tested until her performance broke down, and they concluded, again, that therefore she must not have understood the earlier task when she performed well. This practice of testing your animals until you get the expected result (fallacious remediation) will at first render the appearance of conformity with theoretical models of human exceptionalism, but on closer scrutiny will undermine such models—theories of human exceptionalism in social cognition are not strengthened by diligent post-hoc inquiry into high performance by apes, with no commensurate investigation of the high performances of human participants. In

Povinelli et al. (1997), not one of the human children who performed well in the second set of experiments was subjected to remedial testing. This is not a valid procedure, to systematically interrogate to the point of failure previously successful representatives of one group, but not the other group. Hence, again, if you're an ape under scrutiny by researchers who are offended by your cleverness, then you are liable to be tested and retested and tested again until you deliver the performance that the researchers want.

6 Fallacious Aggregations

These studies also demonstrate another type of sophistry. Recall that the only empirical measure of comprehension in these kinds of studies is the organisms' performances. Researchers compare groups of human children with groups of apes who have been put at a disadvantage (typically through isolation from human communicative conventions, in tasks in which good performance requires an understanding of human communicative conventions). Then, the researchers report that, on average, the humans outperform the apes. If there are individual apes in the comparison group who clearly do understand the tasks (by definition, high performance is the operationalised measure of comprehension), this evidence for ape comprehension is buried in an aggregate measure of performance for the group (a group mean). Leavens (2014) noted that when some of the apes have been enculturated, then this aggregation is dubious: if there are apes that have been raised in human homes or subject to language training, typically these animals far outperform the institutionalised or wild-reared apes. The correct interpretation when enculturated apes outperform non-enculturated apes is that there seems to be an effect of socialisation on the comprehension of human communicative cues. Instead, what researchers have repeatedly done is to average across both enculturated and non-enculturated animals to achieve a group average that is lower than that of the human children in the comparison group (Itakura et al. 1999; Tomasello et al. 1997, among others). In Itakura et al. (1999), Peony—a home-raised chimpanzee—correctly followed human gaze cues in 96 of 96 consecutive trials (a sustained, perfect performance level), but because her data were averaged with the performance data of non-enculturated animals (fallacious aggregation), these researchers concluded that apes have difficulty in following human gaze cues in object-choice tasks (see Leavens 2014 for elaboration).

As noted by Leavens and his colleagues (2015, 2017; Bard and Leavens 2014), there is not one published claim of a species difference in social cognition between apes and human children that has matched the organisms on task-relevant pre-experimental histories, matched the organisms on age, or administered the same procedure to the two groups. To put this claim a little more bluntly, as of this writing, in the year 2018, there is not one iota of evidence that preverbal humans and other organisms point with different psychological processes. When researchers routinely handicap their ape subjects by isolating them from early and intensive exposure to human communicative practices, then test them on their understanding of human communicative cues, and then compare these animals' understanding with human children who are not so handicapped, then is the conclusion that it is the apes who lack cleverness when the animals perform relatively poorly in these kinds of assessments? I think not.

To be fair, however, it is not practically or ethically possible to systematically cross foster apes by human caregivers. For the animals' survival, particularly in zoo environments, sometimes infant apes are home-raised by humans when they are rejected or otherwise neglected by their mothers, but there is no systematic, contemporary cross-fostering project, anywhere in the world, to my knowledge, outside the language-training projects. Because young apes grow up to become lethally dangerous animals, any home-raised ape must eventually be encaged for the safety of their human caregivers. This seems to me to be a significant ethical barrier to the a priori matching of apes and humans on pre-experimental experience, which is required by good experimental design. However, there are a number of great apes who have been cross-fostered, either in zoos for their survival or in the course of ape language-training projects, and the scientific importance of these individuals ought not to be, in my opinion, hidden under the bushel (as it were) of group averages that include institutionalised animals (Leavens 2014). Cross-fostered apes manifest skill sets not characteristic of either wild apes or apes housed in more institutional settings (Bard and Leavens 2014; Call and Tomasello 1994; Leavens and Bard 2011; Leavens et al. 2010).

7 Conclusions

I think a better approach is to catalogue the interplay between life experience and performance on the assays used to assess social competence. For example, whereas pointing by wild apes is very rare (Hobaiter et al. 2014; Veà and Sabater-Pi 1998; but see Douglas and Moscovice 2015), approximately half of institutionalised apes will point in experimentally administered triadic tasks (Leavens and Hopkins 1998; Leavens et al. 2004a); and essentially all home-raised or language-trained apes point (Call and Tomasello 1994; Krause and Fouts 1997; Pedersen et al. 2009; Roberts et al. 2014; Savage-Rumbaugh et al. 1985). This pattern of results clearly indicates that pointing is phenotypically variable in great apes and therefore extremely malleable in the face of different ecological factors influencing their pointing behaviour (Leavens et al. 2005b, 2010). The extreme rareness of pointing by wild apes does not constitute evidence that wild apes lack a cognitive capacity for social understanding that is displayed by captive apes; rather, the obvious implication is that pointing is not as useful for free-ranging, highly mobile apes in their natural habitats as it is for apes living in cages in captivity, or as useful as it is for helpless, motorically inefficacious human babies who are also subject to chronic physical restraint in car seats, cribs, feeding chairs, slings, backpacks, and the like (Leavens et al. 2005b). Every subject in a study of social cognition, human or otherwise, brings to the study a learning history. Every published claim of human superiority over apes in social cognition, manifest in our species before the infants can even speak, is predicated on an implicit assumption that the human children have learned nothing relevant to these tests, in the many months of life experiences they bring to the study. So, to believe that human babies evince superior social understanding to usually very much older great apes, one has to believe something to be true that is patently false: that human babies do not learn anything useful in their production or comprehension of pointing and other social cues, prior to being tested. As soon as it is admitted that human children might learn something useful about triadic communication in their first months of life, then the absurdity of failing to match apes with humans on relevant learning experiences becomes too obvious to ignore (see also Boesch 2012; Leavens and Bard 2011; Leavens et al. 2017).

In conclusion, the state of contemporary research into comparative social cognition is dire, from methodological and logical perspectives. Perhaps it is to be expected that humans see psychological continuity between themselves and their children, but not between themselves and our nearest living relatives, the great apes. This perceptual bias permeates every aspect of comparative social cognition, from the administration of unfair tests, to the selection of psychologically handicapped comparison animals, to the flat refusal to see superior performance by apes as evidence of superior competency, to the lambasting of clever apes with repeated testing until they fail, to the dissolution of superior individual animal results into group averages of ape subjects that have incommensurate life histories—there is no shortage of contemporary, perverted rules of demonstration in common use by researchers to bolster theoretical commitments to human cognitive exceptionalism. Like the fabled Queen who, hearing from her mirror that she might not be the most beautiful, and sets about poisoning her rival, we are systematically extorting the answers we want to hear from our species comparisons (our mirrors) until we get the answers that we want. Great apes and humans share a profound similarity in their intentional communication, when subjected to the same ecological constraints, and there is no objective evidence for the kinds of mental state reasoning by humans in their infancy and toddlerhood that many contemporary researchers claim differentiate humans from their nearest living relatives, notwithstanding the obvious species differences that ensue with the onset of speech in humans. Contemporary claims to the contrary are all smoke and mirrors.

References

Austin JL (1962) How to do things with words. Clarendon Press, Oxford

Bacon F (1620) Novum organum. P. F. Collier & Son, New York (1902 edition)

Bard KA (1992) Intentional behavior and intentional communication in young free-ranging orangutans. Child Dev 62:1186–1197

Bard KA, Leavens DA (2014) The importance of development for comparative primatology. Annu Rev Anthropol 43:183–200

Bard KA, Bakeman R, Boysen ST, Leavens DA (2014) Emotional engagements predict and enhance social cognition in young chimpanzees. Dev Sci 17:682–696

Baron-Cohen S (1989) Perceptual role taking and protodeclarative pointing in autism. Br J Dev Psychol 7:113–127

Bates E, Camaioni L, Volterra V (1975) Performatives prior to speech. Merrill-Palmer Q 21: 205–226

Bates E, Benigni L, Bretherton I, Camaioni L, Volterra V (1977) From gesture to the first word: on the nature of cognitive and social prerequisites. In: Lewis M, Rosenblum L (eds) Interaction, conversation and the development of language. Wiley, New York

Boesch C (2012) Wild cultures: a comparison between chimpanzee and human cultures. Cambridge University Press, Cambridge

Burkart JM, Hrdy SB, van Schaik CP (2009) Cooperative breeding and human cognitive evolution. Evol Anthropol 18:175–186

Butterworth G, Grover L (1988) The origins of referential communication in human infancy. In: Weiskrantz L (ed) Thought without language. Clarendon Press, Oxford

Butterworth G (2003) Pointing is the royal road to language for babies. In: Kita S (ed) Pointing: where language, culture, and cognition meet. Lawrence Erlbaum Associates, Mahwah, NJ

Call J, Tomasello M (1994) Production and comprehension of referential pointing by orangutans (*Pongo pygmaeus*). J Comp Psychol 108:307–317

Carpenter M, Call J (2013) How joint is the joint attention of apes and human infants? In: Metcalf J, Terrace HS (eds) Agency and joint attention. Oxford University Press, Oxford

Carpenter M, Liebal K (2012) Joint attention, communication, and knowing together in infancy. In: Seemann A (ed) Joint attention: new developments in psychology, philosophy of mind, and social neuroscience. The MIT Press, Cambridge, MA

Cartmill EA, Byrne RW (2007) Orangutans modify their gestural signaling according to their audience's comprehension. Curr Biol 17:1345–1348

Douglas PH, Moscovice LR (2015) Pointing and pantomime in wild apes? Female bonobos use referential and iconic gestures to request genito-genital rubbing. Sci Rep 5:13999

Golinkoff RM (1986) 'I beg your pardon': the preverbal negotiation of failed messages. J Child Lang 13:455–476

Gómez J-C (2007) Pointing behaviors in apes and human infants: a balanced interpretation. Child Dev 78:729–734

Gómez J-C (2009) Embodying meaning: insights from primates, autism, and Brentano. Neural Netw 22:190–196

Herrmann E, Call J, Hernandez-Lloreda MV, Hare B, Tomasello M (2007) Humans have evolved specialized skills of social cognition: the cultural intelligence hypothesis. Science 317: 1360–1366

Hobaiter C, Leavens DA, Byrne RW (2014) Deictic gesturing in wild chimpanzees? Some possible cases. J Comp Psychol 128:82–87

Hopkins WD, Taglialatela JP, Leavens DA (2007) Chimpanzees differentially produce novel vocalizations to capture the attention of a human. Anim Behav 73:281–286

Hopkins WD, Taglialatela JP, Leavens DA (2011) Do chimpanzees have voluntary control of their facial expressions and vocalizations? In: Vilain A, Schwartz J-L, Abry C, Vauclair J (eds) Primate communication and human language: vocalisation, gestures, imitation and deixis in humans and non-humans. John Benjamins, Amsterdam, pp 71–88

Itakura S, Agnetta B, Hare B, Tomasello M (1999) Chimpanzee use of human and conspecific social cues to locate hidden food. Dev Sci 2:448–456

Kirchhofer KC, Zimmermann F, Kaminski J, Tomasello M (2012) Dogs (*Canis familiaris*), but not chimpanzees (*Pan troglodytes*), understand imperative pointing. PLoS One 7(2):e30913

Krause MA, Fouts RS (1997) Chimpanzee (*Pan troglodytes*) pointing: hand shapes, accuracy, and the role of eye gaze. J Comp Psychol 111:330–336

Leavens DA (2012a) Pointing: contexts and instrumentality. In: Pika S, Liebal K (eds) Current developments in primate gesture research. John Benjamins, Amsterdam

Leavens DA (2012b) Joint attention: twelve myths. In: Seemann A (ed) Joint attention: new developments in psychology, philosophy of mind, and social neuroscience. The MIT Press, Cambridge, MA
Leavens DA (2014) The plight of the sense-making ape. In: Cappuccio M, Froese T (eds) Enactive cognition at the edge of sense-making. Palgrave Macmillan, Basingstoke
Leavens DA, Bard KA (2011) Environmental influences on joint attention in great apes: implications for human cognition. J Cogn Educ Psychol 10:9–31
Leavens DA, Hopkins WD (1998) Intentional communication by chimpanzees: a cross-sectional study of the use of referential gestures. Dev Psychol 34:813–822
Leavens DA, Hopkins WD, Thomas RK (2004a) Referential communication by chimpanzees (*Pan troglodytes*). J Comp Psychol 118:48–57
Leavens DA, Hostetter AB, Wesley MJ, Hopkins WD (2004b) Tactical use of unimodal and bimodal communication by chimpanzees, *Pan troglodytes*. Anim Behav 67:467–476
Leavens DA, Russell JL, Hopkins WD (2005a) Intentionality as measured in the persistence and elaboration of communication by chimpanzees (*Pan troglodytes*). Child Dev 76:291–306
Leavens DA, Hopkins WD, Bard KA (2005b) Understanding the point of chimpanzee pointing: epigenesis and ecological validity. Curr Dir Psychol Sci 14:185–189
Leavens DA, Hopkins WD, Bard KA (2008) The heterochronic origins of explicit reference. In: Zlatev J, Racine T, Sinha C, Itkonen E (eds) The shared mind: perspectives on intersubjectivity. John Benjamins, Amsterdam, pp 187–214
Leavens DA, Bard KA, Hopkins WD (2010) BIZARRE chimpanzees do not represent "the chimpanzee". Behav Brain Sci 33:100–101
Leavens DA, Sansone J, Burfield AR, Lightfoot S, O'Hara S, Todd BK (2014) Putting the 'joy' in joint attention: affective-gestural synchrony by parents who point for their babies. Front Psychol 5(879):1–7
Leavens DA, Reamer LA, Mareno MC, Russell JL, Wilson D, Schapiro SJ, Hopkins WD (2015) Distal communication by chimpanzees (Pan troglodytes): evidence for common ground? Child Dev 86:1623–1638
Leavens DA, Bard KA, Hopkins WD (2017) The mismeasure of ape social cognition. Anim Cogn. doi:https://doi.org/10.1007/s10071-017-1119-1 (Published online in advance of print)
Liszkowski U, Carpenter M, Henning A, Striano T, Tomasello M (2004) Twelve-month-olds point to share attention and interest. Dev Sci 7:297–307
Lock A (2001) Preverbal communication. In: Bremner JG, Fogel A (eds) Blackwell handbook of infant development. Blackwell, Oxford
Lyn H, Russell JL, Hopkins WD (2010) The impact of environment on the comprehension of declarative communication in apes. Psychol Sci 21:360–365
MacCorquodale K, Meehl PE (1948) On a distinction between hypothetical constructs and intervening variables. Psychol Rev 55:95–107
Masur EF (1983) Gestural development, dual-directional signaling, and the transition to words. J Psycholinguist Res 12:93–109
Menzel CR (1999) Unprompted recall and reporting of hidden objects by a chimpanzee (*Pan troglodytes*) after extended delays. J Comp Psychol 113:426–434
Miklósi Á, Soproni K (2006) A comparative analysis of animals' understanding of the human pointing gesture. Anim Cogn 9:81–93
Moore R (2015) Meaning and ostension in great ape gestural communication. Anim Cogn 19: 223–231
Moore R (2017) Gricean communication and cognitive development. Philos Q 267:303–326
Moore C, Corkum V (1994) Social understanding at the end of the first year of life. Dev Rev 14:349–372
Mundy P, Sigman M, Ungerer J, Sherman T (1986) Defining the social deficits of autism: the contribution of non-verbal communication measures. J Child Psychol Psychiatry Allied Discip 27:657–669

Nisbett RE, Wilson TD (1977) Telling more than we can know: verbal reports on mental processes. Psychol Rev 84:231–259

Pedersen J, Segerdahl P, Fields WM (2009) Why apes point: pointing gestures in spontaneous conversation of language-competent Pan/Homo bonobos. In: Potocki E, Krasinski J (eds) Primatology: theories, methods, and research. Nova, Hauppauge, NY

Povinelli DJ, Eddy TJ (1996) What young chimpanzees know about seeing. Monogr Soc Res Child Dev 61(3):i–vi, 1–152

Povinelli DJ, Reaux JE, Bierschwale DT, Allain AD, Simon BB (1997) Exploitation of pointing as a referential gesture in young children, but not adolescent chimpanzees. Cogn Dev 12:423–461

Povinelli DJ, Bierschwale DT, Čech CG (1999) Comprehension of seeing as a referential act in young children but not juvenile chimpanzees. Br J Dev Psychol 17:37–60

Racine TP (2012) Getting beyond rich and lean views of joint attention. In: Pika S, Liebal K (eds) Current developments in primate gesture research. John Benjamins, Amsterdam

Racine TP, Carpendale JIM (2007) The role of shared practice in joint attention. Br J Dev Psychol 26:3–25

Rheingold HL, Hay DF, West MJ (1976) Sharing in the second year of life. Child Dev 47: 1148–1158

Roberts AI, Vick S-J, Roberts SGB, Menzel CR (2014) Chimpanzees modify intentional gestures to coordinate a search for hidden food. Nat Commun 5:3088

Savage-Rumbaugh ES, Rumbaugh DM, MacDonald K (1985) Language learning in two species of apes. Neurosci Biobehav Rev 9:653–665

Sehon SR (2000) An argument against the causal theory of action explanation. Philos Phenomenol Res 60:67–85

Thomas E, Murphy M, Pitt R, Rivers A, Leavens DA (2008) Understanding of visual attention by adult humans (*Homo sapiens*): a partial replication of Povinelli, Bierschwale, and Čech (1999). J Comp Psychol 122:428–436

Tomasello M (1995) Joint attention as social cognition. In: Moore C, Dunham P (eds) Joint attention: its origins and role in development. Lawrence Erlbaum Associates, Mahwah, NJ

Tomasello M, Call J, Gluckman A (1997) Comprehension of novel communicative signs by apes and human children. Child Dev 68:1067–1080

Tomasello M, Carpenter M, Liszkowski U (2007) A new look at infant pointing. Child Dev 78: 705–722

van der Goot MH, Tomasello M, Liszkowski U (2014) Differences in the nonverbal requests of Great Apes and human infants. Child Dev 85:444–455

Veà JJ, Sabater-Pi J (1998) Spontaneous pointing behaviour in the wild pygmy chimpanzee (*Pan paniscus*). Folia Primatol 69:289–290

Woodruff G, Premack D (1979) Intentional communication in the chimpanzee: the development of deception. Cognition 7:333–362

A Comparison of the Socio-communicative Behavior in Chimpanzees and Bonobos

Jared P. Taglialatela, Scott C. Milne, and Robert E. Evans

Abstract Studying the similarities and differences in socio-communicative behavior between chimpanzees and bonobos is critical to increasing our understanding of the evolution of human sociality and communication. Both species rely heavily on the use of vocalizations during communicative interactions, although the form and function of these signals may vary between the two ape species. For example, bonobo vocalizations seem to be structurally more complex than those produced by chimpanzees, and calls seem to be directed to individuals not in immediate physical proximity. Both species, however, make use of communicative signals from different modalities concurrently, particularly vocalizations and manual gestures. However, this multimodal communication is more commonly observed in chimpanzees when compared to bonobos, who more frequently use vocalizations exclusively, without signals from additional modalities. In addition, there are a number of marked differences in social characteristics between the two species. Though both species exhibit fission–fusion behavior, chimpanzees do so more often, potentially as a result of their habitat profile and foraging strategies. Differences also exist in terms of dominance and aggression. Chimpanzees live in largely patriarchal societies with strong male–male bonds, whereas bonobos tend to be matriarchal with strong female–female bonds. These differences in communicative and social characteristics are thought to be linked to the ecology of the respective habitats of the two apes and their strategies for resource exploitation. In all likelihood, similarities in feeding ecology played a crucial role in selecting for the advanced cognitive abilities of both species—e.g., producing meaningful communicative signals, regulating competition and group cohesion, and making and utilizing tools to aid in foraging.

J. P. Taglialatela (✉)
Department of Ecology, Evolution, and Organismal Biology, Kennesaw State University, Kennesaw, GA, USA

Ape Cognition and Conservation Initiative, Des Moines, IA, USA
e-mail: jtaglial@kennesaw.edu

S. C. Milne · R. E. Evans
Department of Ecology, Evolution, and Organismal Biology, Kennesaw State University, Kennesaw, GA, USA

L. D. Di Paolo et al. (eds.), *Evolution of Primate Social Cognition*, Interdisciplinary Evolution Research 5, https://doi.org/10.1007/978-3-319-93776-2_6

Similarly, differences in their respective habitats may have led to selection for characteristics resulting in the behavioral differences observed today between the two species. Therefore, a clearer understanding of the similarities and differences between the two species most closely related to humans will provide valuable information into our own evolution by elucidating those characteristics shared among humans and our ape ancestors and those derived in the hominin lineage.

Keywords Chimpanzee · Bonobo · Language evolution · Social behavior · Primate communication

1 Introduction and Background

The evolutionary origins and adaptive value of living in large and complex social groups, such as those seen in primates, including humans, have placed particular emphasis on the creation and maintenance of social bonds. Although these bonds can be maintained through a variety of mechanisms, in our own species, language is perhaps the most important implement used to maintain these social bonds. Indeed, human language is unique in the animal kingdom in both its complexity and in the amount of information contained within its signals. Although the vocal domain plays a predominant role, humans actually use multiple signaling modalities concurrently when communicating. For example, people frequently produce manual gestures while speaking (McNeill 1992; Kimura 1973), and the appearance of a speaker's lips and mouth articulating a certain phoneme influences the speech sound that is perceived by a listener (McGurk and MacDonald 1976). Therefore, in terms of both production and perception, human language utilizes more than just the auditory modality.

Given that human language does not leave direct or indelible marks in the fossil record, the study of communication in the extant great apes is critical for understanding the evolutionary origins of human language and speech. Specifically, it is necessary to distinguish ancestral characteristics (i.e., those shared by humans and their closest extant relatives, chimpanzees, and bonobos) from those that are derived in the hominin lineage. To accomplish this, direct investigations and comparisons of the communicative repertoires of the two great ape species most closely related to humans, chimpanzees (*Pan troglodytes*), and bonobos (*Pan paniscus*) are required.

2 Continuities Between the Communicative Behaviors of Extant Apes and Human Language

Recent data indicate that like human language, the communicative signals of apes are learned, used flexibly, and are intentionally produced (Call et al. 2004; Hostetter et al. 2001; Krause and Fouts 1997; Leavens et al. 2004, 2005; Tomasello et al.

1994; Hopkins et al. 2007; Taglialatela et al. 2012; Russell et al. 2013). For example, captive chimpanzees, and other great apes, produce manual communicative gestures only when a human is present and visually oriented toward them (Call et al. 2004; Hostetter et al. 2001; Krause and Fouts 1997; Leavens et al. 1996, 2004; Poss et al. 2006; Tomasello et al. 1994). In such situations, chimpanzees alternate their gaze between a referent (food) and a social agent while gesturing (Leavens and Hopkins 1998) and "repair" these communicative attempts when they have failed (Leavens et al. 2005). Furthermore, previous data indicate that the production of some chimpanzee calls in conjunction with manual gestures selectively activates the Broca's area homologue in chimpanzees, a region critical for speech and language production in humans (Taglialatela et al. 2008, 2011). Collectively, these results suggest that competencies that are fundamental to human language were present in the common ancestor of chimpanzees/bonobos and humans approximately 6 million years ago (MYA). However, the selection pressures that led to the evolutionarily unprecedented capacity for full-blown language and speech in humans still elude our understanding. By closely examining the communicative systems of our closest extant phylogenetic relatives, we are afforded an opportunity to (1) identify those characteristics that were likely present in the most recent common ancestor of chimpanzees/bonobos and humans and (2) hypothesize regarding the selection pressures that may have led early hominins down an unprecedented evolutionary path resulting in the emergence of human language.

3 Comparing Chimpanzee and Bonobo Socio-communicative Behavior

Before comparing the communicative behaviors of chimpanzees and bonobos, we must first have an understanding of the natural history of both species and if and to what extent these factors may account for the observed socio-communicative differences between the two species. Though both species belong to the genus *Pan*, chimpanzees and bonobos diverged from the *Homo* lineage approximately 6 MYA and subsequently diverged from one other as recently as 1 MYA (Becquet et al. 2007; Hey 2010). Both *Pan* species are found in sub-Saharan Africa; however they do not experience habitat overlap in any part of their range. Instead, bonobos live allopatrically to the South of the Congo River, where they are confined to a relatively small but densely forested area within the Democratic Republic of Congo (DRC) (Wrangham et al. 1996; Yamagiwa 1999; Yamakoshi 2004). Here, bonobos are free from competition with the phylogenetically close, and morphologically similar, gorillas (Malenky et al. 1994). Within their range, bonobos have access to swampy habitats as well as primary and secondary forest, though the majority of foraging and daily activity is conducted within the drier areas of their habitat (Hashimoto et al. 1998). Chimpanzees, in contrast, are found to the North of the Congo River where they occupy a wide variety of habitat types, ranging from dry, grassy savannah in the western extremes of their distribution to dense rainforest near the equator (Goodall 1986). Unlike bonobos, chimpanzees experience significant habitat overlap with gorillas in large portions of their range (Stanford and Nkurunungi 2003; Head et al. 2011). This overlap with gorillas, a phylogenetically and morphological similar species, is thought to have played a role in the divergence of socio-communicative traits in chimpanzees and bonobos (Tutin et al. 1991; Malenky and Wrangham 1994).

In recent decades, hypotheses aimed to explain the differences in gestural, and particularly vocal, communication between the two species have focused on differences in how each species utilizes their respective habitats and on contrasts in patterns of social behavior within groups of conspecifics (Head et al. 2011; Malenky and Wrangham 1994; Malenky et al. 1994). However, limited data currently exist directly comparing the socio-communicative behavior of chimpanzees and bonobos, despite the abundant evidence that there are considerable differences between the two species in aspects of social tolerance, social structure, and pro-social behavior (de Waal 1988; Pollick and de Waal 2007). Although both species exhibit a fission–fusion social organization and foraging behavior, bonobos tend to exhibit this behavior less often, and foraging subgroups are much more likely to be age and sex diverse (Watts et al. 2012; White and Wrangham 1988). In addition, chimpanzees are male dominated and tend to exhibit male–male bonding around food sites, have relatively weak alliances among females within a group, and have a highly promiscuous mating system (White 1996; Wrangham et al. 1996; Yamagiwa 1999; Malenky et al. 1994; Yamakoshi 2004; Sakura 1994). Perhaps as a result, chimpanzees are also described as aggressive and highly territorial with relatively frequent intergroup violence. In contrast, the available data indicate that bonobos show higher degrees of female–female bonding than chimpanzees, especially around feeding sites, have matriarchal social groups, and tend to be less aggressive and highly gregarious (White 1996; Malenky et al. 1994; Hare et al. 2012; Parish et al. 2000; Wobber et al. 2010).

Despite these differences, chimpanzees and bonobos both rely heavily on species-specific signals that include a rich repertoire of vocalizations, facial expressions, body postures, and manual gestures that they deploy tactically during communicative interactions (Clay and Zuberbuhler 2009; Hopkins et al. 2007; Taglialatela et al. 2015). As a result, studying the production and usage of vocalizations and gestures, as well as their neural correlates, in both species of *Pan*, has become a common tool used to understand the evolutionary origins of our own communication (Fedurek and Slocombe 2011). Differences have been identified in the structure, tonality, pitch, length, and function, among other things, of calls in bonobos and chimpanzees (Clay and Zuberbuhler 2009; de Waal 1988; Pollick and de Waal 2007). For example, in contrast to the low-frequency and noisy *grunts* and *barks* of chimpanzees, bonobos produce relatively high-frequency tonal *peeps* and *yelps* (de Waal 1988; Pollick and de Waal 2007). Although, it is unknown what accounts for the acoustic differences between the calls of chimpanzees and those produced by bonobos, it is possible that they may have been driven by differences in their feeding ecology. For example, both species utilize calls that are thought to function as a method of long-distance communication between individuals of a group. However, the specific information contained in these calls has been hypothesized to differ between chimpanzees and bonobos (Mitani 1996). Specifically, significant contrasts seem to exist when the function of each species' vocalizations, as they relate to group structure and social cohesion, is considered (Hohmann and Fruth 1994).

Since group composition is highly variable in chimpanzees as well as in bonobos, and both species exhibit high degrees of fission–fusion behavior, it is likely that communicative signals directed to conspecifics outside of visual range are highly important (Hohmann and Fruth 1994). In chimpanzees, long-distance calls are thought to play a major role in regulating group spacing around feeding sites and in the maintenance of dominance hierarchies within groups (White 1996; Hohmann and Fruth 1994). Studies of captive and wild bonobos, however, show that in both settings, they produce high-frequency vocalizations that are tonal, frequency modulated, and often produced in a series (de Waal 1988; Bermejo and Omedes 1999; Hohmann and Fruth 1994). These characteristics suggest that relatively more complex information may be carried within these signals and over relatively longer distances as compared to the noisy grunts and barks produced by chimpanzees (Goodall 1986). As a result, bonobos may use vocalizations as a way to coordinate group movement over large areas and to localize individuals of a group. In fact, some studies have suggested that certain calls in bonobos (e.g., high hoots) may be used to elicit similar calls from other groups and, therefore, likely function to maintain social cohesion between subgroups (Hohmann and Fruth 1994).

Not surprisingly, the function of vocalizations in bonobos and chimpanzees seems to reflect differences in the sociality of the two species. Among captive bonobos, call types do not seem to be bound to a single socio-communicative context and are often not directed to a specific individual. Captive chimpanzee vocalizations, on the other hand, seem to be used in more restricted contexts and also tend to be directed toward individual conspecifics (Moore 2014). However, neither species relies exclusively on vocal utterances to communicate. Chimpanzees and bonobos also utilize manual gestures, body postures, and facial expressions to enhance or modify their communicative messages. Communication of this sort can range from simply clapping the hands, presumably to gain the attention of a communicative partner, to bearing the teeth and swaying the body, signals most commonly associated with aggression (Slocombe et al. 2011).

Although vocalizations and gestures alone are crucial to chimpanzee and bonobo communication, both species are also known to use communicative signals from different modalities concurrently (Moore 2014; Slocombe et al. 2011; Taglialatela et al. 2015). Similar to the way that humans often pair their manual gestures, facial expressions, and changes in body postures with spoken words to provide the listener with more information, chimpanzees and bonobos often also use a signal from another modality along with a vocalization. The utilization of manual gestures concomitantly with vocal utterances in *Pan* suggests that this form of multimodal communication was likely present prior to the development of spoken language in early hominins (Taglialatela et al. 2011, 2015). By using multiple signals from different modalities concurrently, the signaler is able to provide potentially more complex or precise information than in a vocalization alone (Pollick and de Waal 2007). Recently, it has been reported that for both chimpanzees and bonobos, gestural communication is less likely to be restricted to a specific behavioral or social context than vocal communication (Pollick and de Waal 2007). Interestingly, in captive populations, it has also recently been shown that chimpanzees are more likely to pair a vocalization with a manual gesture than bonobos (Moore 2014). Preliminary data suggest that approximately 51% of the vocal utterances produced by chimpanzees are paired with a signal from a different communicative modality (e.g., visual and/or tactile), whereas only 15% of bonobo vocalizations are produced in conjunction with a signal from another modality (Moore 2014). These data are consistent with the hypothesis that bonobos use complex vocalizations autonomously in a wider variety of contexts than chimpanzees. If bonobos utilize vocalizations in the absence of signals from other modalities more often than chimpanzees, and chimpanzees are more likely than bonobos to utilize contextually restricted vocalizations paired with flexible gestures, then it is possible that the ecological profiles and habitat-use patterns of the two species may help provide an explanation for the divergence of these traits.

4 Connecting the Ecology of the *Pan* Species to Putative Differences in Their Socio-communicative Behavior

The differences in the socio-communicative behaviors of chimpanzees and bonobos are likely, at least in part, the result of different ecological conditions present in the habitats of the two ape species, as well as how each species utilizes their respective habitats, over the course of their recent and independent evolutionary histories. Although both species are found in equatorial Africa, their habitats are separated by the Congo River, which forms a natural boundary between the ranges of the two species. Interestingly, the phylogenetic divergence of the *Gorilla/Pan* lineages seems to correlate well with the formation of the Congo River as a natural barrier. Molecular data using mitochondrial DNA sequencing have placed the divergence of the *Gorilla/Pan* lineages at somewhere between 7 and 10 MYA (Sibley and Ahlquist 1984; Patterson et al. 2006; Zihlman et al. 1978). Similarly, these same data have estimated that the *Pan* lineage diverged from the *Homo* lineage around 6–8 MYA and that of chimpanzees and bonobos subsequently diverged from each other somewhere in the range of 1–2 MYA (Sibley and Ahlquist 1984). The divergence of the *Pan* lineage from the *Gorilla* lineage correlates well with the Tertiary and Quaternary Periods of the Miocene epoch, which occurred approximately 5–10 MYA. The ecological changes associated with these periods are thought to have led to niche convergence between chimpanzees and gorillas and divergence of bonobos and may have provided selective pressure for the development of different foraging strategies, leading to the presence of the differences in socio-communicative characteristics seen today in bonobos and chimpanzees (Yamagiwa et al. 1996).

To understand the connection between communication and feeding ecology in both chimpanzees and bonobos, it is important to understand the differences that exist in the foraging behavior of these two species. Most notable among these differences is the tendency of bonobos to use terrestrial herbaceous vegetation (THV) as a fallback food in times of fruit scarcity. This is often contrasted with the maintenance of relatively higher levels of year-round ripe fruit foraging and consumption by chimpanzees regardless of its seasonal abundance (White 1998; White and Wrangham 1988; White and Chapman 1994). Differences also exist in home and day range sizes as well as in the diversity of foods eaten (White 1998; Tutin 1996). For example, whereas chimpanzees and bonobos are both considered ripe fruit specialists, bonobos have a much broader range of staple fruits in their diet as well as higher rates of consumption of pith, leaves, and other THV compared to chimpanzees. The tendency for bonobos to supplement fruit with THV suggests a more opportunistic foraging behavior that is thought to result in the maintenance of social bonds and likely facilitates the large, diverse group sizes of bonobos (White and Chapman 1994; Head et al. 2011; Yamagiwa et al. 1996). These distinctions form the basis of the THV hypothesis which suggests that chimpanzees travel long distances to maintain a consistent uptake of ripe fruit year-round. This is likely due to the fact that chimpanzees live sympatrically with gorillas in so much of their range, whereas bonobos do not (Wrangham 1986). This overlap in habitats among gorillas and chimpanzees suggests that the two species have undergone relatively recent competitive exclusion. Specifically, gorillas tend to consume THV more often than chimpanzees (Stanford and Nkurunungi 2003). It is thought that this competition may explain some of the social characteristics seen in modern chimpanzees that distinguish them from bonobos. These characteristics can be seen as adaptations that emerged as a result of the presence of gorillas and allow chimpanzees to maintain an uptake of essential food sources (Yamakoshi 2004). The THV hypothesis also suggests that close bonds in female bonobos and the gregariousness of bonobo groups are likely a result of the availability of widely dispersed, but generally low-quality, THV that is not contested by gorillas (Yamagiwa et al. 2012; Wrangham et al. 1996; Wrangham 1986). In fact, it has been suggested that West African chimpanzees, a subspecies living free of habitat overlap with gorillas, may actually exhibit social characteristics and levels of group cohesion that seem to parallel bonobos more than other chimpanzee subspecies (Yamakoshi 2004).

The THV hypothesis can be used to formulate a number of specific predictions with regard to shared and distinct social characteristics among chimpanzees, bonobos, and gorillas. For example, both bonobos and gorillas preferentially feed on the fruit but utilize THV regularly as a fallback food throughout the year. The THV hypothesis suggests that the wide dispersal and relatively low quality of the plant materials that make up THV help to facilitate large foraging parties that have a low potential for competition and can harbor strong bonds between conspecifics (Wrangham 1986). As a result of these similarities in their feeding strategies, one can predict that gorillas and bonobos should show similarities in social behavior, specifically in their tolerance of conspecifics while in a context with a high potential for competition, such as feeding (Milne 2015). Similarly, if chimpanzee foraging behavior and feeding ecology contrast with that of bonobos and gorillas, then their social behavior should similarly reflect this difference. Indeed, careful consideration of studies that have examined the social and foraging behavior of bonobos, chimpanzees, and/or gorillas has revealed parallels in the socioecology of gorillas and bonobos in terms of gregariousness among group members and social bonding around food sites. Similarly, contrasts in these same characteristics have been found in chimpanzee groups. For example, foraging groups in chimpanzees are often smaller, aggressive, and feeding on fruit (Stanford and Nkurunungi 2003; White and Wrangham 1988). In fact, some of these data, along with comparisons of communicative behavior in the *Gorilla* and *Pan* species, suggest that bonobos may actually exhibit socio-communicative behavior that is at somewhat of an intermediate point between chimpanzees and gorillas. For example, bonobos have communicative behavior more similar to chimpanzees than gorillas (Hohmann and Fruth 1994; Pollick and de Waal 2007) but social and foraging behavior more similar to gorillas than chimpanzees in many ways (Wrangham 1986; Tutin et al. 1991; Yamakoshi 2004). However, the majority of these studies are from the field and in study groups living in habitats where the visibility of the focal individual is an issue and data collection is often difficult and sometimes inconclusive as a result (White and Wrangham 1988; Yamakoshi 2004). Moreover, several important variables are completely uncontrollable in the wild, such as the distribution of resources or the quality of a food source. While this provides an advantage to comparing the natural ecology of chimpanzees and bonobos, it makes direct comparisons of their social behavior difficult. As a result, some of the claims of the THV hypothesis, as they relate to the connection between the availability of plant material to the African apes and their subsequent social behaviors, have recently been debated (Yamakoshi 2004). But, to fully understand the connection between ecology, sociality, and ultimately communication, a direct comparison of these categories for all of the African Ape species, bonobo, chimpanzee, and gorilla, is necessary (Tutin et al. 1991; Yamakoshi 2004). Though these types of comparisons are relatively scarce at this point, data from our lab suggest that there may not be the parallels and differences in social behavior of the African Apes that their respective ecologies would suggest.

For example, we found that, in feeding contexts, when the potential for competition and aggressive interactions is presumably highest, bonobos were the most social of the three species, as expected, but that gorillas were the least social. Measures of sociality were based on how much time a focal individual spent alone (i.e., greater than 1.5 m from a conspecific), in a given context. In fact, we found, that in captive apes, no significant similarities exist in sociality between gorillas and bonobos in any context, and, though both species spent similar amounts of time foraging, bonobos spent only 68% of their time alone in this context versus gorillas spending 81% of their time alone in feeding contexts (Milne 2015). This contrast is odd given that the THV hypothesis suggests that the availability of low-quality, highly dispersed THV, a resource that gorillas and bonobos both have regular access to, facilitates large, gregarious groups with strong bonds and low instances of aggression. However, comparing fundamental aspects of the socioecology of the African Great apes in ways such as this can help to reveal possible instances of divergence or convergence of social and communicative traits, as a result of feeding ecology or not, and significantly more research into this putative connection is needed (Tutin et al. 1991; Yamakoshi 2004).

The ecological characteristics and foraging profiles of chimpanzees and bonobos, however, have likely influenced the socio-communicative strategies unique to each species. Since both species occupy a variety of habitats with differing degrees of visibility, it is probable that the signal modality that each species uses to communicate with one another are linked to the characteristics of those habitats as well as the social organization and communication that they facilitate.

Chimpanzees often travel through multiple habitat types in a given day, and these habitats have varying levels of visibility (Goodall 1986). It is therefore possible that chimpanzees use seemingly less complex vocal communication than bonobos, but often pair these auditory signals with visual gestures. It may be efficient in these types of habitats to use a relatively rigid vocal repertoire, combine flexible gestures with these vocalizations when feeding in close contact with conspecifics as a way to regulate competition and communicate social roles. However, given that bonobos feed more often than chimpanzees on dispersed food sources like THV, and often have very large and diverse groups, it may be more beneficial to rely heavily on information that can be carried in a complex vocalization alone, rather than a gesture that would require visual contact with a conspecific. In this way, bonobos are able to maintain group cohesion in densely forested habitats and locate individuals when they are spread among large patches of herbaceous foods.

5 Comparative Genomics

In recent years, genomic tools have become increasingly available enabling behavioral comparisons between chimpanzees and bonobos to be supplemented with genetic analyses at both the individual and species levels. Noninvasive or minimally invasive techniques are readily available to collect DNA from individuals and conduct genetic analysis to determine if, and to what extent, observed behavioral differences can be explained by genetic variation. Recently, several genes including the arginine vasopressin receptor 1a gene (AVPR1a), the monoamine oxidase A gene (MAOA), and the oxytocin receptor gene (OXTR) have been identified as candidates for contributing to observed differences in social behavioral phenotypes (Garai et al. 2014; Donaldson and Young 2008). One gene of particular interest regarding chimpanzee and bonobo comparisons is AVPR1a.

The vasopressin neuropeptide itself has a conserved role in regulating social behaviors in mammals, and the nucleotide sequence within the coding region of AVPR1a is highly conserved among mammals. However, there is significant variability in the 5′ microsatellites of AVPR1a (Hammock and Young 2004), and there is considerable evidence that these microsatellites in the 5′ noncoding region may alter individual or species-level expression patterns of the receptor for this neuropeptide (Donaldson et al. 2008). Thus, interspecies behavioral patterns may be modulated by the modified expression of vasopressin receptors across the brains of different species.

A polymorphism 5′ of the AVPR1a gene in prairie voles (*Microtus ochrogaster*) has been associated with monogamy, paternal care, and social interest (Hammock and Young 2005). Young and colleagues have also shown that this regulatory polymorphism is associated with individual differences in patterns of central vasopressin receptor distribution (Hammock et al. 2005). The identification of this polymorphism as well as its ability to profoundly modify social behaviors in rodents sparked interest in determining if polymorphisms 5′ of AVPR1a have similar effects on behavioral phenotypes in humans. In fact, variations in length of a specific microsatellite (RS3) are associated with altruism, male pair-bonding behaviors, incidence of autism, and amygdala activation during a facial recognition task (Knafo et al. 2008; Walum et al. 2008; Meyer-Lindenberg et al. 2009; Tansey et al. 2011).

While investigating the phylogenetic history of microsatellites 5′ of AVPR1a, Donaldson et al. (2008) identified a polymorphic deletion of ~360 bp in the 5′ region of chimpanzees referred to as the DupB microsatellite. The DupB microsatellite contains the RS3 allele that has been shown in humans to have an impact on social behavior. Furthermore, and perhaps most relevant for this discussion, it has been shown that chimpanzees are the only great apes that are polymorphic for a complete deletion of this microsatellite (Donaldson et al. 2008; Staes et al. 2014). In addition, among the subspecies of chimpanzees, the DupB deletion is more common in chimpanzees of West African origin. Specifically, among captive chimpanzees of West African origin, approximately 70% have a complete deletion of the DupB microsatellite, whereas approximately 62% of eastern African chimpanzees living in the wild retain the DupB microsatellite (Anestis et al. 2014).

As is the case with many other species that occupy varied habitats, chimpanzee behavior is closely associated with habitat (Boesch 1994). For example, the likelihood of participating in cooperative hunting and/or sharing meat is much more common among forest chimpanzees compared to savanna-woodland chimpanzees (Boesch and Boesch 1989). Although considerably more data are needed, variation in behavior among chimpanzees inhabiting different habitats is likely related to the different ecological conditions present in these environments. If the presence of the DupB microsatellite is more common in some chimpanzee subspecies and there are observed behavioral differences between those species, modifications to the noncoding *cis* portion of genes involved in social cognition may have provided a substrate for rapid evolution of behavioral phenotypes to fit variations in ecological conditions (King et al. 1997; Hammock and Young 2007).

There continues to be considerable interest if and to what extent the DupB polymorphism of AVPR1a in chimpanzees is associated with social behavior or is predictive of a particular behavioral phenotype. The majority of these studies have relied on quantifying "personality styles" either through handler questionnaires or observational data (Anestis et al. 2014; Hopkins et al. 2012). Anestis and colleagues (2014) demonstrated an association between DupB+ genotype and a "smart" personality type, defined as those individuals that often used coalitions in aggressive encounters, received more grooming than they gave, and successfully initiated play with social partners (Anestis et al. 2014). Hopkins and colleagues (2012) identified a sex difference in personality—males with the DupB+ genotype had higher scores of "dominance" and lower scores of "conscientiousness" than those with the DupB-genotype. However, there was no difference in dominance and contentiousness found between males and females with the DupB-genotype (Hopkins et al. 2012). More recently, Staes and colleagues (2015) found that males with the homozygous DupB+/+ genotype were more "sociable" as determined by a personality analysis despite a small sample size of three DupB+/+ males (Staes et al. 2015). These data provide additional support to the hypothesis that variation at this allele is associated with observable differences in social behavior (Staes et al. 2014). Taken collectively, the current literature suggests that this microsatellite likely exerts some control on socio-behavioral phenotypes in chimpanzees.

In contrast to chimpanzees, bonobos are not polymorphic for the deletion of the DupB locus (Donaldson et al. 2008; Staes et al. 2014). Chimpanzees and bonobos shared a common ancestor approximately 1–2 MYA and exhibit very similar patterns of complex socio-communicative behavior that can be compared directly (Prüfer et al. 2012). However, and as discussed above, they show notable differences in socio-communicative behavior as well. In addition, these two species are the closest extant relatives of our own species, sharing a common ancestor with humans approximately 6 MYA. Therefore, chimpanzees and bonobos present an excellent model to determine if, and to what extent, variation in microsatellites in the non-coding region of genes (notably, AVPR1a) impact complex socio-communicative behavior. Comparisons between closely related, behaviorally similar species with divergent social behaviors will allow us to understand how variations in noncoding genetic elements can contribute to the rise of specific social behaviors. Additionally, the close evolutionary relationship chimpanzees and bonobos share with humans makes this research directly applicable to understanding the genetic factors underlying differential socio-behavioral phenotypes in humans.

6 Conclusions

Studies on the variability of socio-communicative behaviors of the two species most closely related to humans, bonobos, and chimpanzees have and continue to provide unprecedented insight into the evolutionary origins of human language and cognition. Specifically, by identifying both similarities and differences in the ways social information is transmitted between individuals, we can identify characteristics of communicative interactions that may have been present in the last common ancestor of hominins and *Pan*. Identifying shared and derived aspects of communicative behavior allows us to hypothesize about the ecological conditions that may have led to the differences in socio-communicative behavior observed among great apes today. Without the ability to make observational comparisons of social behaviors among extant species that are closely related to humans, it would be impossible to identify the conditions that lead to the evolutionary origin of the complex socio-communicative behavior in our own species.

By integrating behavioral observations of chimpanzees and bonobos with comparative genomics, we will be able to not only hypothesize about the ecological conditions that contributed to differences in socio-communicative behavior but identify specific genetic elements associated with specific socio-communicative behaviors. In this way, the study of socio-communicative behavior in bonobos and chimpanzees will provide unprecedented information about the evolutionary origins of human language.

References

Anestis SF, Webster TH, Kamilar JM, Fontenot MB, Watts DP, Bradley BJ (2014) AVPR1A variation in chimpanzees (Pan troglodytes): population differences and association with behavioral style. Int J Primatol 35(1):305–324

Becquet C, Patterson N, Stone AC, Przeworski M, Reich D (2007) Genetic structure of chimpanzee populations. PLoS Genet 3(4):e66. https://doi.org/10.1371/journal.pgen.0030066

Bermejo M, Omedes A (1999) Preliminary vocal repertoire and vocal communication of wild bonobos (*Pan paniscus*) at Lilungu (Democratic Republic of Congo). Folia Primatol 70(6):328–357

Boesch C (1994) Cooperative hunting in wild chimpanzees. Anim Behav 48(3):653–667

Boesch C, Boesch H (1989) Hunting behavior of wild chimpanzees in the Tai National Park. Am J Phys Anthropol 78:547–573

Call J, Hare B, Carpenter M, Tomasello M (2004) 'Unwilling' versus 'unable': chimpanzees' understanding of human intentional action. Dev Sci 7(4):488–498

Clay Z, Zuberbuhler K (2009) Food-associated calling sequences in bonobos Anim Behav. https://doi.org/10.1016/j.anbehav.2009.02.016

de Waal FBM (1988) The communicative repertoire of captive bonobos (Pan paniscus), compared to that of chimpanzees. Behaviour 106(3):183–251

Donaldson ZR, Young LJ (2008) Oxytocin, vasopressin and the neurogenetics of sociality. Science 322:900–904

Donaldson ZR, Bai Y, Kondrashov FA, Stoinski TL, Hammock EAD, Young LJ (2008) Evolution of a behavior-linked microsatellite-containing element of the 5′ flanking region of the primate avpr1a gene. BMC Evol Biol 8:180–188

Fedurek P, Slocombe KE (2011) Primate vocal communication: a useful tool for understanding human speech and language evolution? Hum Biol 83(2):153–173

Garai C, Furuichi T, Kawamoto Y, Ryu H, Inoue-Murayama M (2014) Androgen receptor and monoamine oxidase polymorphism in wild bonobos. Meta Gene 2:831–843

Goodall J (1986) The chimpanzees of Gombe: patterns of behavior. Harvard University Press, Cambridge

Hammock EA, Young LY (2004) Functional microsatellite polymorphisms associated with divergent social structure in vole species. Mol Biol Evol 21:1057–1063

Hammock EAD, Young LJ (2005) Microsatellite instability generates diversity in brain and sociobehavioral traits. Science 308:1630–1634

Hammock EA, Young LJ (2007) Neuropeptide systems and social behavior: noncoding repeats as a genetic mechanism for rapid evolution of social behavior. Evol Nerv Syst 3:361–371

Hammock EA, Lim MM, Nair HP, Young LJ (2005) Assocation of vasopressin 1a receptor levels with a regulatory microsattelite and behavior. Genes Brain Behav 4(5):289–301

Hare B, Wobber T, Wrangham R (2012) The self-domestication hypothesis: bonobo psychology evolved due to selection against male aggression. Anim Behav 83:573–585

Hashimoto C, Tashiro Y, Kimura D, Enomoto T, Ingmanson EJ, Idani G, Furuichi T (1998) Habitat use and ranging of wild bonobos (*Pan paniscus*) at Wamba. Int J Primatol 19: 1045–1060

Head JS, Boesch C, Makaga L, Robbins MM (2011) Sympatric chimpanzees (*Pan troglodytes* troglodytes) and gorillas (Gorilla gorilla gorilla) in Loango National Park, Gabon: dietary composition, seasonality, and intersite comparisons. Int J Primatol 32(3):755–775

Hey J (2010) The divergence of chimpanzee species and subspecies as revealed in multi population isolation-with-migration analyses. Mol Biol Evol 27:921–933

Hohmann G, Fruth B (1994) Structure and use of distance calls in wild bonobos (*Pan paniscus*). Int J Primatol 15(5):767–782

Hopkins WD, Taglialatela JP, Leavens DA (2007) Chimpanzees differentially produce novel vocalizations to capture the attention of a human. Anim Behav 73:281–286

Hopkins WD, Donaldson ZR, Young LY (2012) A polymorphic indel containing the RS3 microsatellitein the 5′ flanking region of the vasopressin V1a receptor gene is associated with chimpanzee (Pan troglodytes) personality. Genes Brain Behav 11:552–558

Hostetter AB, Cantero M, Hopkins WD (2001) Differential use of vocal and gestural communication by chimpanzees (*Pan troglodytes*) in response to the attentional status of a human (Homo sapiens). J Comp Psychol 115(4):337–343

Kimura D (1973) Manual activity during speaking— I. Right-handers. Neuropsychologia 11:45–50

King DG, Soller M, Kashi Y (1997) Evolutionary tuning knobs. Endeavour 21(1):36–40

Knafo A, Israel S, Darvasi A, Bachner-Melman R, Uzefovsky F, Cohen L, Feldman E, Lerer E, Laiba E, Raz Y, Nemanov L, Gritsenko I, Dina C, Agam G, Dean B, Bornstein G, Ebstein RP (2008) Individual differences in allocation of funds in the dictator game and post-mortem hippocampal mRNA levels are correlated with length of the arginine vasopressin 1 a receptor (AVPR1A) RS3 promotor region repeat. Genes Brain Behav 7:266–275

Krause MA, Fouts RA (1997) Chimpanzee (*Pan troglodytes*) pointing: hand shapes, accuracy, and the role of eye gaze. J Comp Psychol 111:330–336

Leavens DA, Hopkins WD (1998) Intentional communication by chimpanzee (*Pan troglodytes*): cross-sectional study of the use of referential gestures. Dev Psychol 34:813–822

Leavens DA, Hopkins WD, Bard KA (1996) Indexical and referential pointing in chimpanzees (*Pan troglodytes*). J Comp Psychol 110(4):346–353

Leavens DA, Hostetter AB, Westley MJ, Hopkins WD (2004) Tactical use of unimodal and bimodal communication by chimpanzees, *Pan troglodytes*. Anim Behav 67:467–476

Leavens DA, Russell JL, Hopkins WD (2005) Intentionality as measured in the persistence and elaboration of communication by chimpanzees (*Pan troglodytes*). Child Dev 76(1):291–306

Malenky RK, Wrangham RW (1994) A quantitative comparison of terrestrial herbaceous food consumption by *Pan paniscus* in the Lomako forest, Zaire, and *Pan troglodytes* in the Kibale forest, Uganda. Am J Primatol 32:1–12

Malenky RK, Kuroda S, Vineberg EO, Wrangham RW (1994) The significance of terrestrial herbaceous foods for bonobos, chimpanzees and gorillas. In: Wrangham RW, McGrew WC, de Waal FB, Heltne PG (eds) Chimpanzee cultures. Harvard University Press, Cambridge, pp 59–75

McGurk H, MacDonald J (1976) Hearing lips and seeing voices. Nature 264:746–748. https://doi.org/10.1038/264746a0

McNeill D (1992) Hand and mind: what gestures reveal about thought. University of Chicago Press, Chicago

Meyer-Lindenberg A, Kolachana B, Gold B, Olsh A, Nicodemus KK, Mattay V, Dean M, Weinberger DR (2009) Genetic variants in AVPR1A linked to autism predict amygdala activation and personality traits in healthy humans. Mol Psychiatry 14(10):968–975

Milne SC (2015) An empirical and theoretical comparison of the socio-ecological behaviors of captive chimpanzees (*Pan troglodytes*), bonobos (*Pan paniscus*), and western lowland gorillas (Gorilla gorilla): social tolerance and behavioral responses to changes in food quality and distribution. Master of Science in Integrative Biology Theses. Paper 8

Mitani JC (1996) Comparative field studies of African ape vocal behavior. In: Great ape societies. Cambridge University Press, Cambridge, pp 241–254

Moore B (2014) Differences in the communicative behavior and neurobiology of chimpanzees (*Pan troglodytes*) and bonobos (*Pan paniscus*) Master of Science in Integrative Biology Theses. Paper 3

Parish AR, de Waal F, Haig D (2000) The other "closest living relative": how bonobos (*Pan paniscus*) challenge traditional assumptions about females, dominance, intra-and intersexual interactions, and hominid evolution. Ann NY Acad Sci 907(1):97–113

Patterson N, Richter DJ, Guerre S, Lander ES, Reich D (2006) Genetic evidence for complex specialization of humans and chimpanzees. Nature 441(7097):1103–1108

Pollick AS, de Waal FMB (2007) Ape gestures and language evolution. Proc Natl Acad Sci 104(19):8184–8189

Poss SR, Kuhar C, Stoinski TS, Hopkins WD (2006) Differential use of attentional and visual communicative signaling by orangutans (Pongo pygmaeus) and gorillas (Gorilla gorilla) in response to the attentional status of a human. Am J Primatol 68:978–992

Prüfer K, Munch K, Hellmann I, Akagi K, Miller JR, Walenz B, Koren S, Sutton G, Kodira C, Winer R, Knight JR, Mullikin JC, Meader SJ, Ponting CP, Lunter G, Higashino S, Hobolth A, Dutheil J, Karakoc E, Alkan C, Sajjadian S, Catacchio CR, Ventura M, Marques-Bonet T, Eichler EE, Andre C, Atencia R, Mugisha L, Junhold J, Patterson N, Siebauer M, Good JM, Fischer A, Ptak SE, Lachmann M, Symer DE, Mailund T, Schierup MH, Andres AM, Kelso J, Paabo S (2012) The bonobo genome compared with the chimpanzee and human genomes. Nature 486(7404):527–531

Russell JL, Mcintyre J, Hopkins WD, Taglialatela JP (2013) Vocal learning of a communicative signal in captive chimpanzees, *Pan troglodytes*. Brain Lang 127:520–525

Sakura O (1994) Factors affecting party size and composition of chimpanzees (*Pan troglodytes* verus) Bossou, Guinea. Int J Primatol 15:167–183

Sibley CG, Ahlquist JE (1984) The phylogeny of the hominoid primates, as indicated by DNA-DNA hybridization. J Mol Evol 20:2–15

Slocombe KE, Waller BM, Liebal K (2011) The language void: the need for multimodality in primate research. Anim Behav 81:919–924

Staes N, Stevens JMG, Helsen P, Hillyer M, Korody M, Eens M (2014) Oxytocin and vasopressin receptor gene variation as a proximate base for Inter- and intraspecific behavioral bifferences in bonobos and chimpanzees. PLoS One 9(11):e113364

Staes N, Koski SE, Helsen P, Fransen E, Eens M, Stevens JMG (2015) Chimpanzee sociability is associated with vasopressin (Avpr1a) but not oxytocin receptor gene (OXTR) variation. Horm Behav 75:84–90

Stanford CB, Nkurunungi JB (2003) Behavioral ecology of sympatric chimpanzees and gorillas in Bwindi Impenetrable National Park, Uganda: Diet. Int J Primatol 24(4):901–918

Taglialatela JP, Russel JL, Schaeffer JA, Hopkins WD (2008) Communicative signaling activates "Broca's" homologue in chimpanzees. Curr Biol 18:343–348

Taglialatela JP, Russell JL, Schaeffer JA, Hopkins WD (2011) Chimpanzee vocal signaling points to a multimodal origin of human language. PLoS One 6(4):e18852. https://doi.org/10.1371/journal.pone.0018852

Taglialatela JP, Reamer L, Schapiro SJ, Hopkins WD (2012) Social learning of a communicative signal in captive chimpanzees. Biol Lett 8:498–501

Taglialatela JP, Russell JL, Pope SM, Morton T, Bogart S, Reamer LA, Schapiro SJ, Hopkins WD (2015) Multimodal communication in chimpanzees. Am J Primatol 77(11):1143–1148

Tomasello M, Call J, Nagell K, Olguin R, Carpenter M (1994) The learning and use of gestural signals by young chimpanzees: a trans-generational study. Primates 35:137–154

Tutin CEG (1996) Ranging and social structure of lowland gorillas in the Lopé Reserve, Gabon. In: WC MG, Marchant LF (eds) Great ape societies. Cambridge University Press, Cambridge, pp 58–70

Tutin CEG, Fernandez M, Elizabeth Rogers M, Williamson EA, McGrew WC, Altmann SA, Southgate DAT (1991) Foraging profiles of sympatric lowland gorillas and chimpanzees in the Lope Reserve, Gabon. Philos Trans R Soc Lond B 334(1270):179–186

Walum H, Westberg L, Henningsson S, Neiderhiser JM, Reiss D, Igl W, Ganiban JM, Spotts EL, Pedersen NL, Eriksson E, Lichenstein P (2008) Genetic variation in the vasopressin receptor 1a gene (avpr1a) associates with pair bonding behavior in humans. Proc Natl Acad Sci 105(37): 14153–14156

Watts DP, Potts KB, Lwanga JS, Mitani JC (2012) Diet of chimpanzees (*Pan troglodytes* schweinfurthii) at Ngogo, Kibale National Park, Uganda, 1. Diet composition and diversity. Am J Primatol 74:114–129

White FJ (1996) Comparative socio-ecology of *Pan paniscus*. In: McGrew WC, Marchant LF (eds) Great ape societies. Cambridge University Press, Cambridge, pp 29–41

White FJ (1998) Seasonality and socioecology: the importance of variation in fruit abundance to bonobo sociality. Int J Primatol 19:1013–1027

White FJ, Chapman CA (1994) Contrasting chimpanzees and bonobos:nearest neighbor distances and choices. Folia Primatol 63(4):181–191

White FJ, Wrangham RW (1988) Feeding competition and patch size in the chimpanzee species *Pan paniscus* and *Pan troglodytes*. Behaviour 105(1/2):148–164

Wobber V, Hare B, Maboto J, Lipson S, Wrangham RW, Ellison PT (2010) Differential changes in steroid hormones before competition in bonobos and chimpanzees. Proc Natl Acad Sci 107 (28):12457–12462

Wrangham RW (1986) Ecology and social relationships in two species of chimpanzees. In: Rubenstein DI, Wrangham RW (eds) Ecological aspects of social evolution: birds and mammals. Princeton University Press, Princeton, pp 352–378

Wrangham RW, Chapman CA, Clark-Arcadi AP, Isabirye-Basuta G (1996) Social ecology of Kanyawara chimpanzees: implications for understanding the costs of great ape groups. In: WC MG, Marchant LF (eds) Great ape societies. Cambridge University Press, Cambridge, pp 45–57

Yamagiwa J (1999) Socioecological factors influencing population structure of gorillas and chimpanzees. Primates 40(1):87–104

Yamagiwa J, Basabose AK, Kahekwa J, Bikaba D, Ando C, Matsubara M, Iwasaki N, Sprague DS (2012) Long-term research on Grauer's gorillas in Kahuzi-Biega National Park, DRC: life history, foraging strategies, and ecological differentiation from sympatric chimpanzees. In: Long-term field studies of primates. Springer, Heidelberg, pp 385–412

Yamagiwa J, Maruhashi T, Yumoto T, Mwanza N (1996) Dietary and ranging overlap in sympatric gorillas and chimpanzees in Kahuzi-Biega National Park, Zaire. In: WC MG, Marchant LF (eds) Great ape societies. Cambridge University Press, Cambridge, pp 82–98

Yamakoshi G (2004) Food seasonality and socioecology in Pan: are West African chimpanzees another bonobo? Afr Stud Monogr 25(1):45–60

Zihlman AL, Cronin JE, Cramer DL, Sarich VM (1978) Pygmy chimpanzees as a possible prototype for the common ancestor of humans, chimpanzees, and gorillas. Nature 275(5682): 744–746

Part II
Studying Primate Social Cognition: Theory, Observation, Experiments, and Modelling

Primate Social Cognition: Evidence from Primate Field Studies

Julia Ostner

Abstract The social complexity hypothesis proposes that primates evolved large brains in order to master the challenges posed by a complex society. The more complex a social group, the more information an individual needs to remember and compute in order to successfully maneuver its social environment, particularly since with increasing group size, the number of possible dyads to keep track of is increasing exponentially. Accumulating evidence from field studies indicates that primates indeed keep track of their own dominance, kin and affiliative relationships with other individuals as well as of relationships between third parties, which indicates the capacity for triadic awareness. Studies are beginning to unravel how monkeys put this knowledge to use for their own benefit, i.e., when choosing reliable partners in cooperative acts like coalition formation. Several mechanisms underlying this social knowledge are currently discussed, as it is not clear whether nonhuman primates indeed possess the computational capacities to remember social interactions of all dyads in their social group. Emotional bookkeeping has been proposed as a cognitively less demanding mechanism to at least keep track of one's own interactions—an idea gaining support by recent neuroendocrinological evidence.

Keywords Social complexity hypothesis · Primate field studies · Social relationships · Social bonds · Coalition formation · Triadic awareness · Emotional bookkeeping · Neurochemical pathways

---

J. Ostner (✉)
Department of Behavioral Ecology, University of Göttingen, Göttingen, Germany

Primate Social Evolution Research Group, German Primate Center, Leibniz Institute for Primate Research, Göttingen, Germany
e-mail: julia.ostner@biologie.uni-goettingen.de

L. D. Di Paolo et al. (eds.), *Evolution of Primate Social Cognition*, Interdisciplinary Evolution Research 5, https://doi.org/10.1007/978-3-319-93776-2_7

1 Introduction

The social complexity hypothesis posits that the main selective driver favoring the evolution of advanced cognitive abilities was the need to successfully maneuver the complicated social world (Humphrey 1976; Jolly 1966). Managing social relationships, anticipating others' actions, and manipulating their social behavior set high demands on an individual's socio-cognitive abilities (Byrne and Whiten 1988). The social complexity hypothesis has been supported in broad comparative analyses associating increasing sociality, frequently indexed by social group size, with increasing brain size in primates (Barton and Dunbar 1997; Lehmann and Dunbar 2009) and in between species comparisons of cognitive differences in corvids and lemurs (Bond et al. 2003; see Holekamp et al. 2015 for inconsistent results in carnivores; MacLean et al. 2008). It is currently debated whether increasing social complexity accounts for general cognitive advances or only for advances in the social domain, in primates and other animals (van Schaik et al. 2012). In phylogenetically controlled comparative analyses variation in ecological factors predicted brain size, thus, supporting the hypothesis that dietary complexity influences brain evolution (MacLean et al. 2014). In addition, it has been shown that animals including humans could only evolve a large brain where the energetic constraints of maintaining the large brain had been lifted (Isler and van Schaik 2013; Navarrete et al. 2011).

Regardless of the specific selective force responsible for the initial evolution of larger brains in primates, it is undisputed that most primates form complex societies, face a multitude of social challenges, leading selection to favor individuals that are capable of solving socio-cognitive problems (Seyfarth and Cheney 2015). In a recent attempt to structure the level of social complexity, Bergman and Beehner (2015) proposed that the number of different relationship types an individual has to maintain and keep track of in its social group may be a key measure of social complexity. In this definition, a social relationship implies consistently different interactions with a social partner. Thus in an anonymous aggregation of conspecifics, the number of differentiated relationships will be zero, whereas in many primate groups where all individuals know each other individually and interact repeatedly, the number of differentiated relationships can be very large particularly if there are different types of relationships between the same two individuals, e.g., alliance relationships and dominance relationships (Bergman and Beehner 2015).

Research from mice to humans produced convincing evidence that by forming and maintaining affiliative relationships or by being well-integrated into the affiliation network, individuals gain fitness benefits, stressing the evolutionary significance of variation in individual sociality (Holt-Lunstad et al. 2010; Massen et al. 2010; Ostner and Schülke 2014; Ostner and Schülke 2018; Silk 2007). For example, female baboons with strong social bonds enjoy increased longevity and infant survival (Silk et al. 2009, 2010); and survival is also increased in macaques and baboons that are well-connected into the same- and opposite-sex social network (Archie et al. 2014; McFarland and Majolo 2013). Long-term social relationships between males and females are predictive of paternity in chimpanzees and macaques (Langergraber et al. 2013; Ostner et al. 2013) and closely bonded dispersing male Assamese macaques enhance their reproductive success via rank changing coalitions (Schülke et al. 2010). In humans and other animals, individuals with close affiliates show an attenuated physiological stress response when facing a stressful situation, which in the long-term may enhance fitness as chronically increased physiological stress levels are detrimental to an individual's health (Hennessy et al. 2009; Young et al. 2014b). Owing to the impact social relationships have on an individual's fitness, selection should put a premium on an individual's ability to monitor and update their own as well as others' social ties and to use this ability to their own advantage.

In this chapter, I will review available evidence of social cognition from field studies of primate behavior. I see field studies as an important complementary approach to captive research and observational approaches as complementary to experimental designs, because combined, those studies provide evidence for the existence, functional significance, actual use, and mechanisms underlying a given ability in a natural setting (de Waal 1991; Schino et al. 2006).

2 Capacity for Social Cognition: Evidence from Primate Field Studies

Remembering past interactions, keeping track of one's own relationships and monitoring the relations of others are aspects of social cognition (Seyfarth and Cheney 2015). Following Hinde (Hinde 1976, 1983), I define a social relationship as the result of a series of dyadic social interactions over time. If two individuals recognize each other, interact repeatedly, and if the nature and quality of each interaction is predictive of future ones, the two individuals are said to have developed a social relationship.

A crucial precondition to track social relationships is the ability to recognize others individually, a capacity favored in many complex (primate) societies. Evidence for only limited individual recognition in a gregarious primate species comes from wild geladas (Bergman 2010; Bergman and Beehner 2015). Geladas live in multilevel societies and in playback experiments failed to recognize individuals beyond their social unit including individuals they regularly encounter (Bergman 2010). This is in striking contrast to other species including the closely related chacma baboons which vocally recognize group members (e.g., Crockford et al. 2007). It remains to be shown whether the limited individual recognition in geladas is due to (a) cognitive constraints (given that the number of regularly encountered geladas in a band exceeds the number of encountered individuals in a chacma baboon society by far) or to (b) lack of motivation in geladas to learn about other individuals, due to insufficiently strong selective pressure on recognition (Bergman 2010). In contrast to the gelada case, individual recognition, a precondition for many aspects of social cognitive processes, is well-documented across a wide range of taxa (Bergman 2010).

Social information on other individuals is gathered via selective attention biased toward more relevant information. Experimental studies in captivity show that nonhuman primates value social information about their group members highly and selectively, even foregoing food rewards to be able to see pictures of high social relevance, e.g., monkeys higher ranking than themselves (Deaner et al. 2005; Watson et al. 2012), and being selectively attentive to gazes of higher-ranking individuals (Shepherd et al. 2008). Building on these findings and transcending dominance rank relations and artificial stimuli, we show that selective attention for naturally occurring dyadic social interactions is targeted toward those agonistic and affiliative interactions that involve a close social affiliate of the observer compared to interactions where both actors in the interaction are non-affiliates of the observer (Schülke, Ostner unpubl data). Individuals of many species, thus, clearly recognize other individuals and pay selective attention to their behavior.

The next step in monitoring others is keeping track of one's own relationships, which is different from the likely more complex monitoring of third-party relationships. Individuals remember the outcome of past agonistic interactions when acting upon the dyadic dominance relationship, which reduces the probability of escalated fights and injury (de Waal 1986). They also keep track of their affiliative relationships and differentiate their social behavior depending on the quality of their relationship to others. Chimpanzee and macaque males form coalitions and recruit allies based on past affiliative interactions (Berghänel et al. 2011b; Boesch et al. 2006; Schülke et al. 2010; Watts 2002; Young et al. 2014a) and across a large data set of nonhuman primate species social grooming and agonistic support are positively related (Schino 2007). Numerous primate and non-primate species engage in post-conflict affiliative behavior, possibly functioning as reconciliation. Interestingly, the tendency to reconcile after a fight varies depending on the bond strength or the value of the social partner, with post-conflict friendly reunions being more likely after a conflict between individuals sharing a close affiliative relationship (Arnold and Aureli 2007; Aureli et al. 2002). These examples illustrate the capacity of nonhuman primates to keep track of their interactions and relationships with other group members.

Cognitively more demanding may be to monitor and integrate social interactions of third parties, i.e., triadic awareness. There is compelling evidence from experimental and observational studies both from the wild and captivity for the capacity to discriminate own and other's kin relations: wild vervet monkeys, when hearing the scream of an infant, are more likely to look at the infant's mother compared to another female (Cheney and Seyfarth 1980). Ability to discriminate own and others' kin relations was later confirmed in experimental match-to-sample tests in longtail macaques (Dasser 1988) and in observational and playback studies of chacma baboons and Japanese macaques (Cheney and Seyfarth 1980, 1999; Schino et al. 2006). Studies on white-faced capuchins, chacma baboons, sooty mangabeys, as well as Japanese and bonnet macaques provide evidence of triadic awareness of dominance rank relations (Bergman et al. 2003; Perry et al. 2004; Range and Noë

2005; Schino et al. 2006; Silk 1999). Matrilineal organized Japanese macaques and chacma baboons even combine both aspects of relationships and simultaneously and hierarchically classify based on rank and kinship as evidenced in experimental and observational studies (Bergman et al. 2003; Schino et al. 2006).

Recognizing the dominance relationships of third parties is particularly important when making decision about intervention and recruitment in potentially risky agonistic conflicts. As dominance rank is often organized in a linear, transitive fashion, and is relatively stable over time, these studies have their limitations. In several studies, it was not possible to rule out that instead of using cognitively demanding decision rules for intervention and recruitment that involve triadic awareness, primates were using simple rules of thumb, such as: "intervene only when you outrank the target of a conflict" or "recruit the highest ranking ally available" (Perry et al. 2004; Range and Noë 2005) (see Silk 1999 for a study where rules of thumb did not explain intervention decisions). While consistent with the ability of triadic awareness, these studies cannot provide firm evidence for it. Interestingly, a majority of decision humans take also seem to be governed by such rules of thumb or "simple heuristics" (Todd and Gigerenzer 2000, 2003).

Dominance rank and kin relations may be linear, transitive, and temporarily relatively stable, facilitating classification by third parties, but affiliative relationships are not. Affiliative relationships of others, thus, cannot be explained easily by the employment of simple rules of thumb. The preferential soliciting of agonistic support from those white-faced capuchins that are more closely bonded to the recruiter than to the opponent indicates the capacity for and use of triadic awareness of others' affiliative relationships (Perry et al. 2004). Male macaques seem to be able to recognize even relatively transient affiliative relationships of others: Tibetan and Barbary macaque males form close relationships with small infants, which form very soon after the birth of an infant and vanish again upon the start of the next mating season. Thus, males have only very limited time to monitor the relationships between other males and infants and to update this knowledge. Still, when approaching a particular male for a friendly male–infant–male interaction, male Tibetan macaques select those infants as social tools that have the strongest relationship with the target male (Ogawa 1995), and male Barbary macaques that carry an infant select those males for triadic male–infant–male interactions that have the closest relationships to the infant the male is carrying (Kuběnová et al. 2017). In a carefully designed playback study, it was shown that male chacma baboons deduce even faster temporal changes in male–female consort relationships (Crockford et al. 2007).

Overall, there is broad evidence across nonhuman primates that individuals are able to recognize others individually and to keep track of their own and others' social relationships based on kinship, dominance rank, or affiliation (Cheney and Seyfarth 1990; Seyfarth and Cheney 2015). The level of recognition has implications for an individual's memory load, as memory load is a linear function of group size if only one's own rank, kin, and bond relationship need to be discriminated, but increases exponentially if relationships of all other group members are recognized (Byrne and Bates 2010; Seyfarth and Cheney 2015). Combined, these studies highlight the primate propensity to monitor social relationships of others begging the question of its evolutionary significance and associated ultimate benefits.

3 Usage and Benefits of Keeping Track of Own and Others' Social Relationships

Keeping track of interactions and relationships of oneself and others has been associated with a number of benefits. Remembering the outcome of past conflicts and knowing one's place in the hierarchy reduce the likelihood of escalated fighting and increase predictability and social control, which, if lacking, are main causes of chronic psychosocial stress and its associated adverse effects on health (Sapolsky 2005).

Males may effectively decrease the risk of aggression by using infants as social tools (Packer 1980; Paul et al. 1996), as described in the previous section, if they pay attention to other males' relationships with infants and select infants for male–infant–male interactions based on the target male's preference. In addition, as these male–infant–male interactions serve to build and maintain strong social ties (Berghänel et al. 2011b; Kalbitz et al. 2017; Ogawa 1995), choosing an infant that is strongly bonded to the target male may work as a specific incentive for the target male to engage in this bond-building and bond-maintaining social interaction, yielding all the benefits provided by strong bonds. Subordinate male chacma baboons eavesdropping on consorting pairs' spatial and temporal whereabouts may be rewarded with opportunities for sneaky matings and possibly increased paternity success (Crockford et al. 2007).

Arguably one of the most important realms of recognition of dominance and affiliative relationships is the formation of coalitions and alliances (Harcourt 1988). Coalitions, if successful, are an adaptive strategy to increase access to valuable resources, food or mates, and to rise in dominance rank, which in many species is associated with increased reproductive success (Chapais 1995; van Schaik et al. 2004). Forming a coalition with an ally, selecting the most suitable out of a group of bystanders, and deciding whether to intervene in an ongoing conflict, on the side of the aggressor or to support the target, are decisions that are largely governed by the relative rank and affiliative relations of oneself to others as well as between the other parties.

In order to maximize feasibility of a coalition, one should choose the highest ranking ally available to optimize the coalition's combined fighting ability and thus the probability of winning (Noë 1994; Noë and Sluijter 1995; van Schaik et al. 2004), a strategy particularly successful in opportunistic coalitions leading to an immediate payoff. Evidence for such a recruitment strategy comes, e.g., from yellow baboons (Noë and Sluijter 1995) and Barbary macaques (Bissonnette et al. 2009; Young et al. 2014a) when forming (leveling) coalitions to gain access to a receptive female. If coalitions do not lead to an immediate payoff and there is a threat of retaliation by the target, as it is the case in rank-changing coalitions (which, if successful, lead to males not being sorted in the hierarchy based on individual fighting ability), then recruitment decisions should be governed by past experience and reliability indexed, for example, as social bond strength and loyalty (Ostner and Schülke 2014). Indeed, in species with rank-changing coalitions, the frequency of coalition formation is linked to dyadic bond strength (Berghänel et al. 2011b; Ostner and Schülke 2014; Watts 2002; Young et al. 2014a). A study on wild Barbary macaque males forming both leveling and rank-changing coalitions provides evidence for a flexible use of recruitment decisions basing supporter selection independently on social bond strength and dominance rank; in a subset limited to opportunistic, leveling coalitions, supporter selection was solely based on the feasibility of maximizing dominance rank (Young et al. 2014a). These results corroborate findings of nonhuman primates categorizing group members by more than one trait, i.e., dominance rank and bond strength, and show that male macaques flexibly use this cognitive ability by implementing one of two selection criteria when selecting a supporter in different coalitionary contexts.

Apart from recruiting the best possible ally, selection has also favored an individual's ability to monitor not only the target's intrinsic (fighting) ability but also its extrinsic power, i.e., its ability to solicit support from others. The higher the extrinsic power of the subordinate males are, the more destabilized are the dyadic dominance relationships between Barbary macaque males, as indicated by increasing rates of counter-aggression, conflicts initiated by the subordinates, and spontaneous submission of the dominant individual (Berghänel et al. 2011a). The fact that males reacted to their opponents' extrinsic power in dyadic conflicts suggests that they knew about the opponent's potential to recruit help from group members.

Individuals may also use their social knowledge strategically to manipulate others' behavior. Chimpanzees impressively demonstrate their knowledge of third-party relationships in agonistic contexts. When attacked, chimpanzees use screams to solicit support from bystanders (de Waal and van Hooff 1981) and modify the acoustic structure of their screams as a function of the severity of the aggression, thereby providing the bystanders with detailed information about the nature of the conflict (Slocombe and Zuberbühler 2007). Yet, the relationship between severity of aggression and scream structure is not perfect, because the signal is not entirely honest. Chimpanzees acoustically exaggerated the actual severity of the attack to increase the likelihood of being supported, but only if the audience includes at least one bystander that matched or exceeded the attacker in rank; this demonstrates that chimpanzees use their knowledge of others' rank relationships for tactical deception and manipulation of others (Slocombe and Zuberbühler 2007). Keeping track of others' affiliative interactions may also become a political strategy to keep others from forming ties that may strengthen alliances against oneself. Chimpanzees and white-faced capuchin use "separating interventions" to break up friendly interactions among subordinates (de Waal 1978; Perry 1998). Similarly, affiliation between corvids that are starting to form a social bond which may subsequently lead to an alliance is likely to be broken up by other individuals. The pattern of intervention suggests that interventions function to keep other individuals from forming new relationships and alliances (Massen et al. 2014).

4 Mechanisms Underlying the Monitoring of One's Own and Others' Relationships

Several mechanisms underlying an individual's ability to keep track of one's own and others' social relationships have been proposed. These are neither mutually exclusive nor necessarily able to explain the entire range of cognitive capacities observed. Clearly, there are interspecific differences in the mechanisms employed, and humans in particular seem to have more elaborate ways of acquiring explicit social knowledge in addition to those mechanisms that they share with nonhuman primates. Still, the above presented evidence stresses that nonhuman primates (and other animals) keep track of their own and others' social interactions and relationships and use this knowledge to their benefit. I will first discuss mechanisms underlying the monitoring of one's own relationships before turning to others' relations.

For tracking one's own relationships, nonhuman primates could rely on a system of computational or calculated reciprocity implying a strict contingency of gives and takes, i.e., mental scorekeeping (de Waal and Luttrell 1988). However, evidence has been accumulating that there is a strong correlation between services given and received on a long-term basis in the absence of short-term contingencies (Schino and Aureli 2009). For example, in Japanese macaques (Schino et al. 2003), chimpanzees (Gomes et al. 2008), olive baboons (Frank and Silk 2009), and capuchins (Schino et al. 2009), there is only a weak time-match in grooming given and received within a bout, but a balanced long-term account. Tolerance of these temporal, short-term imbalances may be particularly strong in closely affiliated dyads, exemplified in a playback study on wild baboons where support was contingent upon previous grooming in weakly bonded dyads but had no effect in dyads of bonded females (Cheney et al. 2010). Emotional bookkeeping, a mechanism similar to attitudinal reciprocity (de Waal 2000), has been suggested as a mechanism linking temporally separated events (Aureli and Schaffner 2002; Schino and Aureli 2009). Emotions created by past events and interactions can be construed as intervening variables affecting subsequent interactions and thus allowing individuals to keep track of social interactions via a "loose accounting mechanism" and to convert different behavioral commodities into a common currency (Schino and Aureli 2009).

Barbary macaque males form coalitions based on affiliative interactions or social bonds formed months earlier, implying that males possess some mental representation of the relationship that they can refer to in decisions of partner choice (Berghänel et al. 2011b). Several neuroendocrinological pathways have been suggested to be involved in the representation of a social relationship. One starting point is the highly conserved mechanism of the oxytocin pathway underlying mother–infant bonding (Broad et al. 2006; Curley and Keverne 2005). Apart from its function during parturition and lactation, oxytocin, a neuropeptide, plays a crucial role in the regulation and promotion of affiliative behavior with feedbacks to the neural reward system, social memory, and social recognition (Lim and Young 2006). Recent studies show that oxytocin release is highly context- and partner-specific (Olff et al. 2013). Grooming interactions in chimpanzees only evoke a rise in oxytocin from baseline when the grooming partner is closely bonded (Crockford et al. 2013), and oxytocin levels in pair-living tamarins increase with increasing strength of the pair-bond (Snowdon et al. 2010). A study on human subjects shows that vocal contact received by a bonded partner is sufficient to decrease cortisol and increase oxytocin levels following a stressful situation (Seltzer et al. 2010). Oxytocin release during interactions with a bonded partner may, thus, positively feed back on the neural reward circuit, enhancing social recognition and memory and promoting the formation and maintenance of individual relationships (Crockford et al. 2013; Gustella and MacLeod 2012).

An alternative or additional neurochemical candidate involved in social contact and the maintenance of social relationships is the brain opioid system, particularly endorphins (Dunbar 2012; Marchin and Dunbar 2011). Close social contact, such as grooming, leads to a release of endorphins, endogenous opioid peptides, in the central nervous system, an activation of the reward system, and an associated feeling of pleasure and contentment (Curley and Keverne 2005; Keverne et al. 1989; Marchin and Dunbar 2011). As endorphin degradation is rapid, individuals are motivated to continue to engage in the rewarding behavior, i.e., close social contact, to avoid symptoms of withdrawal (Marchin and Dunbar 2011). The blockage of opioid receptors consequently increases an individual's motivation to seek close social contact (Keverne et al. 1989). The oxytocin/vasopressin system is crucial in the formation of social relationships, whereas endorphins may be more responsible for their maintenance, as social withdrawal leads to a decrease in endorphins which in turn lead to an urge to seek social contact, to avoid the pain of isolation and consequently be rewarded for social contact (Marchin and Dunbar 2011). The two neurochemical systems involved in social behavior, the oxytocin/vasopressin and the opioid system, have similar effects, such as the activation of the brain reward system and analgesia, and are part of complex feedback mechanisms making their relative roles in social behavior difficult to disentangle. Contrasting to most other mammalian taxa primates, particularly old world primates, have been released from the strict hormonal and sensory control underlying social bonding, and higher order cognitive functioning facilitated social interactions in the absence of hormonal triggering (Curley and Keverne 2005; Marchin and Dunbar 2011). Consequently, it has been argued that opioids may be the neurochemical "glue" that in concert with other cognitive mechanisms, allow nonhuman primates to maintain their numerous and complex social relationships over extended periods of time periods decoupled from hormonal control (Curley and Keverne 2005; Marchin and Dunbar 2011).

How do individuals master the cognitive demands of monitoring other individuals' social relationships? Individuals could recognize rank and kin relationships of third parties by closely monitoring their social interactions. Although theoretically possible, it seems more likely that individuals use transitive inferences to classify other relationships (Cheney and Seyfarth 1990; Schino et al. 2006). If an individual has observed A to be dominant over B and B over C, then it does not need to observe interactions between A and C directly to conclude that A is also dominant over C. Such a system of transitive inferences works particularly well in species with a linear, transitive dominance hierarchy characteristic for many primate species, and experiments suggest that corvids use it to make inferences about unknown relations (Paz-y-Mino et al. 2004).

Predicting third-party kin relationships may also be based on a system of transitive inferences, for example, in species where kin is organized in matrilines, as in many cercopithecines. In this case, observing a female with an infant share affiliative interactions with a juvenile may lead to the accurate prediction that the infant and the juveniles also share a bond (or kin) relationship (Seyfarth and Cheney 2015). This may explain the findings showing that redirection of aggression is more likely targeted toward an individual belonging to the same matriline as the initial aggressor or even more complex that an individual A is more likely to aggress another individual B if a member of the B matriline earlier that day aggressed an A matriline member (Cheney et al. 1986).

Transitive inference, however, is not an appropriate strategy to predict social relationships between the sexes or in individualistic, i.e., non-nepotistic, societies. Close between-sex relationships are frequently formed in macaques and baboons (Haunhorst et al. 2016; Palombit et al. 1997). If a male maintains close relationships with two females, the bond between the male and either female is not predictive of the relationships between the two females. Thus, different cognitive strategies must have been applied in these cases, begging the question whether individuals rely on one system (transitive inferences) in one context, i.e., rank relationships, and dismiss it in another, i.e., social bonding, or whether they rely on a different strategy altogether (Seyfarth and Cheney 2015). It has been argued that if an individual nonhuman primate is able to monitor and classify its own relationships based on social memory and emotions (see above), this "concept" or "mental representation" of a social relationship may enable the same individual to detect such relationships also in other individuals by observing just a subset of their social interactions (Seyfarth and Cheney 2015).

In sum, there is broad evidence from observational and experimental studies on wild and free-ranging nonhuman primate populations that primates possess the capacity to monitor their own and others' social relationships and use this ability flexibly and strategically for their own benefit. The rapid advancement of noninvasive methodological approaches promises to close gaps in our understanding of the underlying cognitive and neuroendocrinological mechanisms.

Acknowledgments I thank the editors, in particular Laura Di Paolo, for inviting me to contribute to this book. I also thank Susan Perry for very helpful comments on the manuscript.

References

Archie EA, Tung J, Clark M, Altmann J, Alberts SC (2014) Social affiliation matters: both same-sex and opposite-sex relationships predict survival in wild female baboons. Proc R Soc B: Biol Sci 281:20141261

Arnold K, Aureli F (2007) Postconflict reconciliation. In: Campbell C et al (eds) Primates in perspective. Oxford University Press, New York, pp 592–608

Aureli F, Schaffner CM (2002) Relationships assessment through emotional mediation. Behaviour 139:393–420

Aureli F, Cords M, van Schaik CP (2002) Conflict resolution following aggression in gregarious animals: a predictive framework. Anim Behav 64:325–343

Barton RA, Dunbar RIM (1997) Evolution of the social brain. In: Whiten A, Byrne RW (eds) Machiavellian intelligence II. Cambridge University Press, Cambridge, pp 240–263

Berghänel A, Ostner J, Schülke O (2011a) Coalitions destabilize dyadic dominance relationships in male Barbary macaques (*Macaca sylvanus*). Behaviour 148:1256–1274

Berghänel A, Ostner J, Schröder U, Schülke O (2011b) Social bonds predict future cooperation in male Barbary macaques, *Macaca sylvanus*. Anim Behav 81:1109–1116

Bergman TJ (2010) Experimental evidence for limited vocal recognition in a wild primate: implications for the social complexity hypothesis. Proc R Soc Lond B: Biol Sci 277:3045–3053

Bergman TJ, Beehner JC (2015) Measuring social complexity. Anim Behav 103:203–209

Bergman TJ, Beehner JC, Cheney DL, Seyfarth RM (2003) Hierarchical classification by rank and kinship in baboons. Science 302:1234–1236

Bissonnette A, de Vries H, van Schaik CP (2009) Coalitions in male Barbary macaques, *Macaca sylvanus*: strength, success and rules of thumb. Anim Behav 78:329–335

Boesch C, Boesch H, Vigilant L (2006) Cooperative hunting in chimpanzees: kinship or mutualism? In: Kappeler P, van Schaik C (eds) Cooperation in primates and humans. Springer, Heidelberg, pp 139–150

Bond AB, Kamil AC, Balda RP (2003) Social complexity and transitive inference in corvids. Anim Behav 65:479–487

Broad KD, Curley JP, Keverne EB (2006) Mother-infant bonding and the evolution of mammalian social relationships. Philos Trans R Soc B 361:2199–2214

Byrne RW, Bates LA (2010) Primate social cognition: uniquely human, uniquely social, or just unique? Neuron 65:815–830

Byrne RW, Whiten A (1988) The machiavellian intelligence hypotheses: editorial. In: Byrne RW, Whiten A (eds) Machiavellian intelligence. Clarendon Press, Oxford, pp 1–9

Chapais B (1995) Alliances as means of competition in primates: evolutionary, developmental and cognitive aspects. Yearb Phys Anthropol 38:115–136

Cheney DL, Seyfarth RM (1980) Vocal recognition in free-ranging vervet monkeys. Anim Behav 28:362–367

Cheney D, Seyfarth R (1990) The representation of social relations by monkeys. Cognition 37:167–196

Cheney DL, Seyfarth RM (1999) Recognition of other individuals' social relationships by female baboons. Anim Behav 58:67–75

Cheney D, Seyfarth R, Smut B (1986) Social Relationships and social cognition in nonhuman primates. Science 234:1361–1366

Cheney DL, Moscovice LR, Heesen M, Mundry R, Seyfarth RM (2010) Contingent cooperation between wild female baboons. Proc Natl Acad Sci 107:9562–9566

Crockford C, Wittig RM, Seyfarth RM, Cheney DL (2007) Baboons eavesdrop to deduce mating opportunities. Anim Behav 73:885–890

Crockford C, Wittig RM, Langergraber K, Ziegler TE, Zuberbühler K, Deschner T (2013) Urinary oxytocin and social bonding in related and unrelated wild chimpanzees. Proc R Soc B: Biol Sci 280:20122765

Curley J, Keverne E (2005) Genes, brains and mammalian social bonds. Trends Ecol Evol 20:561–567
Dasser V (1988) A social concept in Java monkeys. Anim Behav 36:225–230
de Waal F (1978) Exploitative and familiarity-dependent support strategies in a colony of semi-free living chimpanzees. Behaviour 66:268–312
de Waal F (1986) The integration of dominance and social bonding in primates. Q Rev Biol 61:459–479
de Waal FBM (1991) Complementary methods and convergent evidence in the study of primate social cognition. Behaviour 118:297–320
de Waal FBM (2000) Attitudinal reciprocity in food sharing among brown capuchin monkeys. Anim Behav 60:253–261
de Waal F, Luttrell L (1988) Mechanisms of social reciprocity in three primate species: symmetrical relationship characteristics or cognition. Ethol Sociobiol 9:101–118
de Waal FBM, van Hooff JARAM (1981) Side-directed communication and agonistic interactions in chimpanzees. Behaviour 77:164–198
Deaner RO, Khera AV, Platt ML (2005) Monkeys pay per view: adaptive valuation of social images by rhesus macaques. Curr Biol 15:543–548
Dunbar RIM (2012) Bridging the bonding gap: the transition from primates to humans. Philos Trans R Soc B 367:1837–1846
Frank RE, Silk JB (2009) Impatient traders or contingent reciprocators? Evidence for the extended time-course of grooming exchanges in baboons. Behaviour 146:1123–1135
Gomes CM, Mundry R, Boesch C (2008) Long-term reciprocation of grooming in wild West African chimpanzees. Proc R Soc B: Biol Sci 276:699–706
Gustella A, MacLeod C (2012) A critical review of the influence of oxytocin nasal spray on social cognition in humans: evidence and future directions. Horm Behav 61:410–418
Harcourt AH (1988) Alliances in contest and social intelligence. In: Byrne RW, Whiten A (eds) Machiavellian intelligence: social expertise and the evolution of intellect in monkeys, apes, and humans. Oxford University Press, Oxford, pp 132–152
Haunhorst C, Schülke O, Ostner J (2016) Opposite-sex social bonding in wild Assamese macaques. Am J Primatol 78:872–882
Hennessy MB, Kaiser S, Sachser N (2009) Social buffering of the stress response: diversity, mechanisms, and functions. Front Neuroendocrinol 30:470–482
Hinde R (1976) On assessing the bases of partner preferences. Behaviour 62:1–9
Hinde RA (1983) Primate social relationships: an integrated approach. Blackwell Scientific, Oxford
Holekamp KE, Dantzer B, Stricker G, Shaw Yoshida KC, Benson-Amram S (2015) Brains, brawn and sociality: a hyaena's tale. Anim Behav 103:237–248
Holt-Lunstad J, Smith TB, Layton JB (2010) Social relationships and mortality risk: a meta-analytic review. PLoS Med 7:e1000316
Humphrey NK (1976) The social function of intellect. In: Bateson PPG, Hinde RA (eds) Growing points in ethology. Cambridge University Press, Cambridge, pp 303–317
Isler K, van Schaik CP (2013) How humans evolved large brains: comparative evidence. Evol Anthropol 23:65–75
Jolly A (1966) Lemur social behavior and primate intelligence. Science 153:501–506
Kalbitz J, Schülke O, Ostner J (2017) Triadic male-infant-male interaction serves in bond maintenance in male Assamese macaques. PLoS One 12:e0183981
Keverne EB, Martensz ND, Tuite B (1989) Beta-endorphin concentrations in cerebrospinal fluid of monkeys are influenced by grooming relationships. Psychoneuroendocrinology 14:155–161
Kuběnová B, Majolo B, Šmilauer P, Konečná M, Ostner J, Schülke O (2017) Triadic awareness predicts partner choice in male-infant-male interactions in Barbary macaques. Anim Cogn 20:221–232
Langergraber K, Mitani J, Watts D, Vigilant L (2013) Male–female socio-spatial relationships and reproduction in wild chimpanzees. Behav Ecol Sociobiol 67:861–873

Lehmann J, Dunbar RIM (2009) Network cohesion, group size and neocortex size in female-bonded old world primates. Proc R Soc B: Biol Sci 276:4417–4422

Lim MM, Young LJ (2006) Neuropeptidergic regulation of affiliative behavior and social bonding in animals. Horm Behav 50:506–517

MacLean EL, Merritt DJ, Brannon EM (2008) Social complexity predicts transitive reasoning in prosimian primates. Anim Behav 76:479–486

MacLean EL, Hare B, Nunn C, Addessi E, Amici F, Anderson R, Aureli F, Baker J, Bania A, Barnard A, Boogert N, Brannon E, Bray E, Bray J, Brent LJN, Burkart J, Call J, Cantlon J, Cheke L, Clayton N, Delgado M, DiVincenti L, Fujita K, Herrmann E, Hiramatsu C, Jacobs L, Jordan K, Laude J, Leimgruber K, Messer E, Moura AdA, Ostojic L, Picard A, Platt ML, Plotnik JM, Range F, Reader SM, Reddy R, Sandel AA, Santos LR, Schumann K, Seed A, Sewall K, Shaw R, Slocombe KE, Su Y, Takimoto A, Tan J, Tao R, Van Schaik C, Viranyi Z, Visalberghi E, Wade J, Watanabe A, Widness J, Young J, Zentall T, Zhao Y (2014) The evolution of self-control. Proc Natl Acad Sci 111:E2140–E2148

Marchin A, Dunbar R (2011) The brain opioid theory of social attachment: a review of the evidence. Behaviour 148:985–1025

Massen J, Sterck E, de Vos H (2010) Close social associations in animals and humans: functions and mechanisms of friendship. Behaviour 147:1379–1412

Massen JJM, Szipl G, Spreafico M, Bugnyar T (2014) Ravens intervene in others' bonding attempts. Curr Biol 24:2733–2736

McFarland R, Majolo B (2013) Coping with the cold: predictors of survival in wild Barbary macaques, *Macaca sylvanus*. Biol Lett 9:20130428

Navarrete A, van Schaik CP, Isler K (2011) Energetics and the evolution of human brain size. Nature 480:91–93

Noë R (1994) A model of coalition formation among male baboons with fighting ability as the crucial parameter. Anim Behav 47:211–213

Noë R, Sluijter AA (1995) Which adult male savanna baboons form coalitions? Int J Primatol 16:77–105

Ogawa H (1995) Bridging behavior and other affiliative interactions among Tibetan macaques (*Macaca thibetana*). Int J Primatol 16:707–727

Olff M, Frijling JL, Kubzansky LD, Bradley B, Ellenbogen MA, Cardoso C, Bartz JA, Yee JR, van Zuiden M (2013) The role of oxytocin in social bonding, stress regulation and mental health: an update on the moderating effects of context and interindividual differences. Psychoneuroendocrinology 38:1883–1894

Ostner J, Schülke O (2014) The evolution of social bonds in primate males. Behaviour 151:871–906

Ostner J, Schülke O (2018) Linking sociality to fitness in primates – a call for mechanisms. Adv Study Behav 50:127–175

Ostner J, Vigilant L, Bhagavatula J, Franz M, Schülke O (2013) Stable heterosexual bonds in a promiscuous primate. Anim Behav 86:623–631

Packer C (1980) Male care and exploitation of infants in Papio anubis. Anim Behav 28:512–520

Palombit RA, Seyfarth RM, Cheney DL (1997) The adaptive value of 'friendships' to female baboons: experimental and observational evidence. Anim Behav 54:599–614

Paul A, Kuester J, Arnemann J (1996) The sociobiology of male-infant interactions in barbary macaques, *Macaca sylvanus*. Anim Behav 51:155–170

Paz-y-Mino CG, Bond AB, Kamil AC, Balda RP (2004) Pinyon jays use transitive inference to predict social dominance. Nature 430:778–781

Perry S (1998) Male–male social relationships in wild white-faced capuchins, *Cebus capucinus*. Behaviour 135:139–172

Perry S, Barrett HC, Manson JH (2004) White-faced capuchin monkeys show triadic awareness in their choice of allies. Anim Behav 67:165–170

Range F, Noë R (2005) Can simple rules account for the pattern of triadic interactions in juvenile and adult female sooty mangabeys? Anim Behav 69:445–452

Sapolsky R (2005) The influence of social hierarchy on primate health. Science 308:648–652

Schino G (2007) Grooming and agonistic support: a meta-analysis of primate reciprocal altruism. Behav Ecol 18:115–120

Schino G, Aureli F (2009) Reciprocal altruism in primates: partner choice, cognition, and emotions. Adv Study Behav 39:45–69

Schino G, Ventura R, Troisi A (2003) Grooming among female Japanese macaques: distinguishing between reciprocation and interchange. Behav Ecol 14:887–891

Schino G, Tiddi B, Polizzi Di Sorrentino E (2006) Simultaneous classification by rank and kinship in Japanese macaques. Anim Behav 71:1069–1074

Schino G, Di Giuseppe F, Visalberghi E (2009) The time frame of partner choice in the grooming reciprocation of *Cebus apella*. Ethology 115:70–76

Schülke O, Bhagavatula J, Vigilant L, Ostner J (2010) Social bonds enhance reproductive success in male macaques. Curr Biol 220:2207–2210

Seltzer LJ, Ziegler TE, Pollak SD (2010) Social vocalizations can release oxytocin in humans. Proc R Soc B: Biol Sci 277:2661–2666

Seyfarth RM, Cheney DL (2015) Social cognition. Anim Behav 103:191–202

Shepherd S, Deaner RO, Platt ML (2008) Social status gates social attention in monkeys. Curr Biol 16:R119–R120

Silk JB (1999) Male bonnet macaques use information about third-party rank relationships to recruit allies. Anim Behav 58:45–51

Silk J (2007) Social components of fitness in primate groups. Science 317:1347–1351

Silk JB, Beehner JC, Bergman TJ, Crockford C, Engh AL, Moscovice LR, Wittig RM, Seyfarth RM, Cheney DL (2009) The benefits of social capital: close social bonds among female baboons enhance offspring survival. Proc R Soc B: Biol Sci 276:3099–3104

Silk JB, Beehner JC, Bergman TJ, Crockford C, Engh AL, Moscovice LR, Wittig RM, Seyfarth RM, Cheney DL (2010) Strong and consistent social bonds enhance the longevity of female baboons. Curr Biol 20:1359–1361

Slocombe KE, Zuberbühler K (2007) Chimpanzees modify recruitment screams as a function of audience composition. Proc Natl Acad Sci 104:17228–17232

Snowdon CT, Pieper BA, Boe CY, Cronin KA, Kurian AV, Ziegler TE (2010) Variation in oxytocin is related to variation in affiliative behavior in monogamous, pair bonded tamarins. Horm Behav 58:614–618

Todd P, Gigerenzer G (2000) Precis of Simple heuristics that make us smart. Behav Brain Sci 23:727–780

Todd P, Gigerenzer G (2003) Bounding rationality to the world. J Econ Psychol 24:143–165

van Schaik CP, Pandit SA, Vogel ER (2004) A model for within-group coalitionary aggression among males. Behav Ecol Sociobiol 57:101–109

van Schaik CP, Isler K, Burkart JM (2012) Explaining brain size variation: from social to cultural brain. Trends Cogn Sci 16:277–284

Watson KK, Ghodasra J, Furlong M, Platt ML (2012) Visual preferences for sex and status in female rhesus macaques. Anim Cogn 15:401–407

Watts DP (2002) Reciprocity and interchange in the social relationships of wild male chimpanzees. Behaviour 139:343–370

Young C, Majolo B, Schülke O, Ostner J (2014a) Male social bonds and rank predict supporter selection in cooperative aggression in wild Barbary macaques. Anim Behav 95:23–32

Young C, Majolo B, Heistermann M, Schülke O, Ostner J (2014b) Responses to social and environmental stress are attenuated by strong male bonds in wild macaques. Proc Natl Acad Sci 111:18195–18200

Contribution of Social Network Analysis and Collective Phenomena to Understanding Social Complexity and Cognition

Denis Boyer and Gabriel Ramos-Fernandez

Abstract The social brain hypothesis postulates the increasing complexity of social interactions as a driving force for the evolution of cognitive abilities. Whereas dyadic and triadic relations play a basic role in defining social behaviours and pose many challenges for the social brain, individuals in animal societies typically belong to relatively large networks. How the structure and dynamics of these networks also contribute to the evolution of cognition, and vice versa, is less understood. Here we review how collective phenomena can occur in systems where social agents do not require sophisticated cognitive skills, and how complex networks can grow from simple probabilistic rules, or even emerge from the interaction between agents and their environment, without explicit social factors. We further show that the analysis of social networks can be used to develop good indicators of social complexity beyond the individual or dyadic level. We also discuss the types of challenges that the social brain must cope within structured groups, such as higher information fluxes, originating from individuals playing different roles in the network, or dyadic contacts of widely varying durations and frequencies. We discuss the relevance of these ideas for primates and other animals' societies.

Keywords Collective phenomena · Complex networks · Agent-based modelling · Group cognition

---

D. Boyer (✉)
Instituto de Física, Universidad Nacional Autónoma de México, Ciudad de México, México

Centro de Ciencias de la Complejidad, Universidad Nacional Autónoma de México, Ciudad de México, México
e-mail: boyer@fisica.unam.mx

G. Ramos-Fernandez
Centro de Ciencias de la Complejidad, Universidad Nacional Autónoma de México, Ciudad de México, México

Centro Interdisciplinario de Investigación para el Desarrollo Integral Regional (CIIDIR) Unidad Oaxaca, Instituto Politécnico Nacional, Oaxaca, México
e-mail: ramosfer@alumni.upenn.edu

L. D. Di Paolo et al. (eds.), *Evolution of Primate Social Cognition*, Interdisciplinary Evolution Research 5, https://doi.org/10.1007/978-3-319-93776-2_8

1 Introduction

The social brain, social complexity or social intelligence hypothesis (SCH; Jolly 1966; Humphrey 1976; Dunbar 1988) postulates that selective pressures originating from the social environment played a major role in the evolution of cognitive abilities of group-living species, particularly in primates. This hypothesis assumes that individuals in complex societies must be able to keep track of moment-to-moment interactions with different individuals, integrate these social interactions into general patterns of social relationships and use this knowledge to predict others' behaviour. Because these abilities are cognitively challenging, it is supposed that an increase in social complexity necessarily entails an increase in cognitive complexity.

The SCH has been criticised for being anthropomorphic (Barrett et al. 2007), attributing human-like cognitive features to other animals without considering alternative ways of dealing with a complex social environment. In addition, there has been some debate on what actually constitutes social complexity (Hemelrijk 2002; Barrett et al. 2007, 2012; Bergman and Beehner 2015). In this chapter we review how models of collective behaviour can provide insights into the cognitive abilities that are minimally required for social structures to emerge out of simple interaction patterns. We also review how analysing social networks using formal metrics can provide useful measures of social complexity which may help to ground the discussion on what is actually challenging about the social environment.

2 What Is Social Complexity?

Traditionally, what has been considered challenging about social behaviour in primates is the existence of triadic relationships: it is not only important that two individuals know about each other's behaviour but also how would each behave in the presence of a third individual (Cheney and Seyfarth 1990). The possibility of coalition formation has been noted as one of the hallmarks of social complexity in primates (Silk et al. 2004) and carnivores (Smith et al. 2010). Proponents of the SCH argue that in order to predict whether another individual will participate in a coalition or not, individuals must use abstractions such as dominance rank or matrilineal relationship. At the core of the debate about whether such abstractions are necessary or not is the issue of which higher-level properties influence lower levels of social organisation: does the dominance hierarchy, as the general pattern of social relationships, influence moment-to-moment interactions? More crucially, do individuals use these general patterns to predict how will others behave? In the case of dominance relationships, is the concept of dominance rank necessary for making decisions by the individuals involved in those interactions?

We argue that theories that focus only on interactions and relationships, even if useful, cannot explain all the complexity of social behaviour in primates. It is tempting to relate social complexity with the number of interactions an individual can process during a short period of time. However, such framework may be difficult to scale up. Exploring higher levels of complexity beyond triadic relationships seems complicated and impractical, since it would require to take into account tetradic, pentadic interactions (and so forth). The presence of many practically simultaneous stimuli is likely to complicate decision-making, without necessarily implying the recognition of new social patterns. A different approach, developed throughout this chapter, consists in determining how a small number of relatively simple interactions between neighbouring individuals can self-organise and generate macroscopic structures at the group level and how this can inform us about the actual complexity faced by individual group members.

Flack (2012) proposes that in order to understand social complexity, we should specify how macroscopic or aggregate properties (such as social structure) arise from microscopic dynamics (such as temporally dynamic social interactions). She suggests that higher organisational levels such as the social structure arise when coarse-grained representations of social interactions at lower levels become useful for animals to make decisions. This paradigm is highly relevant for social cognition: if individual behaviour is indeed influenced by higher levels of organisation (so-called slow variables that change more slowly than day-to-day interactions), it means that individuals should have the computational capacity to consider them. In other words, any component of the system, if it is to be influenced by the statistical regularities of the system's behaviour, should be able to record and process these regularities in order to make decisions about its behaviour.

Barrett et al. (2007) suggest that social complexity theory could benefit from an embodied approach, where cognition is the engagement with the world in terms of perception and action mechanisms, not only in terms of abstract mental constructs (like dominance rank in the example above). This engagement allows individuals to exploit the structure of the environment. The discovery of mirror neurons (Di Pellegrino et al. 1992), which activate when performing an action but also when others perform it, suggests that monkeys could have an automatic (and likely unconscious) understanding of others as goal-directed agents (Gallese and Goldman 1998). This could actually constitute a form of pattern recognition involving active perception, which implies a social understanding but not requiring social concepts. In fact, this pattern recognition could be what individuals, according to Flack (2012), need to respond to the statistical regularities of the slow variables. Pattern recognition of social features could also be updated as the social environment changes, using the most familiar configuration to act on the world. This experientially informed pattern recognition could be expensive in terms of neuronal tissue and connectivity (which would give support to the social brain hypothesis as an explanation for large brains).

3 Emerging Complexity in Collective Phenomena

Some aspects of social behaviour can be complex for collective reasons rather than because of the sophistication of the cognitive tasks to be performed by individuals. Viewing groups as systems of interacting elements, tools borrowed from physics can be useful to elaborate simple models that help to understand how other forms of complexity, not reduced to that of the individual brain, can emerge in social groups. We focus on two classes of problems, which are often interrelated. One is emergence itself, namely, the birth of collective patterns which cannot be predicted from the study of individuals or their pair-wise (or triadic) interactions taken in isolation. The second class aims to understand the functions of collective structures, such as networks. We will discuss their stability and resilience and, more generally, how collective aspects can orient individual behaviour itself and give rise to certain dynamics in response to internal or external changes.

3.1 From Small-Scale Interactions to Macroscopic Patterns in Systems of Agents

A distinctive feature of collective patterns is that they exhibit some kind of order, for instance, coordination, that can emerge without a leader nor strong environmental stimuli acting on the individuals.

Coordination often occurs in animal groups as the result of some kind of imitation in a broad sense, that is, when individuals take similar decisions than other group members. In the simplest scenarios, no specific social links need to be specified, and a few simple assumptions can be made regarding the cognitive abilities of the organisms. For instance, in ant colonies efficient foraging decisions are taken collectively, while each individual follows pheromone trails left by others (Deneubourg and Goss 1989; Bonabeau et al. 1997). On the one hand, trails are reinforced by the attraction exerted by pheromone and, on the other hand, pheromone evaporation prevents the emergence of useless paths, such as those leading to areas without resources or that are unnecessarily long. Likewise, the analysis of three-dimensional coordinated movements of starling flocks (Fig. 1) reconstructed from video recordings reveals relatively simple interindividual interactions (Ballerini et al. 2008; Bialek et al. 2014). As shown by the idealised model proposed by Vicsek et al. (1995), moving in the same direction can be achieved in very large groups provided that each individual tends to orient its velocity along the average flight direction of its immediate neighbours. This alignment rule can also be made stochastic by introducing some random angular noise in the orientation decisions, which may represent either the inability of each individual to estimate exactly the heading directions of neighbours or some type of "free will". If this individual noise exceeds a critical threshold, or if the density of neighbours becomes too small, collective motion can disappear abruptly and be replaced by disordered states

Fig. 1 Starling murmuration at the Royal Society for the Protection of Birds (RSPB) Nature Reserve at Minsmere, Suffolk, UK. Picture by Airwolfhound, Creative Commons License CC BY-SA 2.0

composed of noncoordinated individual movements. The larger the total number of individuals in the group, the more abrupt is the transition (Vicsek et al. 1995). In order to strengthen spatial cohesion, Couzin et al. (2002) introduced an agent-based model that incorporates attraction as well as short-range repulsion, in addition to the alignment rule. Depending on the interaction parameters, this model displays rich dynamical behaviours that resemble those observed in real groups, such as swarms, torus or dynamical parallel groups (Couzin et al. 2002).

Real flocks are not only an illustration of coordinated, cohesive movement taking place in groups, they also exhibit complex internal dynamics that allow a fast transfer of information among individuals, without loss and keeping the variance of individual velocities small (Bialek et al. 2014). Groups can respond quickly to external perturbations (for instance, the presence of a predator), even if only a small fraction of individuals are actually informed of such perturbations. Using the type of individual movement rules described above, simulations show that a small number of informed individuals is sufficient to accurately orient the whole group towards a preferred direction and with a low probability of fragmentation (Couzin et al. 2005). This type of information transfer does not require signalling or that group members know which individuals have information. In addition, the presence of long-range spatial correlations between individual velocities, observed in experiments with starlings, suggests that these systems are poised to respond maximally to environmental perturbations (Cavagna et al. 2010).

The cohesion of large groups can fluctuate in time, as individuals interact with their environment. Sheep herds (Ginelli et al. 2015) or troops of baboons (Strandburg-Peshkin et al. 2015) alternate slow foraging phases during which they spread out, with fast aggregation events that maintain cohesion and are triggered by the movements of initiators. Field observations in sheep can be explained by an individual based-model inspired from Couzin et al. (2005) that includes resting, walking and running phases (Ginelli et al. 2015), and where the attractive force increases sharply only when the distance between two neighbours exceeds a certain value. The individual rules generate a collective intermittent dynamics of fast aggregation events spaced in time, which possibly resolves the dilemma of having enough foraging space to avoid interindividual competition, combined with the imperative of protection against predators offered by cohesion. Despite the fact that they belong to more socially stratified societies, baboons on the move exhibit similar spatial dynamics and, like sheep, are prone to follow democratically the largest subgroup of initiators at any time during their foraging trips, rather than dominant individuals (Strandburg-Peshkin et al. 2015). Similarly, Dostie et al. (2016) showed that a group of baboons traversing habitat with different kinds of resource distributions modified its shape adaptively for efficient foraging while still constrained by the predation risk perceived by the individuals. These shape changes seem to be a partial outcome of individuals modifying their traveling trajectory, with those in the core of the group having more influence on the positions of those who lie in the periphery of the group than vice versa (Bonnell et al. 2017).

An idea suggested from the studies above is that groups can behave as entities equipped with some "collective cognition" that emerges from interactions and that can sense the environment. This principle is not restricted to animal behaviour and is reminiscent, in a fairly different context, of the collective guidance of cells. Evidence actually shows that neural crest cells can aggregate and follow very weak chemical gradients cohesively, under conditions when single cells cannot sense gradients (Camley et al. 2016).

Spatially explicit simulation models can be useful to further understand the emergence of social organisations, for instance, social differentiation and task division. In the model of Hemelrijk (1999, 2002, 2013), a set of individuals form a group and compete by pairs. After winning a dominance interaction, an individual chases away the other. All individuals start with the same probability of winning. After a fight, the dominance parameter of the winner increases and that of the loser decreases, affecting their probability of winning the next fight. If fights are frequent, the model produces a spatial segregation of individuals according to their social rank, with the dominant individuals occupying the centre of the group and the subordinates the periphery (acting as shields against eventual predators, as was observed in the baboon study by Bonnell et al. 2017 cited above). In addition, this model exhibits a phase transition from equalitarian to despotic organisation, as a parameter representing the propensity to fight is varied. In this view, social organisation is an emergent collective phenomenon.

A similar model based on reinforcement of past dominance interactions can explain task division in bumblebee nests (Hogeweg and Hesper 1983). If the simulated nest is divided in a central part and a periphery (as in real nests), the dominance value of an agent will affect its position in space and therefore its activity. Namely, peripheral low ranked individuals will be more likely to forage and high-ranked ones to feed the brood. Therefore, stable specialised classes can emerge from the individual experiences of initially identical agents, hence without the need of assuming that such classes reflect genetic differences. Similar principles may be applicable to primate societies: in chacma baboons, for instance, under predation risk individual placements are determined by competitive abilities, with the dominant animals occupying the centre of the group (Dostie et al. 2016).

As mentioned earlier, environmental factors are key to group formation, cohesion and structure. In particular, grouping patterns in primates and other animals can occur as side effects of feeding on the same food patches, rather than from direct attractive interactions (Sugardjito et al. 1987; Mitani et al. 1991; te Boekhorst and Hogeweg 1994). Spatially explicit agent-based models are useful to study the effects of heterogeneity in resource distribution on the formation of temporary aggregations in a population of foraging individuals (Wilson and Richards 2000; Getz and Saltz 2008). Aggregation mediated by resources is one of the mechanisms that could lead to high levels of fission-fusion dynamics in primate societies, a property that has been interpreted as socially complex and thus as a selective pressure for cognitive abilities (Aureli et al. 2008). Based on observations of spider monkeys, Ramos-Fernandez et al. (2006) developed a simple foraging model where noninteracting individuals visit fruit trees of varying sizes, following mental maps and a maximum efficiency principle. The individual foraging rules can generate frequent fusion and fission events that resemble the association patterns observed in spider monkeys, in particular, for intermediate levels of environmental heterogeneity. In this regime, the model predicts that the average subgroup size reaches a maximum. A network analysis is also carried out, where the nodes represent agents and the edges connect pairs of individuals that have been in contact at least once (occupying the same tree) during the simulation period. In heterogeneous environments, most of the individuals of the group belong to the same "giant" cluster of interconnected individuals. This is due to the fact that many individuals may feed at the same time during their trajectory on one of the few trees that are very rich in fruits. If the edges are weighted in proportion to the time spent together by the nodes they connect, further analysis shows that the giant cluster is itself composed of small "cliques", i.e. substructures composed of individuals that are more densely connected to each other than to others in the group. In comparison, very scarce resource distributions induce much fewer associations, resulting in fragmented networks composed of small isolated clusters. Very abundant resources have a similar effect, as individuals may cluster around different large trees. This model shows that a complex grouping pattern can emerge from relatively simple rules guiding individual foraging decisions, without necessarily involving explicit rules modifying grouping tendencies or dyadic interactions, and that social structures are sensitive to the spatial distribution of resources.

Strong effects of environmental factors on association networks have been reported using empirical data. A network analysis performed by Mokross et al. (2014) quantified how the structure of mixed-species bird flocks was affected by habitat modifications in the Amazonian rain forest. It was found that habitat type had a strong effect on the structure of the networks of non-trophic interactions among bird species, as well as on behavioural interactions within flocks. Frequency of associations among species declined with increasing levels of forest fragmentation due to human activities. Flock cohesiveness and the density of cliques were also positively correlated with the mean vegetation height. These results suggest that social structures are comparatively more sensitive to environmental changes than other ecological networks.

3.2 Emerging Structures in Network Models

The above studies illustrate the convenience of network analysis for describing systems composed of individuals that are not necessarily cohesive and have different activities taking place simultaneously. We further review a few representative network models in nonspatial contexts, as they give useful insights on the relationships between individual connectivity and emerging macroscopic properties. Precise definitions of the network measures mentioned here can be found in Newman (2003).

One of the most studied collective phenomena in networks is percolation. The percolation property is characterised by the presence of a giant cluster connecting a large fraction of a set of nodes (Bollobás and Riordan 2006). In the presence of a giant cluster, it is likely that a path connecting a pair of nodes chosen at random exists. This property is well understood in random networks (Fig. 2a), which are constructed by adding connections one by one and at random among a set of initially unconnected nodes. Although random, this process leads to the emergence of a giant component when the number of connections per individual reaches a critical value (Newman 2003).

Generally, complex networks are characterised by heterogeneous connections, namely, the number of connections (or the degree) of a node can vary largely from one node to the other (Fig. 2b). The presence of a few highly connected nodes, or "hubs", characterised by a degree significantly larger than the average degree of the network, is very rare in random networks but commonly observed in technological, biological or social networks (Barabási and Albert 1999). In animal social networks, node heterogeneity can play an important role in information flow (Lusseau and Newman 2004; Pinter-Wollman et al. 2014).

Since random networks have homogeneous (Poissonian) degree distributions, other rules must be implemented to account for the observed heterogeneity of many real-world networks. For instance, networks can be grown by adding nodes one by one and by connecting them to previously existing nodes (Krapivsky et al. 2000). If each new node is connected to a randomly chosen existing node, a homogeneous network with exponential degree distribution is obtained. The preferential attachment rule, in turn, consists in connecting the new node to an existing node which is chosen with a probability proportional to its degree. Therefore, highly connected

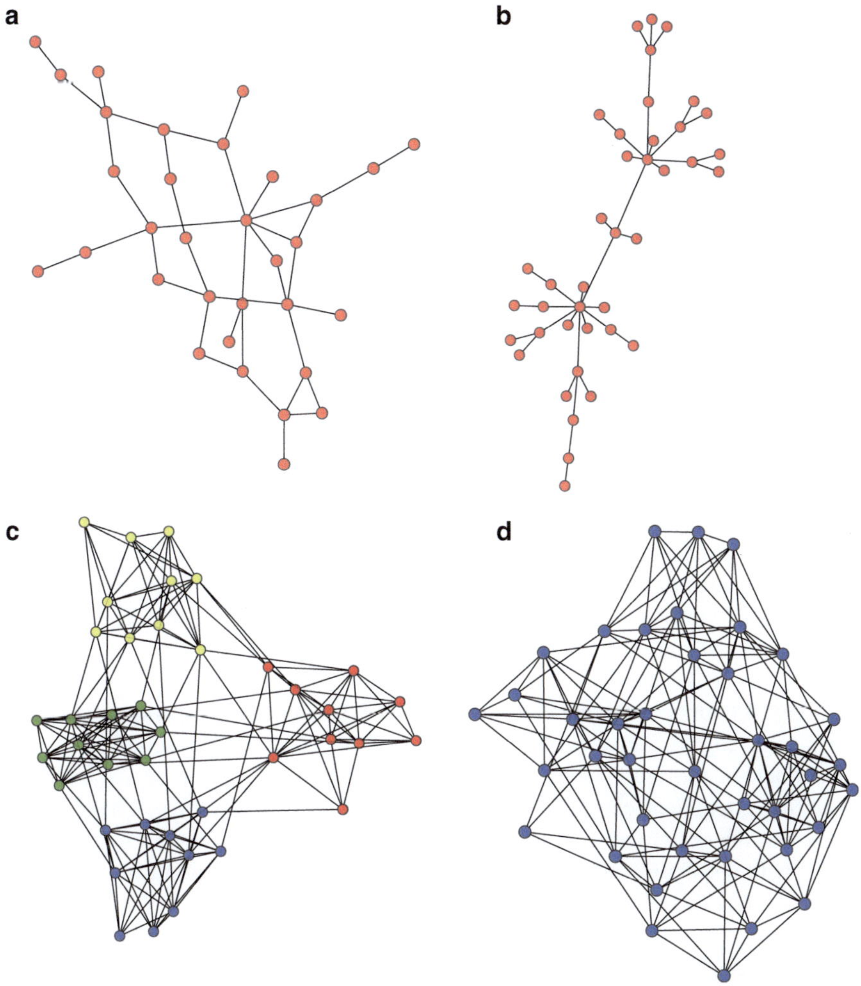

Fig. 2 Types of network architecture. (**a**) A connected component with 40 nodes and 39 edges extracted from a random network, where every pair of nodes has the same probability of being linked. The degree distribution (or the number of connections per node) would be Poissonian; (**b**) a network of 40 nodes and 39 edges, established with preferential attachment, i.e. nodes with more edges have a higher probability of having more edges and becoming "hubs". The degree distribution of such a network would be non-Poissonian (and for very large networks, scale-free); (**c**) a network with high modularity or cliquishness, where 40 nodes are clustered into 4 components (which themselves are random networks with a probability of an intra-module edge being 0.8 and only 5 edges joining each module to others). A clear separation into four modules can be observed, and some nodes linking different modules would be relevant as "brokers"; (**d**) a network with low modularity, composed of the same number of nodes and edges as the network in (**c**) but with a probability of 0.66 of an intra-module edge and 10 edges joining modules. No clear separation into modules can be observed

nodes tend to attract the new connections. This process produces non-exponential networks, characterised instead by power-law degree distributions that entail the presence of highly connected nodes (Barabási and Albert 1999). These "hubs" are usually the older ones, and the network as a whole grows hierarchically, as the degree of a node is highly correlated with its order of appearance. Primate social networks do not seem to be scale-free, but those of other species like bottlenose dolphins or Columbian ground squirrels exhibit such structure (see Kanngiesser et al. 2011 for a discussion).

A peculiarity of many social networks, compared to other types of networks, is their cliquishness (Newman and Park 2003). The clustering coefficient is the likelihood that two nodes connected to a same node are themselves connected to each other. When connections are transitive, this coefficient is high (Fig. 2c–d). In random networks or networks constructed according to preferential attachment, cliques are practically absent and the clustering coefficient very small. Other models that generate dense networks have helped to understand how cliquishness can confer resilience to a system. In the model of Marsili et al. (2004), a network is constantly perturbed by random deletion of links. In parallel, the nodes seek to establish new connections either randomly in the population or via "friends". In the latter case, a node can connect to a neighbour of one of its own neighbours. The dynamics of this model can be viewed as a continuous struggle against volatility (link deletion) with connections arising from new opportunities partially mediated by the existing network. In an interval of values for the rate of the connections via friends, the resulting networks are both densely connected and with a very high clustering coefficient. Surprisingly, these structures are not the unique solutions, since, depending on the initial conditions, networks with low average degree and low clustering coefficient can also form for the same parameter values. These results imply that dense networks, in which short-range reconnections are allowed, can be robust with respect to node deletion and may even resist under conditions in which dense networks would not form.

A commonly studied local property, giving information on the position of a node in its network, is centrality. The betweenness centrality of a node i is the number of shortest paths joining other nodes (geodesics) that run through i. Central nodes are not necessarily the most connected ones, but their removal may considerably increase the length of the shortest path for going from one node to another (see Newman 2003 and refs. therein). Eigenvalue centrality is defined from the eigenvalue spectrum of the network adjacency matrix and it is related to diffusion processes in the network. These two measures are similar and tell about network resilience, namely, the effect that node removal will have on path lengths and therefore network topology and function.

3.3 *From Structure to Interactions: Functions of Complex Networks*

Simulation models are useful to illustrate how macroscopic structures can be the outcomes of individual behaviours and to predict unobserved but possible features by varying parameters. Note that in the vicinity of a phase transition, mild changes in behaviours produce drastic differences at larger scales. Conversely, as shown below, macroscopic features crucially contribute to orient individual behaviour, even when agents are not aware of them. One may then view social dynamics as a constant interplay between the evolution in time of large-scale structures and that of individuals with local (e.g. dyadic level) interactions. An ongoing challenge in social network analysis is to understand how a given architecture may facilitate certain social behaviours or certain types of dynamics.

Robustness is a property of many complex systems, and certain network architectures are known to be robust under attacks. Structural robustness can be probed theoretically by node (or link) deletion. If nodes are removed at random in a scale-free network, for instance, the mean topological distance between any pair of nodes remains practically unaffected, in sharp contrast with the case of homogeneous networks (Albert et al. 2000; Newman 2006). In this sense, scale-free networks are robust. However, the situation is quite different if attacks are not random but targeted, namely, if the most connected nodes are progressively deleted. In this case, the giant cluster component does not remain cohesive, and the mean topological distance between pairs of nodes increases sharply with the fraction of deleted nodes. The removal of very few nodes may even fragment a network into two or more components, thus preventing communication in the whole system. In empirical networks, these strategic nodes, or "brokers", are not always the most connected actually, but they are often located at the boundary between modules or communities (Newman 2006). These nodes have a high betweenness centrality, i.e. they frequently lie on the shortest path that connects a pair of nodes randomly chosen in the network. By using centrality measures, Lusseau and Newman (2004) have identified brokers in the association network of a group of bottlenose dolphins. The removal of the most central individual was predicted to substantially affect the mean distance between pairs of individuals in the network. A fission of the group was actually observed when this individual happened to temporarily leave the group. Similarly, the resilience of the grooming network of a captive chimpanzee group was probed by removing central individuals, highlighting their key role on group cohesiveness (Kanngiesser et al. 2011).

Centrality measures can thus help to understand the role that individuals play in a social network and how collective decisions are taken. King et al. (2011) showed with the help of a network analysis that each departure from the sleeping site in chacma baboons was a self-organising process, initiated by an individual that was followed by the rest of the group. The initiators tended to be more sociable individuals, namely, those that were central in the grooming and spatial association networks, rather than dominant individuals. These findings add support to the idea that collective behaviour in animal groups can follow relatively simple cognitive heuristics. Brent et al. (2015) further provided evidence that during the collective foraging trips of killer whales, leadership was taken by the older, more experienced individuals, especially in difficult times of low prey abundance. Such leadership also provides a simple mechanism by which ecological knowledge can be transferred within the group. Whether being a broker or a leader requires more sophisticated cognitive abilities remains elusive.

The structure of a social network can also be important for the way in which collective decision-making (and therefore group cognition) can occur. Dense networks, or networks with long-distance connections or with particular individuals functioning as hubs, could be more efficient in exchanging information about the environment and responding appropriately. For example, in an experiment with human subjects, networks of cooperators that were structured in spatially based cliques were more efficient in solving problems that required broad exploration, whereas networks with greater long-range connectivity were better at solving problems that required less exploration (Mason et al. 2008).

From the above examples, one sees that networks are not just static objects made of nodes and links but also structures on which many dynamical processes can take place. Two well-studied processes are epidemic spreading (Pastor-Satorras et al. 2015) and synchronisation of activities (Arenas et al. 2008). Understanding how the structure of a network determines the evolution of its elements is of primary importance. In this context, resilience, a key feature of complex systems, refers to the ability of a system to readjust its activity in response of environmental or internal changes. In principle, it is determined both by the structure of the network and the dynamics that describe the evolution of the states of the nodes. Inspired by ecological and regulatory systems, Gao et al. (2016) have considered a general framework where a set of coupled dynamical elements are located on a fixed network of interactions. The study of the stationary states and their stability shows that global states of high activity can undergo sudden transitions to undesired states of much lower activity, when environmental conditions or interaction parameters are changed. A system is said resilient if the states of high activity remain stable over a broad range of parameter values. Networks that are dense and heterogeneous are found to be more resilient (Gao et al. 2016).

The resilience of animal social networks undergoing natural "knockouts" can be studied in the field. Barrett et al. (2012) observed the effects of the death of dominant females in a group of baboons. The resulting instability in the social hierarchy was counterbalanced by restructuring into more cliquish groups. The results suggest that adjustments are aimed at reducing the uncertainty of relationships, with no need for specific mechanisms of conflict management (Barrett et al. 2012; see also Flack et al. 2006 and Sect. 4 below).

As mentioned earlier, a property of social networks is the presence of subcommunities or modules (Newman 2006). One might expect that any social network would respond to a mild perturbation by readjustments that preserve modularity. Modularity is a particularly important network property, as it can orient individual behaviour towards cooperation. Marcoux and Lusseau (2013) simulated prisoner's dilemma games on static networks of varying modularity index and found that modular networks significantly promoted the evolution of cooperation (Nowak and Sigmund 1998). Therefore, cooperation could evolve in social groups without complicated mechanisms such as genetic similarity, provided that the underlying social networks were sufficiently modular. This conclusion has to be contrasted with Darwinian theories that predict the evolution of selfish individuals when interactions take place randomly in a population. In fact, in an analysis of 70 interaction networks of 30 different primate species, Voelkl and Kasper (2009) found that the structure was precisely what favoured the fixation of cooperative traits, compared to well-mixed populations of similar size or randomly connected networks. Similarly, cooperation happens to be the dominating trait in evolutionary games that take place on scale-free networks (Santos and Pacheco 2005). This latter result differs completely from the outcomes of binary games or games on regular networks, where under the same rules, cooperators are unable to resist the invasion by defectors.

4 Can Social Network Metrics Be Used as an Indicator of Social and Cognitive Complexity?

It is tempting to use metrics derived from social network analysis as measures of social complexity. The density of links, the heterogeneity in the degree distribution or the modularity of a network could be considered as measures of its complexity (Costa et al. 2007). In keeping with the most commonly used definition of social complexity (Dunbar 1988), one could use the number of different interactions (e.g. grooming, proximity, co-feeding, etc.) that would need to be included in a multilayered network that effectively described the social structure of a particular species (Barrett et al. 2012).

However, our review shows that many structural and functional properties of groups and their social networks are an emerging result of individual interactions, without necessarily implying that the complexity at the collective level rests on complex cognitive processes. If we do not know which aspects of social networks are actually taken into account by individuals in their decision-making, it is not clear which metrics of social network analysis could be useful for defining social complexity in a way that is relevant for the study of social cognition.

Brent (2015) argues that while primates do seem capable of understanding third-party relationships, it is not known whether they consider other more distant individuals in their social network, to whom they are connected only indirectly, in their decision-making. An intriguing aspect of real-world networks is the existence of recurring substructures or "motifs" (Milo et al. 2002). These appear to be the result of constraints during the evolution of a network and would function as the elementary building blocks or computational circuits through which information flow and other processes occur in a network. The assembly of these relatively simple building blocks would naturally lead to large functional structures such as social networks. It is intriguing to think that social interactions also lead to the occurrence of these motifs (such as triadic closure in affiliative relationships or alliance formation in agonistic contexts) and that the whole network would grow as a result of these local processes coming together. Moreover, in terms of cognition, motifs could be recognised by individuals using relatively simple pattern recognition mechanisms, which would also allow them to act in appropriate ways without necessarily taking the whole network into account (Barrett et al. 2007).

In the case of network growth, as we review in Sect. 3.2, a process that can lead to a heterogenous (and therefore complex) distribution of network degrees is preferential attachment, i.e. new nodes joining those nodes that are already well-connected. If animals recognise these highly connected nodes when establishing their social relationships, it would imply that they have some way of recognising the connections that go to others beyond their own connections or at least some proxies for the highly connected nodes.

Several studies have shown that animals can modify the structure of their network in predictable ways. For example, the ground-breaking study of Flack et al. (2006) showed experimentally that particular individual macaques with "policing" roles in their social network can affect the structure of the network as a whole. Removing these particular individuals resulted in a change in all interactions, producing less diverse and more fragmented networks. Clearly, from a purely functional standpoint, policing stabilises and promotes cohesiveness within the social network. However, we know little about the cognitive mechanisms underlying this policing role or the effect that the absence of a policer has on the decision-making by others.

Building on Flack et al. (2006), Barrett et al. (2012) have proposed that social animals develop multilayered interaction networks precisely in order to reduce the uncertainty about their social relationships (Fig. 3). In other words, more interaction contexts (and thus more social complexity) reduce uncertainty in predicting the behaviour of social companions. A perturbation of the network in one context (as in the knockout experiments by Flack et al. 2006 or the natural observations by Barrett et al. 2012) leads to adjustments in other contexts, which reduce the uncertainty (measured by a decrease in Shannon's entropy) of the whole network. This supports the idea that a multilayered social network is the best representation of

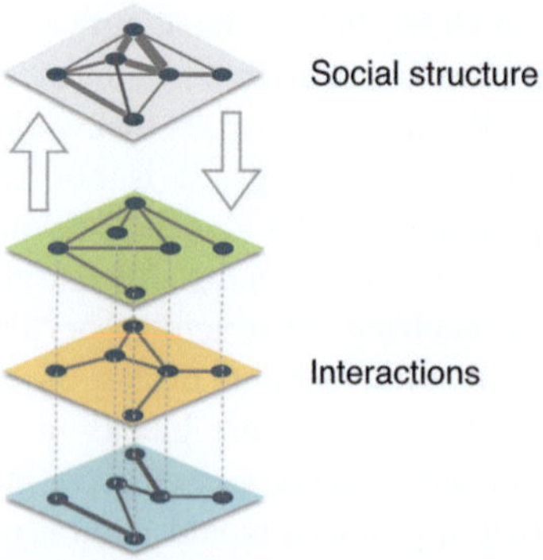

Fig. 3 A conceptual model of social structure by means of a multilayered network, with each layer corresponding to a different interaction and the whole social network being the object containing these different layers. Modified from Barrett et al. (2012)

social structure and that different layers (or interactions) are interrelated. It remains to be shown to what extent are animals actively making these adjustments as a result of their understanding of their social network.

Acknowledgements We acknowledge financial support from DGPA-PAPIIT grant IN105015, CONACYT grant 157656 and Instituto Politecnico Nacional. We thank Louise Barrett for fruitful comments on the manuscript.

References

Albert R, Jeong H, Barabási A-L (2000) Attack and error tolerance of complex networks. Nature 406:378–382

Arenas A, Díaz-Guilera A, Kurths J, Moreno Y, Zhou C (2008) Synchronization in complex networks. Phys Rep 469(3):93–153

Aureli F, Schaffner CM, Boesch C, Bearder SK, Call J, Chapman CA, Connor R, Di Fiore A, Dunbar RIM, Peter Henzi S, Holekamp K, Korstjens AH, Layton R, Lee P, Lehmann J, Manson JH, Ramos-Fernandez G, Strier KB, van Schaik CP (2008) Fission-fusion dynamics: new research frameworks [with comments]. Curr Anthropol 49(4):627

Ballerini M, Cabibbo N, Candelier R, Cavagna A et al (2008) Interaction ruling animal collective behavior depends on topological rather than metric distance: evidence from a field study. Proc Natl Acad Sci USA 105(4):1232–1237

Barabási A-L, Albert R (1999) Emergence of scaling in random networks. Science 286(5439): 509–512

Barrett L, Henzi P, Rendall D (2007) Social brains, simple minds: does social complexity really require cognitive complexity? Philos Trans R Soc B 362:561–575

Barrett L, Henzi SP, Lusseau D (2012) Taking sociality seriously: the structure of multi-dimensional social networks as a source of information for individuals. Philos Trans R Soc Lond Ser B Biol Sci 367(1599):2108–2118

Bergman TJ, Beehner JC (2015) Measuring social complexity. Anim Behav 103:203–209

Bialek W, Cavagna A, Giardina I, Mora T et al (2014) Social interactions dominate speed control in poising natural flocks near criticality. Proc Natl Acad Sci USA 111(20):7212–7217

Bollobás B, Riordan O (2006) Percolation. Cambridge University Press, Cambridge

Bonabeau E, Theraulaz G, Deneubourg J-L, Aron S, Camazine S (1997) Self-organization in social insects. Trends Ecol Evol 12(5):188–193

Bonnell TR, Clarke PM, Henzi SP, Barrett L (2017) Individual-level movement bias leads to the formation of higher-order social structure in a mobile group of baboons. PeerJ Preprints 5: e2808v1

Brent LJ (2015) Friends of friends: are indirect connections in social networks important to animal behaviour? Anim Behav 103:211–222

Brent LJN, Franks DW, Foster EA, Balcomb KC, Cant MA, Croft DP (2015) Ecological knowledge, leadership, and the evolution of menopause in killer whales. Curr Biol 25:746–750

Camley BA, Zimmermann J, Levine H, Rappel W-J (2016) Emergent collective chemotaxis without single-cell gradient sensing. Phys Rev Lett 116:098101

Cavagna A, Cimarelli A, Giardina I, Parisi G, Santagati R, Stefanini F, Viale M (2010) Scale-free correlations in starling flocks. Proc Natl Acad Sci USA 107(26):11865–11870

Cheney DL, Seyfarth RM (1990) How monkeys see the world: inside the mind of another species. University of Chicago Press, Chicago

Costa LDF, Rodrigues FA, Travieso G, Villas Boas PR (2007) Characterization of complex networks: a survey of measurements. Adv Phys 56(1):167–242

Couzin ID, Krause J, James R, Ruxtony GD, Franks NR (2002) Collective memory and spatial sorting in animal groups. J Theor Biol 218:1–11
Couzin ID, Krause J, Franks NR, Levin SA (2005) Effective leadership and decision-making in animal groups on the move. Nature 433:513–516
Deneubourg JL, Goss S (1989) Collective patterns and decision-making. Ethol Ecol Evol 1: 295–311
Di Pellegrino G, Fadiga L, Fogassi L, Gallese V, Rizzolatti G (1992) Understanding motor events: a neurophysiological study. Exp Brain Res 91:176–180
Dostie MJ, Lusseau D, Bonnell T, Clarke PMR, Chaplin G, Kienzle S, Barrett L, Henzi SP (2016) Proof of principle: the adaptive geometry of social foragers. Anim Behav 119:173–178
Dunbar RIM (1988) Primate social systems. Chapman & Hall, London, UK
Flack JC (2012) Multiple time-scales and the developmental dynamics of social systems. Philos Trans R Soc B 367:1802–1810
Flack JC, Girvan M, de Waal FBM, Krakauer DC (2006) Policing stabilizes construction of social niches in primates. Nature 439:426–429
Gallese V, Goldman A (1998) Mirror neurons and the simulation theory of mind-reading. Trends Cogn Sci 2:493–501
Gao J, Barzel B, Barabási AL (2016) Universal resilience patterns in complex networks. Nature 530:307–312
Getz WM, Saltz D (2008) A framework for generating and analyzing movement paths on ecological landscapes. Proc Natl Acad Sci USA 105(49):19066–19071
Ginelli F, Peruani F, Pillot M-H, Chaté H, Theraulaz T, Bon R (2015) Intermittent collective dynamics emerge from conflicting imperatives in sheep herds. Proc Natl Acad Sci USA 112:12729–12734
Hemelrijk CK (1999) An individual-orientated model of the emergence of despotic and egalitarian societies. Proc R Soc B Biol Sci 266(1417):361–369
Hemelrijk CK (2002) Understanding social behaviour with the help of complexity science (invited article). Ethology 108(8):655–671
Hemelrijk CK (2013) Simulating complexity of animal social behaviour. In: Edmonds B, Meyer R (eds) Simulating social complexity: understanding complex systems. Springer-Verlag, Berlin, pp 581–615
Hogeweg P, Hesper B (1983) The ontogeny of the interaction structure in bumble bee colonies: a MIRROR model. Behav Ecol Sociobiol 12:271–283
Humphrey N (1976) The social function of intellect. In: Bateson PPG, Hinde RA (eds) Growing points in ethology. Cambridge University Press, Cambridge, MA, pp 303–317
Jolly A (1966) Lemur social behaviour and primate intelligence. Science 153:501–506
Kanngiesser P, Sueur C, Riedl K, Grossmann J, Call J (2011) Grooming network cohesion and the role of individuals in a captive Chimpanzee group. Am J Primatol 73:758–767
King AJ, Sueur C, Huchard E, Cowlishaw G (2011) A rule-of-thumb based on social affiliation explains collective movements in desert baboons. Anim Behav 82:1337–1345
Krapivsky PL, Redner S, Leyvraz F (2000) Connectivity of growing random networks. Phys Rev Lett 85:4629–4632
Lusseau D, Newman MEJ (2004) Identifying the role that animals play in their social networks. Proc R Soc Lond B 271:S477–S481
Marcoux M, Lusseau D (2013) Network modularity promotes cooperation. J Theor Biol 324: 103–108
Marsili M, Vega-Redondo F, Frantisek Slanina F (2004) The rise and fall of a networked society: a formal model. Proc Natl Acad Sci USA 101(6):1439–1442
Mason WA, Jones A, Goldstone RL (2008) Propagation of innovations in networked groups. J Exp Psychol 137(3):422–433
Milo R, Shen-Orr S, Itzkovitz S, Kashtan N, Chklovskii D, Alon U (2002) Network motifs: simple building blocks of complex networks. Science 298(5594):824–827

Mitani JC, Grether GF, Rodman PS, Priatna D (1991) Association among wild orang-utans: sociality, passive aggregations or chance? Anim Behav 42:33–46
Mokross K, Ryder TB, Correa Cortes M, Wolfe JD, Stouffer PC (2014) Decay of interspecific avian flock networks along a disturbance gradient in Amazonia. Proc R Soc B 281:20132599
Newman MEJ (2003) The structure and function of complex networks. SIAM Rev 45(2):167–256
Newman MEJ (2006) Modularity and community structure in networks. Proc Natl Acad Sci USA 103:8577–8582
Newman MEJ, Park J (2003) Why social networks are different from other types of networks. Phys Rev E 68:036122
Nowak MA, Sigmund K (1998) Evolution of indirect reciprocity by image scoring. Nature 393: 573–577
Pastor-Satorras R, Castellano C, Van Mieghem P, Vespignani A (2015) Epidemic processes in complex networks. Rev Mod Phys 87:925–979
Pinter-Wollman N, Hobson EA, Smith JE, Edelman AJ et al (2014) The dynamics of animal social networks: analytical, conceptual, and theoretical advances. Behav Ecol 25(2):242–255
Ramos-Fernandez G, Boyer D, Gomez VP (2006) A complex social structure with fission-fusion properties can emerge from a simple foraging model. Behav Ecol Sociobiol 60:536–549
Santos FC, Pacheco JM (2005) Scale-free networks provide a unifying framework for the emergence of cooperation. Phys Rev Lett 95:098104
Silk JB, Alberts SC, Altmann J (2004) Patterns of coalition formation by adult female baboons in Amboseli, Kenya. Anim Behav 67:573–582
Smith JE, Van Horn RC, Powning KS, Cole AR, Graham KE, Memenis SK, Holekamp KE (2010) Evolutionary forces favoring intragroup coalitions among spotted hyenas and other animals. Behav Ecol 21:284–303
Strandburg-Peshkin A, Farine DR, Couzin ID, Crofoot MC (2015) Shared decision-making drives collective movement in wild baboons. Science 348:1358–1361
Sugardjito J, Te Boekhorst IJA, Van Hooff JARAM (1987) Ecological constraints on the grouping of wild orang-utans (Pongo pygmaeus) in the Gunung Leuser National Park, Sumatra, Indonesia. Int J Primatol 8:17–41
te Boekhorst IJ, Hogeweg P (1994) Effects of tree size on travelband formation in orang-utans: data analysis suggested by a model study. In: Brooks R, Maes P (eds) Artificial life IV. MIT Press, Cambridge, pp 119–129
Vicsek T, Czirok A, Ben-Jacob E, Cohen I, Shochet O (1995) Novel type of phase transition in a system of self-driven particles. Phys Rev Lett 75(6):1226–1229
Voelkl B, Kasper C (2009) Social structure of primate interaction networks facilitates the emergence of cooperation. Biol Lett 5(4):462–464
Wilson WG, Richards SA (2000) Consuming and grouping: resource-mediated aggregation. Ecol Lett 3:175–180

Comparative Economics: Using Experimental Economic Paradigms to Understand Primate Social Decision-Making

Julia Watzek, Mackenzie F. Smith, and Sarah F. Brosnan

Abstract One of the goals of studying primate social cognition is to better understand how decision-making is similar or different across primate species, including humans. Recently, researchers have begun to use paradigms from experimental economics that allow for direct comparisons across species using identical or highly similar experimental approaches. In many cases, paradigms used extensively in humans, such as the Assurance Game, Matching Pennies Game, and Prisoner's Dilemma, have been adapted for other species in order to understand how different payoff structures influence decision-making. This approach has been utilized to explore not only individual differences and the influence of the social environment on behavior, but also the ecological relevance of such paradigms to the species in question. This comprehensive exploration allows for the most robust understanding of the evolution and development of social decision-making.

Keywords Game theory · Cooperation · Coordination · Competition · Inequity · Comparative · Social cognition

1 Introduction

Primates' decisions are embedded within a social matrix, and both social behavior and decision-making have been studied in a variety of primate species. Most data on decision-making, however, has been derived from research testing individuals by

J. Watzek · M. F. Smith
Department of Psychology, Language Research Center, Georgia State University, Atlanta, GA, USA

S. F. Brosnan (✉)
Department of Psychology, Language Research Center, Georgia State University, Atlanta, GA, USA

Department of Philosophy, Neuroscience Institute, Center for Behavioral Neuroscience, Georgia State University, Atlanta, GA, USA
e-mail: sbrosnan@gsu.edu

L. D. Di Paolo et al. (eds.), *Evolution of Primate Social Cognition*, Interdisciplinary Evolution Research 5, https://doi.org/10.1007/978-3-319-93776-2_9

themselves. Few studies have explored how social influences affect decision-making, and even fewer have systematically compared social decision-making across taxa using equivalent experimental approaches. Comparative studies on how different primate species, including humans, solve social problems can provide insight into the evolution of social decision-making in ways that cannot be addressed by studying any one species alone (Brosnan et al. 2013; Santos and Rosati 2015). Similarities and differences between species can illuminate the phylogeny of social decision-making and may shed light on the evolutionary pressures that led to the emergence of both unique and shared decision-making strategies and biases. In addition, comparative studies can inform hypotheses about the underlying cognitive mechanisms that are necessary for such strategies to emerge. This can tell us not only when decision-making strategies are shared, indicating either a common origin (in closely related species) or shared selective pressures, but also when mechanisms are distinct, indicating that the decision outcome was sufficiently important to have risen separately in different species.

A promising new line of research addresses the theoretical questions behind the nature of social decision-making across a range of species using experimental economic games. Game theoretic research has been conducted extensively in humans and focuses on how humans make decisions as they interact in social settings, particularly in cooperative and competitive scenarios. For example, an individual's actions may change depending on their partner's identity, or when their rewards are valued differently depending on what their partner receives. Game theory can reduce complex ecological decision-making situations to basic action-payoff structures that are often based on a single dichotomous choice. It can be used to model strategic interactions, compare results between different payoff distributions and contexts, and predict behavioral outcomes. While designing a comparable task that works across different species presents a challenge (Smith et al. 2018), these games are simple and easily adapted to different procedures; thus, the game theoretic approach lends itself particularly well to the comparative study of social decision-making.

In this chapter, we discuss how experimental economic games can be applied to a variety of animal species and validated against existing results from human research on game theoretical paradigms. We then consider the ways in which comparative studies utilizing game theory have already begun to shed light on the evolution of social decision-making in primates and their underlying mechanisms.

2 Experimental Economic Paradigms

2.1 Coordination: Assurance Game

The Assurance Game, or Stag Hunt Game, is an interactive strategic paradigm in which both players need to coordinate on one of two options, Stag or Hare. Players receive the largest payoff only when they both choose Stag (the payoff-dominant outcome). Choosing Hare yields a consistent, albeit smaller reward regardless of their partner's choice (the risk-dominant outcome), while choosing Stag when their partner chooses Hare results in no payout at all. This game is easy to assess because coordination on Stag is the obvious choice for any subject who understands the payoffs (and indeed is the dominant choice for humans to whom the game has been explained). The Assurance Game nicely captures how cooperation can be achieved and maintained through mutualism, because both players are better off by cooperating than by acting alone. However, both individuals need to trust their partner to cooperate in order to achieve the most beneficial outcome [see also Engelmann et al. (2015) for an investigation of such strategic trust in chimpanzees]. Otherwise, defecting to Hare is the less risky decision.

We have tested four primate species representing a range of primate taxa on the Assurance Game: humans and chimpanzees (apes), rhesus monkeys (Old World monkeys), and capuchin monkeys (New World monkeys) (Brosnan et al. 2011, 2012; Parrish et al. 2014; see also Bullinger et al. 2011). We took particular care to ensure a common protocol to the highest degree possible. Pairs were tested sitting immediately next to each other and were not anonymous. Neither animal nor human subjects received instructions, but they were not prevented from communicating with their partners during the experiment. Like the animal subjects, humans received tangible (though monetary rather than edible) rewards at the end of each trial and had previously participated in behavioral testing at the same lab. One notable consequence of this procedure was that results for our humans deviated substantially from those seen in more human-typical procedures that include instruction and pretesting.

We ran two different versions of the task, a manual exchange task based on procedures common in animal research and a computerized analogue based on procedures more common in human research. In the manual study, players in each pair chose between the two options by returning one of two tokens to the experimenter (Brosnan et al. 2011). In the computerized study, players indicated their choice by selecting one of two icons on a computer screen that they shared with their partner (Brosnan et al. 2012; Parrish et al. 2014). An advantage to the computerized procedure is that we could more tightly control subjects' knowledge. Specifically, we had an asynchronous version in which players' choices were revealed as they were made, thus providing information about the first player's response to the second player (although this did not mean that the latter recognized that this information was available, or chose to use it). In the synchronous condition, neither partners' choices were displayed until after both had made a decision, which allowed us to test whether the subjects benefitted from the information present in the asynchronous condition.

Some pairs in all four species settled on the coordinated strategy that resulted in the payoff-dominant outcome, i.e., predominantly played Stag/Stag. However, the species differed in the number of pairs that found this outcome and in the way that they reached it. For capuchin monkeys, only one of six pairs found the payoff-dominant outcome in the token exchange paradigm (Brosnan et al. 2011). However, in the computerized version that allowed for more trials, all pairs found the payoff-dominant outcome in the asynchronous procedure, although none did in the synchronous condition (Brosnan et al. 2012). These results indicate that capuchin monkeys used the available information about their partner's choice to match it with their own decision. While response matching would also favor coordination on

Hare/Hare, it pays less than coordinating on Stag/Stag, and even a slight tendency to prefer the (sometimes) higher-paying Stag option could explain the strong bias toward playing Stag.

Chimpanzees' strategies in the Assurance Game were strongly associated with their experience with cognitive tasks and behavioral testing (Brosnan et al. 2011). Pairs that had participated in daily cognitive testing since they were infants found the payoff-dominant outcome both with and without knowledge of their partner's choice, and even generalized the response strategy to a novel token set, indicating that they understood the strategy behind their choices. Subjects with little to no such experience, however, were unable to find the Stag/Stag outcome, although most of these subjects developed a consistent response pattern (matching their partner).

All five pairs of rhesus monkeys succeeded in finding the payoff-dominant outcome in both versions of the computerized task (Brosnan et al. 2012). Humans were also able to do so; however, their performance was not nearly as good as in traditional economic games. In the manual task, only 5 out of 26 pairs found the payoff-dominant Stag/Stag strategy, while 10 pairs settled on the risk-dominant Hare/Hare strategy (Brosnan et al. 2011). Notably, none of the latter pairs ever played Stag/Stag, indicating that those pairs failed to explore alternate strategies once they had found a mutually beneficial one. In the computerized task, all 22 of 27 pairs that spoke about the task during the experiment found the payoff-dominant Stag/Stag strategy, while the five pairs that did not talk about the task settled on the risk-dominant Hare/Hare strategy, again without having played Stag/Stag. Our interpretation is that humans were much less likely to explore the parameter space and, because they never experienced the Stag/Stag outcome, did not realize that there was a better option in the manual task. In the computer game, however, they used language to verbally consider options, leading to a higher incidence of the payoff-dominant strategy (humans in the manual task were not told not to talk, but, likely due to the presence of the experimenter, did not do so). In an alternate condition in which subjects could not talk, pairs who played the asynchronous game did better than those who played the synchronous one, indicating that humans, like other primates, use other cues to improve performance when language is unavailable.

A remaining question, however, was whether humans and rhesus monkeys used the same strategy to find the Stag/Stag outcome. Both species' performance in the synchronous task rules out a simple response-matching strategy. When subjects were presented with simulated players who varied in their proportion of Stag responses, rhesus monkeys showed a strong Stag bias, regardless of the strategy played by the simulation, whereas humans matched their response proportions to those displayed by the simulated players (Parrish et al. 2014). Due to the payoff structure, the two species ended up with almost exactly the same average payoffs. These results nicely illustrate how different species can reach identical outcomes using very different cognitive mechanisms and emphasize the importance of distinguishing between outcomes and mechanisms. Indeed, rhesus monkeys also show probability matching in other studies (Lau and Glimcher 2005), indicating that even when two species can use a cognitive mechanism, we cannot assume that they both use it in any given situation.

Ideally, comparative research programs involve both identical methodologies across species and methodologies designed with species-specific considerations in mind (e.g., Salwiczek et al. 2012; Prétôt et al. 2016). Two additional studies have assessed the behavior of chimpanzees and 4-year-old children in a potentially more ecologically relevant version of the Assurance Game. In both studies, chimpanzees and humans cooperated at high levels to reach the payoff-dominant outcome (Bullinger et al. 2011; Duguid et al. 2014). Interestingly, coordination among chimpanzees was only preferred when the benefits exceeded those gained from individual efforts. Chimpanzee coordination on Stag decreased when the value of the Hare option was increased, particularly when subjects could not see what their partner chose before making a decision themselves. Children, however, maintained high cooperation rates through increased communication and visual monitoring prior to making a decision (Duguid et al. 2014). Thus, similar to adults, children as young as 4 years old use communication as a way to find the best strategy for coordination.

Although all primate species were able to find the payoff-dominant outcome in the Assurance Game, they differed considerably in how they did so. Conceivably, different ecological pressures could have led to the selection for different skills. Capuchin monkeys are known to cooperate extensively both in the field (e.g., Perry 2003; Crofoot et al. 2011) and in the laboratory (e.g., de Waal and Berger 2000; de Waal 2000; Mendres and de Waal 2000; de Waal and Davis 2003; Brosnan et al. 2006). One possible explanation for their inability to coordinate without information about their partner's behavior is that capuchins live in small, cohesive groups and are rarely out of sight of each other. Thus, they may never need to coordinate their behavior with a partner whose choices they cannot observe directly. Chimpanzees, however, participate in activities such as highly coordinated group hunting, which requires cooperation even when group members are out of sight, which may explain why they could maintain coordination, albeit less well, when they could not see their partner's choice. Chimpanzees also coordinate more when there is a greater benefit, and indeed, group hunts have been argued to predominantly occur in habitats in which prey is more difficult to catch alone and coordination is required (Boesch and Boesch 1989).

2.2 *Competition: Matching Pennies*

The Assurance Game has a solution that is mutually beneficial for both partners with respect to overall payout (Stag/Stag). Once found and settled on, the choices of both players are highly predictable in every round of the game. This is likely similar to the ways in which many species cooperate, and indeed, such coordination is presumably a necessary prerequisite for more complex forms of cooperation. However, behaving predictably can be detrimental in other social situations, such as in predator-prey interactions or intragroup competition. Recently, Martin et al. (2014) used a standardized procedure to present chimpanzees and humans with a nonverbal, computerized version of the Matching Pennies Game. In this competitive strategic game, joint actions always result in one winner and one loser. Each player chooses between two options; the matcher earns a payoff when the two players' choices match, whereas the mismatcher earns a payoff when the two players' choices mismatch. Thus, accurately predicting the other player's behavior while acting unpredictably yourself is crucial to maximize individual gain, and there is no possibility of mutual gain. Unlike the Assurance Game, the Nash equilibrium in this game is a mixed-strategy equilibrium. Players should choose between options with a certain probability in order to maximize their gains given a specific payout matrix. Switching between both options is necessary because consistently choosing only one or the other option could easily be exploited by the opponent.

Across three versions of the Matching Pennies Game, chimpanzees deviated from game theoretic predictions significantly less than did humans (Martin et al. 2014). Specifically, chimpanzees recognized patterns in their opponents' choice history at a higher rate than humans. This allowed them to more accurately predict and optimally respond to their opponents' behavior, which is crucial to succeeding in the competitive settings presented here and, presumably, in intragroup competition in the wild.

While it may seem odd to find a decision-making task in which a nonhuman primate's ability surpasses that of humans, the authors argue that these results are consistent with the cognitive tradeoff hypothesis. This hypothesis states that cortical specialization for distinct human cognitive capacities (e.g., language or a complex sense of self) reduced more basic capacities, such as the rapid reaction times and pattern recognition capabilities evidenced by the chimpanzees in this study. One caveat to this interpretation is that this population of chimpanzees has had extensive previous experience with cognitive tasks while the humans did not; experience improves performance of both chimpanzees (Brosnan et al. 2011) and humans (Silberberg and Kearns 2009) on cognitive tasks such as this.

Due to the payout structure of the game, a specific predicted choice by the opponent calls for a different best response from each player; one tries to match it while the other tries to mismatch it. This raises the question of whether different mechanisms underlie the strategies that emerge for the two players. In this study, matcher reaction times were faster than mismatcher reaction times (Martin et al. 2014). This difference in reaction times could reflect more difficult calculations prior to making a decision for the mismatcher. Alternatively, matching may be an automatic motor response, while mismatching requires an override of this default response. The latter interpretation is supported by recent studies indicating that humans automatically match their opponents' physical actions, an effect that holds against significant incentives to not match but disappears when both players are blindfolded (Cook et al. 2012; Belot et al. 2013). This "automatic imitation" effect parallels capuchin monkeys' performance in the Assurance Game. Thus, future research should investigate the potential role of an involuntary, automatic response as an underlying mechanism of cooperation. Presenting subjects with both an asynchronous and a synchronous version of the Matching Pennies Game could disentangle the effect of an automatic matching response from that of seeing the other player's choice and responding to it in order to exploit the given payout structure.

2.3 Cooperation and Competition: Prisoner's Dilemma

The benefits of cooperation range widely: Alarm calls warn other group members of a predator, breeding pairs reduce offspring care and territory defense by sharing with helpers, and alliances increase the chances of winning a fight. However, these benefits also come at a fitness cost: Alarm callers are more likely to get eaten, helpers often do not reproduce themselves or delay reproduction, and intervening in a fight risks injury. So how can cooperation emerge and be maintained when, in the short term, each individual would be better off by being selfish and letting others do the work?

The Prisoner's Dilemma is a classic paradigm among the dyadic experimental economic games that captures both of these cooperative and competitive aspects. It is used to model situations in which two individuals could gain benefits from cooperating, but there is a temptation to earn more, at least in the short term, by defecting. Critically, the cooperative outcome is only expected in repeated interactions with known individuals, which frequently occur in many primate species, indicating that they might be able to achieve the mutually beneficial outcome. Players can either cooperate or defect, and the payoffs follow this structure: defecting on a cooperator > mutual cooperation > mutual defection > cooperating and being defected upon. The conundrum is that defection maximizes each player's payout in a given round regardless of their opponent's choice; however, mutual defection results in less overall gain than mutual cooperation. The Prisoner's Dilemma is thus a mathematical formulation of the evolutionary puzzle that cooperation presents, in that benefits to the group may be costly to the individual.

The paradigm has been widely explored in several animal species, including wild primates (Dugatkin 1997), and has been used extensively to assess social decision-making strategies in humans within a game theoretic framework (Camerer 2003). However, experimental studies with nonhuman primates are lacking. In one recent exception, Haroush and Williams (2015) adapted the paradigm for use with rhesus monkeys in a joint computer task. Overall, subjects were more likely to defect than to cooperate, suggesting either that the short-term temptation to defect outweighed the long-term benefit of cooperating or that the monkeys did not expect their partners to cooperate consistently and settled for the less risky outcome of mutual defection. In particular, subjects were less likely to cooperate if their partner defected in the previous round, but they were more likely to cooperate after mutual cooperation from their partner, indicating that they took their partner's past action into account when making their own decision. Critically, subjects did not only respond to their individual payouts but were sensitive to the social context of the experiment. When subjects played against the computer or played each other in separate rooms (as opposed to sitting side-by-side), the likelihood of cooperation and that of reciprocation after previous cooperation decreased. This result indicates that subjects recognized the role of a nearby conspecific in determining the outcome and suggests that mechanisms triggered by the presence of a partner in a social context may override an inherent tendency to defect.

The contingent reciprocity ("I'll be nice to you if you were nice to me") needed for these games can develop in a number of ways. In situations with repeated interactions with known individuals, individual recognition, and the cognitive ability to remember relationships, trust may underlie reciprocity (e.g., Engelmann et al. 2015). This is the case for humans, and seems likely for other highly social, intelligent species, but may be more challenging for others due to psychological constraints such as limited memory, self-control, and numerical skills (Stevens and Hauser 2004). Another less cognitively demanding way to promote cooperation is to rely on kinship or close affiliative bonds because cheating might be beneficial economically, but costly socially (Melis et al. 2008; Engelmann and Herrmann 2016). Finally, reciprocity may also develop when individuals are simply predisposed to help others who help them, e.g., through attitudinal reciprocity (Brosnan and de Waal 2002) or emotional bookkeeping (Schino 2007). Future work including other decision situations and other species are critical for understanding when and how complex forms of cooperation have evolved and what social structures and cognitive abilities underpin or limit it.

2.4 Inequity: Ultimatum Game

Inhibiting immediately self-rewarding behaviors may result in a greater long-term benefit in many social situations. We know that many primates are able to cooperate to achieve mutual rewards when it aligns with their own self-interest (the Assurance Game), and can even avoid the temptation to defect in order to maintain a beneficial outcome (the Prisoner's Dilemma), but will individuals choose to forego absolute gains in order to protest relatively unequal outcomes? We and others have tested this using the Ultimatum Game, a bargaining game in which one player (the "proposer") is asked to split a sum of money with their partner (the "responder"), who can either accept or reject the offer. If the respondent accepts, the money is split as proposed; if the respondent rejects, then neither player receives anything (which equalizes the outcomes, but at a cost to both proposer and responder).

Traditional economic thinking predicts that the "rational" behavior is for the proposer to offer the lowest possible amount and for the respondent to accept any non-zero amount offered, because any value presents a gain over the alternative of receiving nothing. This, however, fails to take into account the importance of relative outcomes. Humans deviate substantially from these economic predictions, a finding that has been widely debated as an indication of the human sense of fairness. Proposers in Western cultures typically offer 40–50% of the available money, and respondents typically reject low offers (those below 20–30%), which is thought to be a form of punishment for selfish behavior (Camerer 2003). There is substantial variation across cultures (Henrich et al. 2001), but in no culture do proposers routinely offer the lowest possible amount. The Ultimatum Game is less easily adapted for comparative study than dichotomous choice games, due to the contingent nature of the second player's decision (i.e., it is only an Ultimatum Game if the responder understands that the proposer is the one who determined the proposed allocation). However, it presents a game theoretic paradigm that may provide insight into the evolution of social decision-making in the context of inequitable outcomes and could shed light on the evolution of the human sense of fairness.

An initial adaptation of the Ultimatum Game for use with nonhuman primates found that, unlike humans, chimpanzees rationally maximized their rewards, with proposers offering the smallest option and responders accepting any offer (Jensen et al. 2007; see also Kaiser et al. 2012). However, when the same procedure that was used with the apes was presented to humans, they also behaved in this self-interested way (Smith and Silberberg 2010). In a more recent study using a token exchange task, both chimpanzees and children tended to split the rewards equally when their partner's cooperation was required after the reward distribution was proposed (the Ultimatum Game), but not when the partner could not reject the offer (similar to the Dictator Game, in which people also offer much lower amounts than in the Ultimatum Game) (Proctor et al. 2013). This result indicates that both chimpanzee and child proposers were sensitive to the role their partner could play and understood that their partner would wish for more rather than less.

Interestingly, we saw no rejections, in either chimpanzees or children. While different from typical one-shot, anonymous Ultimatum Games, this behavior makes sense in the context of normal interactions with a social partner, which are neither one-shot nor anonymous. Refusing an offer leads to short-term equity but low long-term benefits, as it essentially shuts down that cooperative relationship. Responders would be better off convincing their partners to choose the other option, which would lead to long-term benefits (Milinski 2013). In fact, both chimpanzees and children showed behavioral evidence of dissatisfaction (e.g., protests), and in human studies, when responders are given the opportunity to express dissatisfaction (through writing notes to the anonymous proposer), rates of acceptance of unequal offers increased, along with the rates of sending very nasty notes (Xiao and Houser 2005).

We also see extensive evidence that animals respond negatively to unequal outcomes (e.g., by refusing to participate when their payoff for completing a task is lower than that of a social partner's) beyond the Ultimatum Game, but interestingly only in species that cooperate outside of mating and kinship bonds, including chimpanzees, capuchin monkeys, several species of macaques, dogs, and corvids (reviewed in Brosnan and de Waal 2014). While more research is needed, this supports the proposal that inequity aversion functions to maintain cooperative behavior, since negative responses to unequal outcomes prevent exploitation and ensure equitable partnerships (Brosnan 2006, 2013; Fehr and Schmidt 2006). Understanding how cooperation and inequity aversion relate will also help us better understand how human ideals, such as our sense of fairness, evolved.

3 Future Directions

Comparative studies using experimental economic paradigms can shed light on the evolution of social decision-making in a number of ways. By studying more than a single species, we can generate hypotheses about the evolutionary roots of a behavior. Systematic comparisons of species that share some features, yet differ in others, allow us to illuminate the phylogeny of social decision-making strategies and biases and to determine which of these features are linked to these strategies. Moreover, comparative research can evaluate the cognitive mechanisms that underlie observed decision-making behavior both within and between species as well as across methodologies. Even in situations in which different species display similar behavioral strategies, the underlying mechanisms that are driving them might be different. Comparative studies therefore allow us to assess the selective pressures that may have shaped social decision-making across a range of species and ecological contexts.

Considering all of the games we discussed, it is clear that different primates are using different cognitive mechanisms to make decisions. Capuchins rely on matching, and their apparent inability to maintain play when they do not have information about their partner's choices indicates that cooperation is limited to direct interpersonal interactions. This puts some rather hard limits on the flexibility we may expect to see in capuchin decision-making. Rhesus macaques, on the other hand, can solve a coordination task without information on their partner's play, although recent evidence indicates that they require that additional information in more complex situations (i.e., a Hawk-Dove Game; Brosnan et al. 2017). This may indicate more cognitive flexibility, which is not unexpected given their large, fluid social groups, which would provide pressure to evolve such strategies, and their noted cognitive abilities (Deaner et al. 2006), which would provide the means to do so. Chimpanzees appear to be the most flexible, able to develop strategies that they can apply in other situations in the Assurance Game and able to accurately predict and respond to patterns in their partners' responses in the Matching Pennies Game. We anticipate that they will be able to find solutions in more complex situations.

Moving forward, a full understanding of the evolution of human behavior and cognition cannot be limited to primates, despite the obvious benefits of their close phylogenetic relationship to humans. We cannot understand the evolution of social cognition more broadly without the inclusion of other more distantly related species. For example, studies of inequity aversion with cooperative nonprimate species like dogs or corvids can clarify how cooperative bonds outside of kinship and mating may tie into the evolution of the human sense of fairness and its role in stabilizing cooperation. However, particularly with these more distantly related species, it is critical to design truly comparable procedures that are appropriate for their ecology, cognitive abilities, and anatomy (e.g., Salwiczek et al. 2012), a task that is made easier using paradigms from experimental economics.

Another aspect that will be key for future development is to include multiplayer games. While dyadic experiments provide the most control, and are therefore essential for early studies, dyadic interactions do not happen in a social vacuum, and, more critically, many ecologically relevant social behaviors, such as group hunting and territorial defense, involve more than two individuals. Recently, studies of social cognition have begun to explore these behaviors at the group level, including studies using games derived from experimental economics (Hall et al. 2015; Watzek et al. 2015). Although results from such studies are more difficult to interpret, our growing understanding of social behavior within dyads will allow us to determine when the social context influences decision-making in ways that we cannot discern when only two individuals are interacting. Such studies can illuminate both how the relationship between partners influences behavior and the factors that determine whether such social interactions between certain partners occur in the first place (e.g., Melis et al. 2006; Burkart and van Schaik 2013; Suchak et al. 2014; Engelmann and Herrmann 2016).

Finally, while carefully controlled studies are critical to teasing apart underlying mechanisms and making true comparisons across species, one drawback is that they lack ecological relevance. This has two challenges. First, we may inadvertently underestimate a species' capacities if we rely only on one simplified model that is used across species. When tasks are divorced from species' natural propensities, subjects may misinterpret them or fail to understand which cues are the relevant ones for solving the task (Lotem and Halpern 2012; Prétôt et al. 2016). Second, we may miss much of the context that is necessary to understand why an animal behaves as it does. Therefore, comparative research programs should first employ direct comparisons using identical methodologies, second test species that do less well on paradigms adapted to take their species' capabilities and ecologies into account, and third use the results from these model systems to make predictions about and then test behavior in more natural environments. This is not something that can be done quickly, but will ultimately provide the clearest understanding of the behavior in question. Future research with a broader scope in the tested species and the social situations they encounter will deepen our understanding of the evolutionary origins of social cognition in both primates and nonprimates.

References

Belot M, Crawford VP, Heyes C (2013) Players of matching pennies automatically imitate opponents' gestures against strong incentives. Proc Natl Acad Sci USA 110:2763–2768

Boesch C, Boesch H (1989) Hunting behavior of wild chimpanzees in the Tai national park. Am J Phys Anthropol 75:547–573

Brosnan SF (2006) Nonhuman species' reactions to inequity and their implications for fairness. Soc Justice Res 19:153–185

Brosnan SF (2013) Justice- and fairness-related behaviors in nonhuman primates. Proc Natl Acad Sci USA 110(Suppl 2):10416–10423

Brosnan SF, de Waal FBM (2002) A proximate perspective on reciprocal altruism. Hum Nat 13: 129–152

Brosnan SF, de Waal FBM (2014) Evolution of responses to (un)fairness. Science 346:1251776

Brosnan SF, Freeman C, de Waal FBM (2006) Partner's behavior, not reward distribution, determines success in an unequal cooperative task in capuchin monkeys. Am J Primatol 68: 713–724

Brosnan SF, Parrish AE, Beran MJ et al (2011) Responses to the assurance game in monkeys, apes, and humans using equivalent procedures. Proc Natl Acad Sci USA 108:3442–3447

Brosnan SF, Wilson BJ, Beran MJ (2012) Old world monkeys are more similar to humans than new world monkeys when playing a coordination game. Proc R Soc B Biol Sci 279:1522–1530

Brosnan SF, Beran MJ, Parrish AE et al (2013) Comparative approaches to studying strategy: towards an evolutionary account of primate decision making. Evol Psychol 11:606–627

Brosnan SF, Price SA, Leverett K, Prétôt L, Beran M, Wilson BJ (2017) Human and monkey responses in a symmetric game of conflict with asymmetric equilibria. J Econ Behav Organ 142: 293–306

Bullinger AF, Wyman E, Melis AP, Tomasello M (2011) Coordination of chimpanzees (*Pan troglodytes*) in a stag hunt game. Int J Primatol 32:1296–1310

Burkart JM, van Schaik C (2013) Group service in macaques (*Macaca fuscata*), capuchins (*Cebus apella*) and marmosets (*Callithrix jacchus*): a comparative approach to identifying proactive prosocial motivations. J Comp Psychol 127:212–225

Camerer CF (2003) Behavioral game theory: experiments in strategic interaction. Princeton University Press, Princeton

Cook R, Bird G, Lunser G et al (2012) Automatic imitation in a strategic context: players of rock-paper-scissors imitate opponents' gestures. Proc R Soc B Biol Sci 279:780–786

Crofoot MC, Rubenstein DI, Maiya AS, Berger-Wolf TY (2011) Aggression, grooming and group-level cooperation in white-faced capuchins (*Cebus capucinus*): insights from social networks. Am J Primatol 73:821–833

de Waal FBM (2000) Attitudinal reciprocity in food sharing among brown capuchin monkeys. Anim Behav 60:253–261

de Waal FBM, Berger ML (2000) Payment for labour in monkeys. Nature 404:563

de Waal FBM, Davis JM (2003) Capuchin cognitive ecology: cooperation based on projected returns. Neuropsychologia 41:221–228

Deaner RO, van Schaik CP, Johnson V (2006) Do some taxa have better domain-general cognition than others? A meta-analysis of nonhuman primate studies. Evol Psychol 4:149–196

Dugatkin LA (1997) Cooperation among animals: an evolutionary perspective. Oxford University Press, New York

Duguid S, Wyman E, Bullinger AF et al (2014) Coordination strategies of chimpanzees and human children in a stag hunt game. Proc R Soc B Biol Sci 281:20141973

Engelmann JM, Herrmann E (2016) Chimpanzees trust their friends. Curr Biol 26(2):252–256

Engelmann JM, Herrmann E, Tomasello M (2015) Chimpanzees trust conspecifics to engage in low-cost reciprocity. Proc R Soc B Biol Sci 282:20142803

Fehr E, Schmidt KM (2006) The economics of fairness, reciprocity and altruism – experimental evidence and new theories. In: Kolm S-C, Ythier JM (eds) Handbook of the economics of giving, altruism and reciprocity. Elsevier, Philadelphia, pp 615–691

Hall K, Lambeth SP, Schapiro SJ, Brosnan SF (2015) Chimpanzees respond differently to inequity based on relative food values. In: Abstracts of the 38th annual meeting of the American society of primatologists, Bend, 17–20 June 2015

Haroush K, Williams ZM (2015) Neuronal prediction of opponent's behavior during cooperative social interchange in primates. Cell 160:1233–1245

Henrich J, Boyd R, Bowles S et al (2001) In search of homo economicus: behavioral experiments in 15 small-scale societies. Am Econ Rev 91:73–78

Jensen K, Tomasello M, Call J (2007) Chimpanzees are rational maximizers in an ultimatum game. Science 318:107–109

Kaiser I, Jensen K, Call J, Tomasello M (2012) Theft in an ultimatum game: chimpanzees and bonobos are insensitive to unfairness. Biol Lett 8:942–945

Lau B, Glimcher PW (2005) Dynamic response-by-response models of matching behavior in rhesus monkeys. J Exp Anal Behav 84:555–579

Lotem A, Halpern JY (2012) Coevolution of learning and data-acquisition mechanisms: a model for cognitive evolution. Philos Trans R Soc B 367:2686–2694

Martin CF, Bhui R, Bossaerts P et al (2014) Chimpanzee choice rates in competitive games match equilibrium game theory predictions. Sci Rep 4:5182

Melis AP, Hare B, Tomasello M (2006) Chimpanzees recruit the best collaborators. Science 311: 1297–1300

Melis AP, Hare B, Tomasello M (2008) Do chimpanzees reciprocate received favours? Anim Behav 76:951–962

Mendres KA, de Waal FBM (2000) Capuchins do cooperate: the advantage of an intuitive task. Anim Behav 60:523–529

Milinski M (2013) Chimps play fair in the ultimatum game. Proc Natl Acad Sci USA 110: 1978–1979

Parrish AE, Brosnan SF, Beran BJ, Würsig MJ (2014) Differential responding by rhesus monkeys (*Macaca mulatta*) and humans (*Homo sapiens*) to variable outcomes in the assurance game. Anim Behav Cogn 1:215

Perry S (2003) Coalitionary aggression in white-faced capuchins, *Cebus capucinus*. In: de FBM W, Tyack P (eds) Animal social complexity: intelligence, culture and individualized societies. Harvard University Press, Cambridge, pp 111–114

Prétôt L, Bshary R, Brosnan SF (2016) Comparing species' decisions in a dichotomous choice task: adjusting task parameters improves performance in monkeys. Anim Cogn 19(4):819–834

Proctor D, Williamson RA, de Waal FBM, Brosnan SF (2013) Chimpanzees play the ultimatum game. Proc Natl Acad Sci USA 110:2070–2075

Salwiczek LH, Prétôt L, Demarta L et al (2012) Adult cleaner wrasse outperform capuchin monkeys, chimpanzees and orangutans in a complex foraging task derived from cleaner – client reef fish cooperation. PLoS One 7:e49068

Santos LR, Rosati AG (2015) The evolutionary roots of human decision making. Annu Rev Psychol 66:321–347

Schino G (2007) Grooming and agonistic support: a meta-analysis of primate reciprocal altruism. Behav Ecol 18:115–120

Silberberg A, Kearns D (2009) Memory for the order of briefly presented numerals in humans as a function of practice. Anim Cogn 12:405–407

Smith P, Silberberg A (2010) Rational maximizing by humans (*Homo sapiens*) in an ultimatum game. Anim Cogn 13:671–677

Smith MF, Watzek J, Brosnan SF (2018) The importance of a truly comparative methodology for comparative psychology. Int J Comp Psychol 31. https://escholarship.org/uc/item/6x91j98x

Stevens JR, Hauser MD (2004) Why be nice? Psychological constraints on the evolution of cooperation. Trends Cogn Sci 8:60–65

Suchak M, Eppley TM, Campbell MW, de Waal FBM (2014) Ape duos and trios: spontaneous cooperation with free partner choice in chimpanzees. PeerJ 2:e417

Watzek J, Rossettie MS, Raines AR, Brosnan SF (2015) More for me, less for you: capuchin monkeys (*Cebus apella*) responses to inequity in a group setting. Paper presented at the 52nd annual conference of the animal behavior society, Anchorage, 10–14 June 2015

Xiao E, Houser D (2005) Emotion expression in human punishment behavior. Proc Natl Acad Sci USA 102:7398–7401

The Special Case of Non-human Primates in Animal Experimentation

Augusto Vitale and Marta Borgi

Abstract Non-human primates (NHP) occupy a special place in the relatively new Directive 2010/63/EU on the protection of animals used for scientific purposes. As a matter of fact, the word "primate" is the most frequently used in the text of the Directive, and some articles are specifically dedicated to these animals. The first part of this chapter will provide a brief history of the normative treatment of NHP, followed by a description of the normative passages in the Directive 2010/63/EU which deal with the use of NHP in Europe. A comparison among the member states which carry on research using primate models and their implementation of the EU Directive in national terms will be presented as well. Furthermore, an overview on the use of NHP in European biomedical studies and toxicology, both in terms of numbers and kind of studies, will be followed by a comparison with data coming from Japan and the USA. Then, the possible reasons why NHP represent a special case, in terms of their use in laboratory research, will be discussed: Are there justifiable reasons why NHP should be considered more valuable in moral and then legislative terms? The last part will offer some considerations on the day-by-day reality of working with NHP in biomedical research, trying to understand whether the relationship between a laboratory monkey and the laboratory staff (researchers, technicians, students, etc.) is different from the one established with, for example, a laboratory rodent.

Keywords Animal experimentation · Directive 2010/63/EU · Ethics of research · Human-animal bond · Non-human primates · Primatology · Principle of the 3Rs

---

A. Vitale (✉) · M. Borgi
Center for Behavioral Sciences and Mental Health, Istituto Superiore di Sanità, Rome, Italy
e-mail: augusto.vitale@iss.it; marta.borgi@iss.it

L. D. Di Paolo et al. (eds.), *Evolution of Primate Social Cognition*, Interdisciplinary Evolution Research 5, https://doi.org/10.1007/978-3-319-93776-2_10

1 Introduction

Non-human primates (NHP) are not found only in nature, but they are present in research laboratories in different countries around the world. Although the use of captive apes in research has significantly decreased in the recent years, NHP are still used both for basic and translation/applied research, as well as for regulatory use and routine production.[1]

The use of NHP in research laboratories raises important ethical and practical problems in terms of meeting their behavioural, environmental and social needs in a laboratory context. These animals are considered to have increased capacity for suffering, and this represents one of the reasons for having received special protection in the European legislation.

In this context, this chapter offers an overview of the use of NHP in research laboratories, with a special focus on the European scenario [for other use on NHP in biomedical research, see also Bontrop (2001), Myers (2012) and Evans and Silvestri (2013)]. A detailed report on the numbers of laboratory animals used in the different countries represents the first important step to better understand the current global situation regarding animal experimentation. Therefore, in this chapter we summarise statistical data regarding the use of NHP for scientific purposes and the legal status of these animals in the European Union (EU), also giving some information on why these species are considered special to some extent. Furthermore, since it appears that a special bond can develop between staff and NHP, in this contribution we will also briefly describe emotional aspects of working with these animals in a captive setting.

2 The Use of Non-human Primates in Laboratory Research

It is difficult to build a global picture of the use of NHP in laboratory research, since many countries do not report the number of animals used for experimental and other scientific purposes. By reviewing studies published in 2001 in peer-reviewed journals, it has been estimated that globally every year 100,000–200,000 NHP of different species are used in biomedical research, mostly in North America, Japan and Europe (Carlsson et al. 2004; Hagelin 2004).

The main reason for their use appears to be their high degree of similarity with humans and their complex behaviour that allow researchers to obtain answers very close to those found in humans themselves. NHP cognitive abilities, as well as motor

[1]Although the categorisation of procedures in "basic" or "applied" research can be blurred and arbitrary, here we refer to the text of the Directive 2010/63/EU in which a distinction is made between "Basic Research" and "Translation or applied research" (see article 5—"Purposes of procedures" and the Commission implementing decision of 14 November 2012 on the common format for the submission of the information pursuant to Directive 2010/63/EU).

skills, are unique in the non-human animal world. They have provided us with important new insights into brain function and prevention of infectious diseases in humans and are considered a mainstay in basic cognitive-behavioural research and in behavioural studies of neurodegenerative diseases (Garbarini 2010; Capitanio and Emborg 2008). A more recent application is the use of transgenic monkeys (see, e.g. Cyranoski 2016).

In this section of the chapter, we will deal specifically with the use of NHP in the EU, and we will briefly mention the situation in other countries, such as the USA and Japan.

All member states (MS) of the EU are legally required to provide statistical data on the number of animals used in scientific research and the purpose of the procedures involved.[2] The latest available report on the number of animals used for experimental and other scientific purposes in EU was published from the European Commission in 2013 and refers to animals used in 2011 (except for a country, France, which reported for the year 2010) (European Commission 2013). Data reveal that almost 11.5 million animals (11,481,521) were subjected to experimental procedures in the 27 countries which were members of the EU at that time (Croatia joined the EU in 2013). Of these, NHP accounted for only the 0.05% of the total which corresponds to a number of 6095 primates of different species. Similarly to the previous report [data relative to the year 2008,[3] European Commission (2010b)], rodents and rabbits accounted for the 80% of the animals used in EU in 2011; among these, mice were by far the most used animals (61%), followed by rats (14%), fish (12.5%) and birds (5.9%).

Only a few of the 27 EU MS used NHP in 2011: France (1810 animals), Germany (1770), the UK (1459), Italy (450), Spain (343), the Netherlands (206), Czech Republic (30), Belgium (20) and Hungary (7) (Table 1). Of the total number of NHP used in that year in the EU, only 1.4% (83 animals) were prosimians and 11.5% (700 animals) New World monkeys, while the vast majority (87.2% which accounted for 5312 animals) were Old World monkeys (e.g. macaques) (Table 2). Since the year 1999, no apes have been used in experimental procedures in the EU. The area of research using by far the highest number of NHP for experimental and other scientific purposes was toxicology. NHP used for toxicological and other safety evaluation (including safety evaluation of products) accounted for more than

[2]"Procedure" means "any use, invasive or non-invasive, of an animal for experimental or other scientific purposes, with known or unknown outcome, or educational purposes, which may cause the animal a level of pain, suffering, distress or lasting harm equivalent to, or higher than, that caused by the introduction of a needle in accordance with good veterinary practice" (Directive 2010/63/EU, Article 3, "Definitions").

[3]These statistical data on the number of animals used for experimental and other scientific purposes in the EU were collected under the provisions of Directive 86/609/EEC, which required to present data collected from the member states every 3 years. This Directive has been replaced by Directive 2010/63/EU, and the submission and publication of data have been completely revised, in particular, Article 54 ("Reporting") of the Directive requires that member states collect and make publicly available, on an annual basis, statistical information on the use of animals in procedures.

Table 1 Amount of NHP used in the nine MS reporting the use of NHP in 2011 (European Commission 2013)

MS	Number of NHP
France[a]	1810
Germany	1770
United Kingdom	1459
Italy	450
Spain	343
Netherlands	206
Czech Republic	30
Belgium	20
Hungary	7

[a]France reported for the year 2010

Table 2 Number and percentage of NHP used for experimental and other scientific purposes in 2011 in the EU (European Commission 2013)

	Prosimians	New world monkeys	Old world monkeys	Total (%)
Biological studies of a fundamental nature	83	244	304	631 (10.3%)
Research and development of products and devices for human medicine and dentistry and for veterinary medicine	0	221	937	1158 (19%)
Production and quality control of products and devices for human medicine and dentistry and for veterinary medicine	0	79	138	217 (3.6%)
Toxicological and other safety evaluations	0	44	3391	3435 (56.4%)
Diagnosis of disease	0	0	0	0 (0%)
Education and training	0	0	209	209 (3.4%)
Other	0	112	333	445 (7.3%)
Total (%)	83 (1.4%)	700 (11.5%)	5312 (87.2%)	6095 (100%)

half of the total (56.4%). The other proportions were used in research and development of products and devices for human medicine and dentistry and for veterinary medicine (19%), production and quality control of the same products and devices (3.6%), biological studies of a fundamental nature (10.3%), education and training (3.4%) and other reasons (7.3%) (Table 2). The main reason why safety testing represents one of the major uses of NHP is that inclusion of a non-rodent species is mandatory in the safety assessment of new pharmaceutical compounds, as requested by safety testing regulations. NHP will be used in those cases in which the particular biological structure targeted by a particular drug presents significant similarities between a monkey and a human: that could be the case of the female reproductive system or of the visual system (Perretta 2011). However, although their genetic similarities to humans make NHP particularly suitable for being used in this field of application, a significant change in emphasis is that the need to use NHP should now be considered on a case-by-case basis (SCHEER 2017). Currently, after the implementation of the Directive 2010/63/EU, MS report the number of procedures applied to animals in research rather than the actual numbers. As a matter of fact, the number of NHP utilised has not varied significantly since 2013 (SCHEER 2017)

On a global scale, there are no known statistics regarding the experimental purposes that NHP are used for, though, based on academic papers reporting the use of NHP, it has been estimated that most studies that use NHP are conducted in the fields of microbiology and neuroscience (Carlsson et al. 2004). NHP play a very significant role in neuroscience research, because the complexity of their brain approaches the complexity of human brain.

It is important to notice that the number of NHP used for experimental and other scientific purposes has been decreasing during the last decade. Possible reasons could be public pressure and costs. In comparison with their use in 2005, a decrease of 8.4% was reported in 2008, and this downward trend continued in 2011 with an average decrease of the 36%. The most notable proportional reduction is in the use of prosimians, a decrease of 93% between the years 2008 and 2011. The number of New World monkeys decreased by 22.6%, while that of Old World monkeys decreased by about 28% (Table 3; see also Rennie and Buchanan-Smith 2005).

Some countries, including Japan, do not have requirements to report numbers of animals used in scientific experiments. Instead, the Japanese Association for Laboratory Animal Science (JALAS) conducts surveys on the number of animals used for scientific purposes. Hence, the numbers of laboratory animals provided rely on voluntary questionnaires and therefore might not accurately reflect the actual number of animals involved in experimentation (Takahashi-Omoe and Omoe 2007). As an example, according to the JALAS survey, more than 11,000 NHP were maintained in the year 2009 although this figure is likely to represent an underestimate of the true number used since only 67.4% of Japanese institutions responded to this survey (Yagami et al. 2010).

The USA uses more NHP per year in research than any other country in the world. Comparative figures from the USA show an average annual use of more than 50,000 NHP over the past years, according to the US Department of Agriculture, whose statistics show the number of animals used by species and by category of pain and distress. According to the latest report available, about 58,000 NHP were used in research in 2014 (more than 100,000 if also animals held by facilities but not used in any research are considered; United States Department of Agriculture 2015).

Table 3 Changes in species number and proportion between 2005 and 2011 (European Commission 2007, 2010b, 2013)

Species	Number of NHP used in EU[a] 2005[c]	Number of NHP used in EU[b] 2008[c]	Number of NHP used in EU[b] 2011[c]	Change 2005–2008	Change 2008–2011	Change 2005–2011	% Change 2005–2008	% Change 2008–2011	% Change 2005–2011
Prosimians (Prosimia)	677	1261	83	584	−1178	−594	86.3	−93.4	−87.74
New world monkeys (Ceboidea)	1564	904	700	−660	−204	−864	−42.2	−22.6	−55.2
Old world monkeys (Cercopithecoidea)	8208	7404	5312	−804	−2092	−2896	−9.8	−28.3	−35.3
Apes (Hominoidea)	0	0	0	0	0	0	0.0	0.00	0
Total	10,449	9569	6095	−880	−3474	−4354	−8.4	−36.3	−41.7

[a]EU-25 (Austria, Belgium, Denmark, Finland, France, Germany, Greece, Ireland, Italy, Luxembourg, Portugal, Spain, Sweden, the Netherlands, the United Kingdom, Cyprus, Czech Republic, Estonia, Hungary, Latvia, Lithuania, Malta, Poland, Slovakia, Slovenia)
[b]EU-27 (Romania and Bulgaria joined the EU in 2007)
[c]France reporting for the previous year

3 Legislation for Non-human Primates in Research: An Overview

One of the main differences, in terms of legal status, between human primates and NHP is that humans are recognised as persons, whereas the other primates are not. This means, among other things, that humans are by law protected following the Universal Declaration of Human Rights, which recognises individual liberty and dignity. No such recognition formally exists for the other animals. Different initiatives motivated by ethical considerations for which we should protect animals that suffer more, such as the Great Ape Project, have proposed to expand the legal and moral consideration humans are object of, to the great apes (chimpanzees, bonobos, gorilla and orang-utans). The idea was to give these species a limited legal stature, focusing on three main rights and interests: the right to live, the protection of individual liberty and the prohibition of torture. The argumentation was presented in a book, edited by Peter Singer and Paola Cavalieri and developed through a series of chapters signed by authoritative writers, biologists and lawyers (Richard Dawkins, Marc Bekoff, Douglas Adams, Jane Goodall among the others) (Singer and Cavalieri 1993).

The Great Ape Project has been (and still is) an important initiative, because it inspired scientists and legislators to think in different perspectives, in relation to animal welfare and animal rights. Its influence has been perceived around the globe. For example, in 2000 New Zealand has included, in its Animal Welfare Act, special norms in relation to the use of "non-human hominids". In Part 6 of the Act ("use of animals in research, testing and teaching"), it reads: "The Director-General must not give approval under subsection (1) unless he or she is satisfied—(a) that the use of the non-human hominid in the research, testing, or teaching is in the best interests of the non-human hominid; or (b) that the use of the non-human hominid in the research, testing, or teaching is in the interests of the species to which the non-human hominid belongs and that the benefits to be derived from the use of the non-human hominid in the research, testing, or teaching are not outweighed by the likely harm to the non-human hominid"). An increasing number of governments have enacted restrictions in the use of great apes in research. In 2006 Austria has introduced such restrictions and so the Netherlands, Great Britain and Sweden. The ban in Sweden does not include non-invasive research such as, for example, cognitive-behavioural studies. A similar case is represented by Germany. On June 25, 2008, Spain became the first country to announce that it will extend rights to great apes, advising the government to propose norms that would give chimpanzees, bonobos, gorillas and orang-utans the right to life, to liberty and not to be used in experiments. It is interesting to notice that in 2014 the court of Cassation in Argentina recognised habeas corpus for a female orang-utan, giving her de facto the recognition of the moral status of a persona, therefore not anymore mere property of someone else, and with individual and personal rights. In Japan, in 2006, an agreement was reached to end invasive chimpanzee research.

In the USA the Animal Welfare Act is the main legislation which defines the standards and rules for the use and care of NHP involved in biomedical research. No specific regulations state that NHP cannot be used in laboratory research, and no indications are specifically provided for what use. However, in 2011 a report was published by the Institute of Medicine (under request from the NIH), which stated that most of the current use of chimpanzees in captive research was not necessary (Altevogt et al. 2011). Furthermore, captive chimpanzees in the USA have recently been granted an endangered species status, which makes them object of the Endangered Species Act. This Act does not allow any invasive research on endangered species (see Grimm 2015).

For what concerns the EU, in 2007 the European Parliament adopted a declaration in which it was urging the European Commission to enforce a ban on the use of great apes and NHP captured in the wild in research laboratories. Furthermore, it was asked to set up a deadline for the ban of all NHP utilised in biomedical and toxicological research, in favour of alternative methods. The aim was to introduce these requests in the text of the forthcoming, at that time, revision of the Directive 86/609/EU on the protection of animals utilised in research. The European Commission called for a panel of experts to deliberate on this matter. After a careful analysis, the Scientific Committee on Health and Environmental Risks (SCHER) reached the conclusions that the ban on the use of NHP in experimental research was premature. As a matter of fact, SCHER declared that different sectors of research, such as the research on HIV/AIDS and the study of neurodegenerative diseases (Parkinson's, Alzheimer's disease), still need NHP models (SCHER 2009; Nature Editorial 2008). However, the SCHER report was also emphasising that, although at the moment the available scientific and technological knowledge was not allowing for the replacement of NHP with alternative methods, the European Commission should have periodically checked the development of alternative methods, in order to arrive in the long run at the ban of these animals from biomedical and toxicological studies. Therefore, in 2017 the now Scientific Committee on Health Environmental and Emerging Risks (SCHEER) was asked to give a new revised opinion on the use of NHP in research. After an online and public consultation with interested parties, the SCHEER committee reached a series of conclusions: "Recognising the high levels of public concern about NHP research, regulatory authorities in some world regions have also adopted ethical limits or boundaries on NHP use. However, the close phylogenetic relationship of NHPs with humans makes them the best available animal models for addressing particular research questions. Therefore, there is a consensus within certain sections of the scientific community that, where alternatives do not exist, appropriate use of NHPs remains essential in some areas of biomedical and biological research and for the safety assessment of pharmaceuticals. For example: NHPs can play a key role in drug development due to their similarity to humans with regard to sensory organs, hormonal systems, reproduction, immune system, etc., to evaluate efficacy and safety, especially for biopharmaceutical compounds. There has been progress in a move towards a non-animal predictive mechanism-based approach, e.g. for testing of drugs for reproductive toxicity, but there are still obstacles to overcome in terms of regulatory acceptance and scientific validity. In spite of this, NHPs should only be used when it is scientifically demonstrated that none of the other non-rodent species commonly used in safety testing is appropriate for the purpose of the study. As long as sufficiently validated alternatives are not available, a total ban would make further progress in such research and some safety studies impossible, at least in Europe. On the other hand, a ban may lead to acceleration of investment into the development of alternative methods as seen for cosmetics. However, a ban in Europe would likely force NHP research outside of Europe, and since animal welfare standards for laboratory NHPs are on average higher in many European countries than in other parts of the world, it follows that there would likely be a net decrease in animal welfare", and "In contrast to current areas of research with great potential to replace current NHP use such as in safety testing of pharmaceuticals, other areas of research, partly new, may require increased NHP use in the future, e.g., emerging infectious diseases. A solid harm-benefit assessment is needed here, too. The majority of NHP use in the EU is for safety assessment studies, performed to meet regulatory requirements. NHPs should only be used when there are no alternatives and it is scientifically demonstrated that none of the other non-rodent species commonly used in safety testing is appropriate for the purpose of the study. In addition, NHP use can be avoided when in vitro preliminary studies demonstrate NHP are not a suitable animal model. It is indeed important to consider the limitations of the NHP when choosing which species to use in drug safety test" (SCHEER 2017).

4 The Non-human Primates and the Directive 2010/63/EU

The use of animals in the EU is regulated under the Directive 2010/63/EU (European Commission 2010a). Although in comparison with other species the use of NHP is very limited (see Sect. 2), these animals are the object of special consideration. The Directive focuses on the use and care of NHP in several passages of its text. New regulations (compared to the previous Directive 86/609/EEC; European Commission 1986) have been introduced concerning acquisition, aims of use, evaluation and authorisation of research protocols employing NHP, including retrospective assessment (a retrospective evaluation by the competent authority, for more ethically sensible projects, that is, the use of NHP, cats and dogs and the use of a severe level of invasiveness). In general, although the new Directive acknowledges that animals, including NHP, are still needed to protect human and animal health and the environment, its ultimate goal is to fully replace the use of animals in biomedical research as soon as it is scientifically possible to do so, as stated in Recital 10. To that end, the Directive, together with seeking to ensure a high level of protection for animals that still need to be used in procedures, promotes the advancement of alternative approaches and provides for a review of the Directive itself examining "the possible replacement of the use of animals, and in particular non-human primates, as a matter of priority where it is possible, taking into account the advancement of science" (Recital 49). For these reasons the new Directive permits the use of NHP "only in those biomedical areas essential for the benefit of human beings, for which no other alternative replacement methods are yet available" (Recital 17). In the next paragraphs, the main elements of change regarding the use of NHP introduced by the Directive 2010/63—in comparison with that of 1986—are listed (see also Hartung 2010).

As regulated in the Directive 2010/63 (Article 8, specifically dedicated to NHP), the use of these species is permitted only when the purpose of the procedure cannot be achieved by the use of other species and only in procedure having one of the following purposes: (1) basic research, (2) research aimed at preservation of the species and (3) translational or applied research for the avoidance, prevention, diagnosis or treatment of debilitating (i.e. reducing a person's normal physical or psychological ability to function) or potentially life-threatening clinical conditions in human beings and for the same aims in the development, manufacture or testing of the quality, effectiveness and safety of drugs, foodstuffs and feedstuffs and other substances or products. For the same reasons, with the exception of basic research, it is also allowed the use specimens of endangered NHP. Notwithstanding these provisions, great apes shall not be used in procedures except when "a Member State has justifiable grounds for believing that action is essential for the preservation of the species or in relation to an unexpected outbreak of a life-threatening or debilitating clinical condition in human beings" (provisional measure as in Article 55 "Safeguard clauses").

In general, since 2010, when the new Directive came into force, MS are not allowed to take animals from the wild, except for exemptions granted by competent authorities "on the basis of scientific justification to the effect that the purpose of the procedure cannot be achieved by the use of an animal which has been bred for use in procedures" (Article 9). MS shall also ensure that all species of NHP (as listed in Annex I) may only be used where those animals have been bred for use in procedure and, from the dates set out in Annex II, shall ensure that NHP may be used only where they are the offspring of NHP which have been bred in captivity or where they are sourced from self-sustaining colonies, i.e. "a colony in which animals are bred only within the colony or sourced from other colonies but not taken from the wild, and where the animals are kept in a way that ensures that they are accustomed to humans", as in Article 10(1), third subparagraph. Member states shall also ensure that breeders of NHP "have a strategy in place for increasing the proportion of animals that are the offspring of NHP that have been bred in captivity" (Article 28).

Another regulation introduced by the Directive 2010/63 deals with the detailed information on each NHP, as well as dog and cats, that all breeders, suppliers and users shall kept, including an individual history file which covers not only relevant reproductive or veterinary but also social information on the individual animal; this file will follow the animal as long as it is kept for the purposes of the Directive and 3 years after its death or rehoming and accompanies the animal in the case of rehoming (Article 31). These animals shall also be provided with a permanent individual identification mark (Article 32). Article 34 of the Directive governs the inspections of all breeders, suppliers and users carried out by competent authorities to verify compliance with the requirements of the Directive. Also, in this case NHP are taken into particular consideration: while inspections shall be carried out on at least one third of the users each year, breeders, suppliers and users of NHP shall be inspected at least once a year.

To conclude, an important element of change introduced by the new Directive deals with the retrospective assessment, i.e. an evaluation carried out by the competent authority to verify the harm inflicted on animals and whether the objectives of the project were achieved. The retrospective assessment also takes into consideration "any elements that may contribute to the further implementation of the requirement of replacement, reduction and refinement" as specified in Article 39(1), letter c. The retrospective assessment has been introduced in order to monitor the efficacy of measures aimed at limiting the discomfort created to animals and has been made mandatory for all projects using NHP and projects involving procedures classified as "severe".

It is important to highlight that the Directive 2010/63 recognises animals as having an intrinsic value. It is stated: "Animals have an intrinsic value which must be respected. There are also the ethical concerns of the general public as regards the use of animals in procedures. Therefore, animals should always be treated as sentient creatures and their use in procedures should be restricted to areas which may ultimately benefit human or animal health, or the environment. The use of animals for scientific or educational purposes should therefore only be considered where a non-animal alternative is unavailable" (Directive 2010/63/EU, Recital 12). According to this vision, overall the Directive 2010/63/EU introduces some major advances for animal welfare, including more restrictions of use of NHP, also considering that their use is of great concern to the public and in the light of their genetic proximity to human beings and their highly developed social skills which raise "specific ethical and practical problems in terms of meeting their behavioural, environmental and social needs in a laboratory environment".

5 Non-human Primates and the Principle of the "3Rs"

Because of their phylogenetic closeness with humans, NHP raise a series of ethical issues, related to their supposed sophisticated ability to experience pain and suffering. Alternatives in the use and to the use of NHP in experimentation are therefore particularly warranted.

The possible alternatives in and to the use of NHP in research have to be dealt with the application of the "principle of the 3Rs". This methodological recipe was first proposed by Rex Burch and William Russell in 1959, and it represents the backbone of a more ethical (or "humane" as the two authors indicated in the title of their book) approach to the use of animals in scientific procedures (Russell and Burch 1959). In brief, three steps are proposed to the researcher when he or she decides to use an animal model for his or her research. The first step would be to, if possible, apply the principle of "replacement", that is, to verify if alternatives to the use of an animal model do exist to answer his or her research question. Secondly, and in the case replacement is not possible, the number of individuals utilised must be reduced ("reduction") as much as possible but within the possibility to carry on appropriate statistical analysis. Thirdly, all efforts must be done to reduce discomfort and sufferance as much as possible to the experimental animal ("refinement").

The "3Rs" principle is the most effective tool for dealing with the dilemma brought forth by the need for scientific and medical progress and the ethical relevance of animal welfare. If correctly understood and applied, the 3Rs principle promotes both ethical and scientifically sound progress. Moreover, each of the 3Rs is driven forward by improvements in research techniques and new scientific and technological developments, and evaluation of research protocols should always take in consideration the multifaceted and dynamic nature of the "3Rs".

Does the "3Rs" principle generally work for NHP? This question is very much related to the knowledge of the principle in the scientific community. As a matter of fact, ignorance of the "3Rs" is no more contemplated because the EU Directive explicitly asks for its application in Article 4 ("principle of replacement, reduction and refinement"): "(1) Member States shall ensure that, wherever possible, a scientifically satisfactory method or testing strategy, not entailing the use of live animals, shall be used instead of a procedure. (2) Member States shall ensure that the number of animals used in projects is reduced to a minimum without compromising the objectives of the project. (3) Member States shall ensure refinement of breeding, accommodation and care, and of methods used in procedures, eliminating or reducing to the minimum any possible pain, suffering, distress or lasting harm to the animals". Furthermore, Annex VI to the Directive asks the researcher to specify how he/she intends to apply the model of Russell and Burch to his/her research protocol. Obviously, these provisions apply to NHP as well: perhaps even more so, having already pointed out the special attention the legislation dedicates to these animals.

We don't have here neither the space nor the expertise to cover all of the fields of application in which NHP are currently used. We therefore focus particularly on NHP used in neuroscience research, which represents our field of expertise.

For what concerns "replacement", the possibilities have to be examined on a case-by-case basis, depending strictly on the type of information which has to be obtained from a particular experimental protocol. As a matter of fact, opportunities for replacing the use of NHP in neuroscience research are currently limited. In silico models to mimic cognitive functions under different conditions and treatments are promising (Geerts et al. 2012; Jellen et al. 2015), and most probably the future will bring increasingly more sophisticated computerised models of cognition and function memory, towards the dismissal of the use of NHP in particular areas of neuroscience. Classical alternative methods, such as the use of tissue and cell in vitro system, as well as the use of human volunteers, are potentially very valid, but the intact monkey provides models which are very attractive because of their complexity. The relatively new creation of transgenic monkeys to mimic human pathologies could be a reason for an increase of the use of monkeys in disease research and behavioural neuroscience (see, e.g. Liu et al. 2016). It is also evident that the number of NHP used in research laboratory is steadily declining, as we have noticed before. We cannot say if this corresponds to the use of alternative models for the same scientific enquiry or just abandons a particular line of research. Further research is required to try to understand the reasons for this decline and what consequences this decline has brought to scientific research.

In the case of "reduction", the number of NHP utilised in neuroscience protocols is already very low. For example, in Italy a 3-year-long project on neurophysiology utilises a maximum of 8–9 individuals in total. In behavioural studies, with limited degree of invasiveness, the number is higher, but the individuals can be reutilised. The reuse of NHP can decrease the number of animals used overall, and the choice to do so can be due to ethical reasons but economic reasons as well. The EU Directive 2010/63, however, restricts the cases in which individual animals can be reused for scientific purposes, as stated in Article 16: "(1) Member States shall ensure that an animal already used in one or more procedures, when a different animal on which no procedure has previously been carried out could also be used, may only be reused in a new procedure provided that the following conditions are met: (a) the actual severity of the previous procedures was 'mild' or 'moderate'; (b) it is demonstrated that the animal's general state of health and well-being has been fully restored; (c) the further procedure is classified as 'mild', 'moderate' or 'non-recovery'; and (d) it is in accordance with veterinary advice, taking into account the lifetime experience of the animal".

"Refinement" refers to any approach which avoids or minimises the actual or potential pain, distress and other adverse effects experienced at any time during the life of the animals involved and which enhances their wellbeing (Buchanan-Smith et al. 2005). Researchers should ensure that every aspect of the lifetime experience of the animals is refined, and there are many ways and circumstances in which this can be achieved. Conditions can be improved to minimise the level of sufferance experienced by experimental subjects, at two levels: (1) during the experiments. In neuroscience this can imply the use of positive training techniques, through positive reinforcement, the use of better techniques to record the activity of more neurons at once, limiting the total time needed for the recording and the use of better techniques to make it easier to fix cranial apparatus, for the use of recording electrodes; (2) during lifetime. Here possibilities are endless: improving transportation, improving social housing, improving physical housing, the use of environmental enrichments and improving relationships between monkeys and staff. In recent times, the technique of positive reinforcement in management routines has become a very effective way to improve the quality of life of captive NHP (Prescott and Buchanan-Smith 2003; Schapiro et al. 2005). Once a particular protocol has been implemented, it should constantly be reviewed in order to identify ways of improvements. The cognitive sophistication of NHP allows the possibility to apply training techniques to reduce the amount of stress during procedures.

6 A Special Bond

Together with animals such as dogs and cats, NHP occupy a special position among laboratory animals and have received special attention and protection in the Directive. In the case of dogs and cats, their special position is mostly due to their status in human society: they are increasingly kept as pet in human houses and considered truly members of our families (Herzog 2011; Downey and Ellis 2008; Serpell 1996). The main reason why they are considered a special case in the legislation is indeed that there is a high level of public concern as to the fate of such animals. In the case of NHP, as stated in the Directive, their use in scientific procedures raises specific ethical and practical problems due to their genetic proximity to human beings and their highly developed social skills, as well as their behavioural, environmental and social needs.

Animals as NHP, dogs and cats appear to receive special attention and concern also in the research community. In a recent survey we conducted on attitudes towards laboratory animals in a sample of Italian researchers, NHP resulted to be a case for special concern, together with animals commonly kept as pets, like dogs. For example, hardly any of the researchers interviewed would, if given the chance, submit a NHP or a dog, a laboratory procedure that would cause severe pain and distress and even more so if the primate was a chimpanzee (data not published).

One question one could ask is whether working with NHP in a research generates a feeling of bonding with the experimental animals, which is unique in animal science (see Vitale 2011). Before that, why a researcher should decide to study NHP? Eugène Marais once wrote that all things must start with an act of love (Marais 1968). Later on Wilson has indicated our tendency to feel emotionally connected to living beings similar to us (Wilson 1984). Could it be then that the ones among us who decide to study monkeys do so because there is an ancestral connection? Is close phylogeny the way to a special bond with these animals? When Linda Fedigan and Shirley Strum asked some primatologists "Why NHP?", the answer was pretty straightforward: "because NHP are the best experimental animal to understand human evolution" (Strum and Fedigan 2000). Alison Jolly affirmed that she was "fascinated by creatures which are different from people, hard to understand, and come in radiation of beautiful species". The beauty and variety of the order primates can also play a role in the fascination exercised by these animals. The mighty mountain gorilla is a primate, but so is the miniscule mouse lemur. Students come to our lab to ask to work with primates, because they have watched some television documentaries and found monkeys very pretty. But it is not just that, the phylogenetic closeness plays a role as well, as rightly pointed out by Wilson. "They are just like us", it is something that can be heard when talking about NHP behaviour, and this kind of argument does not belong only to laypersons outside the scientific community.

Such feeling of closeness can generate empathic responses towards the study animal, and this can explain why, in years of experience with a colony of common marmosets (*Callithrix jacchus*), we have noticed that many students and caretakers, as well as researchers, develop a relationship with the individual they observe that appears sentimentally strong: they exchange stories on named subjects, and they tend to anthropomorphise their behaviour more often than people working with mice. However, in our survey on attitudes towards laboratory animals in Italian researchers, many participants have reported feeling of respect and ethical concern towards the animals for which they are responsible, including fish or rodents.

Relationships between research personnel and their animals may greatly differ, although some common elements appear to affect such bonds, including the status of the animal in the human society (as in the case of dogs and cats), the person's attitude to and experience with that particular species (and this can be the case again of dogs and cats) and the time spent with the individual animal, as well as the behaviour of the animal (e.g. its capacity to recognise and distinguish different persons, as in the case of dogs) (Bayne 2002; Chang and Hard 2002; Davis 2002; Vitale and Pollo 2011). Another aspect is related to the number of animals kept in a laboratory, commonly high in the case of animals such as fish or rodents and low in the case of animals such as dogs and NHP, an aspect that can have an influence on the staff's concern for the animal welfare. As an example, giving names to all animals in a colony of fish or mice might not be practical, while it is more common when working with dogs or NHP.

For sure, if a bond between research people and the animals exists, this relationship may come at considerable cost. The psychological burdens of these relationships include feelings of affected persons who have to use and kill animals in research. Animal care staff and researchers may experience moral ambivalence associated with their bond with laboratory animals. As pointed out by Herzog (2002), research institutions can help individuals cope with the ethical consequences of relationships with research animals by, for example, giving animal care personnel an ethical voice through involvement in the institutional animal care and using committee decision process or by publicly acknowledging the emotional and moral costs of human-laboratory animal relationships (Herzog 2002).

During our work we have repeatedly ask ourselves whether a researcher who performs invasive research on animals, particularly NHP, can maintain and justify a feeling of love and respect for his/her study subjects. We know of researchers studying neurophysiology of macaques who declare bona fide feelings of affection for the animals they place in restraining chairs day after day, to study how their brains work. We suggest that at the heart of the matter is the assertion of a hierarchy of moral status. If different living beings vary in moral status, can we consider caring for human patients more valuable than caring for NHPs [it seems so, see the concept of "affinity for people", de Cock Buning (1995)]? However, these considerations do not necessarily mean that the researcher who carries out invasive research does not feel for his/her subjects. Coleman reports that in Japanese primate centres, on annual basis, Buddhist ceremonies are held to remember individual macaques sacrificed at the end of experimental procedures. Furthermore, in other institutions, special rooms are dedicated to the memories of past monkeys, with their portraits hanging from the room walls, where personnel can meet and exchange "monkey stories" (Iliff 2002; Coleman 2011).

Our personal experience with a colony of common marmosets has thought us that the identification with a suffering monkey is a very powerful phenomenon. The monkey who is sick becomes like a person, and when we ask about his/her health, it would be difficult to understand for an outsider that we are actually talking about a non-human animal. A prominent researcher working on non-human primate behaviour once said to us that looking at pictures of past monkeys is, nothing more or less, like looking at a family album.

Whether researchers can and do form human-animal bonds and the day-to-day ways they interact with the animals for which they are responsible is still far to be completely understood. This is of particular importance in the field of animal experimentation, since experiencing the human-animal bond in the laboratory context can potentially improve the welfare of animals and the quality of the research, as well as increase job satisfaction for animal caregivers.

What we propose here is to consider that the scientific and experimental work carried on with NHP could be potentially enriched in some cases by particular attitudes of the researchers towards their subjects, perhaps more than in the case of other laboratory species. The question would be whether this scenario is detrimental to scientific knowledge, as a sort of "sentimental pollutant", or it is actually a form of enrichment of the scientific practice. These could be questions worth exploring.

7 Conclusions

In the present contribution, we presented an overview on the use of NHP for scientific purposes. The use of NHP in research laboratories represents a very interesting and stimulating case study for animal experimentation in general. Although these animals represent a small percentage of the animals used in research in the MS of the EU, they occupy a special niche in the normative text. They can raise powerful ethical considerations on the topic of pain and suffering inflicted to experimental animals and can be a source of empathic responses by researchers working with these species. They may thus represent a "special case" to investigate the multiple factors intervening in our relationship with the animals we observe and study and how this multifactorial scenario can influence the scientific enterprise.

Acknowledgements We would like to thank Simone Pollo for the stimulating discussions on the use of animals in research. The collection of data that led to the writing of part of this chapter has received funding from the European Union Seventh Framework Programme (FP7-HEALTH-2013-INNOVATION-1) under grant agreement no 602616 (Project ANIMPACT).

References

Altevogt BM, Pankevich DE, Shelton-Davenport MK, Kahn JP (eds) (2011) Chimpanzees in biomedical and behavioural research: assessing the necessity. National Academic Press, Washington

Bayne K (2002) Development of the human-research animal bond and its impact on animal well-being. ILAR J 43(1):4–9

Bontrop RE (2001) Nonhuman primates: essential partners in biomedical research. Immunol Rev 183:5–9. https://doi.org/10.1034/j.1600-065x.2001.1830101.x

Buchanan-Smith HM, Rennie AE, Vitale A, Pollo S, Prescott MJ, Morton DB (2005) Harmonising the definition of refinement. Anim Welf 14:379–384

Capitanio JP, Emborg ME (2008) Contributions of non-human primates to neuroscience research. Lancet 371(9618):1126–1135. https://doi.org/10.1016/S0140-6736(08)60489-4

Carlsson HE, Schapiro SJ, Farah I, Hau J (2004) Use of primates in research: a global overview. Am J Primatol 63:225–237. https://doi.org/10.1002/ajp.20054

Chang FT, Hard LA (2002) Human-animal bonds in the laboratory: how animal behavior affects the perspective of caregivers. ILAR J 43(1):10–18

Coleman K (2011) Caring for non-human primates in biomedical research facilities: scientific, moral and emotional considerations. Am J Primatol 73:220–225. https://doi.org/10.1002/ajp.20855

Cyranoski D (2016) Monkeys genetically modified to show autism symptoms. Nature 529:449

Davis H (2002) Prediction and preparation: Pavlovian implications of research animals discriminating among humans. ILAR J 43(1):19–26

de Cock Buning T (1995) What apes teach us about ethics. In: Corbey R, Theunissen B (eds) Ape, man, apeman: changing views since 1600. Leiden University Publishers, Leiden, pp 385–394

Downey H, Ellis S (2008) Tails of animal attraction: incorporating the feline into the family. J Bus Res 61:434–441

European Commission (1986) Council directive 86/609/EEC of 24 November 1986 on the approximation of laws, regulations and administrative provisions of the member states regarding the protection of animals used for experimental and other scientific purposes. Off J L 358, 18/12/1986, p 1–28. http://eur-lex.europa.eu/LexUriServ/LexUriServ.do?uri=CELEX:31986L0609:EN:HTML. Accessed 10 April 2016

European Commission (2007) Report from the commission to the council and the European Parliament. Fifth report on the statistics on the number of animals used for experimental and other scientific purposes in the member states of the European Union. Commission of the European Communities Brussels, 5/11/2007. http://eur-lex.europa.eu/legal-content/EN/TXT/PDF/?uri=CELEX:52007DC0675&from=EN. Accessed 10 April 2016

European Commission (2010a) Directive 2010/63/EU of the European Parliament and of the council of 22 September 2010 on the protection of animals used for scientific purposes. Off J Eur Union L 276, 20/10/2010, p 33–79. http://eur-lex.europa.eu/legal-content/EN/TXT/HTML/?uri=CELEX:32010L0063&from=EN. Accessed 10 April 2016

European Commission (2010b) Report from the commission to the council and the European Parliament. Sixth report on the statistics on the number of animals used for experimental and other scientific purposes in the member states of the European Union. European Commission, Brussels, 8/12/2010. http://eur-lex.europa.eu/legal-content/EN/TXT/PDF/?uri=CELEX:52010DC0511R(01)&from=EN. Accessed 10 April 2016

European Commission (2013) Report from the commission to the council and the European Parliament. Seventh report on the statistics on the number of animals used for experimental and other scientific purposes in the member states of the European Union. European Commission, Brussels, 5/12/2013. http://eur-lex.europa.eu/legal-content/EN/TXT/PDF/?uri=CELEX:52013DC0859&from=EN. Accessed 10 April 2016

Evans DT, Silvestri G (2013) Nonhuman primate models in AIDS research. Curr Opin HIV AIDS 8:255–261. https://doi.org/10.1097/COH.0b013e328361cee8

Garbarini N (2010) Primates as a model for research. Dis Model Mech 3:15–19. https://doi.org/10.1242/dmm.004861
Geerts H, Spiros A, Roberts P, Twyman R, Alphs L, Grace AA (2012) Blinded prospective evaluation of computer-based mechanistic schizophrenia disease model for predicting drug response. PLoS One 7(12):e49732. https://doi.org/10.1371/journal.pone.0049732
Grimm D (2015) Research chimps to be listed as 'endangered'. http://www.sciencemag.org/news/2015/06/research-chimps-be-listed-endangered. https://doi.org/10.1126/science.aac6831
Hagelin J (2004) Use of live nonhuman primates in research in Asia. J Postgrad Med 50(4):253–256
Hartung T (2010) Comparative analysis of the revised directive 2010/63/EU for the protection of laboratory animals with its predecessor 86/609/EEC – a t4 report. ALTEX 27(4):285–303
Herzog H (2002) Ethical aspects of relationships between humans and research animals. ILAR J 43 (1):27–32
Herzog H (2011) Some we love, some we hate, some we eat: why it's so hard to think straight about animals. Harper Perennial, New York
Iliff SA (2002) An additional "R": remembering the animals. ILAR J 43(1):38–47
Jellen LC, Aliper A, Budzin A, Zhavoronkov A (2015) Screening and personalizing nootropic drugs and cognitive modulator regimens in silico. Front Syst Neurosci 9:4. https://doi.org/10.3389/fnsys.2015.00004
Liu Z, Nie YH, Zhang CC, Cai YJ, Wang Y, Lu HP, Li YZ, Cheng C, Qiu ZL, Sun Q (2016) Generation of macaques with sperm derived from juvenile monkey testicular xenografts. Cell Res 26(1):139. https://doi.org/10.1038/cr.2015.112
Marais E (1968) The soul of the ape. Atheneum, New York
Myers DD (2012) Nonhuman primate models of thrombosis. Thromb Res 129(2):S65–S69. https://doi.org/10.1016/j.thromres.2012.02.037
Nature Editorial (2008) Call to action: European scientists who support neuroscience research on primates should tell their politicians why. Nature 456:281–282
Perretta G (2011) Uso dei primati nella ricerca scientifica alla luce della nuova Direttiva. In: Manciocco A, Romano E, Zoratto F, Branchi I, Berry A (eds) Sperimentazione animale: aspetti teorici, normativi e applicativi della nuova Direttiva europea 2010/63. Rapporti ISTISAN 11/39, pp 20–26
Prescott MJ, Buchanan-Smith HM (2003) Training nonhuman primates using positive reinforcement techniques. J Appl Anim Welf Sci 6(3):157–161. https://doi.org/10.1207/S15327604JAWS0603_01
Rennie AE, Buchanan-Smith HM (2005) Report on the extent and character of primate use in scientific procedures across Europe in 2001. Lab Primate Newsl 44(2):6–12
Russell WMS, Burch RL (1959, reprinted 1992) The principles of humane experimental technique. Universities Federation for Animal Welfare, Wheathampstaed
SCHER (Scientific Committee on Health and Environmental Risks) (2009) Scientific opinion on the need for non-human primates in biomedical research, production and testing of products and devices. http://ec.europa.eu/environment/chemicals/lab_animals/pdf/scher_o_110.pdf
SCHEER (Scientific Committee on Health, Environmental and Emerging Risks) (2017) Final opinion on 'the need for non-human primates in biomedical research, production and testing of products and devices (update 2017)'. https://ec.europa.eu/health/sites/health/files/scientific_committees/scheer/docs/scheer_o_004.pdf
Serpell J (1996) In the company of animals: a study of human-animal relationships. Cambridge University Press, Cambridge
Schapiro SJ, Perlman JE, Thiele E, Lambeth S (2005) Training non-human primates to perform behaviours useful in biomedical research. Lab Anim 34:37–42. https://doi.org/10.1038/laban0505-37
Singer P, Cavalieri P (eds) (1993) The great ape project: equality beyond humanity. Fourth Estate, London
Strum SC, Fedigan LM (eds) (2000) Primate encounters: models of science, gender, and society. The University of Chicago Press, Chicago

Takahashi-Omoe H, Omoe K (2007) Animal experimentation in Japan: regulatory processes and application for microbiological studies. Comp Immunol Microbiol Infect Dis 30(4):225–246. https://doi.org/10.1016/j.cimid.2007.02.001

United States Department of Agriculture (2015) United States department of agriculture, animal and plant health inspection service. Annual report animal usage by fiscal year (2014). https://www.aphis.usda.gov/animal_welfare/downloads/7023/Animals%20Used%20In%20Research%202014.pdf

Vitale A (2011) Primatology between feelings and science: a personal experience perspective. Am J Primatol 73:214–219. https://doi.org/10.1002/ajp.20863

Vitale A, Pollo S (2011) Introduction to the special section: "The effects of bonds between human and nonhuman primates on primatological research and practice". Am J Primatol 73(3):211–213

Wilson EO (1984) Biophilia. Harvard University Press, Cambridge

Yagami K, Mashimo T, Sekiguchi F, Sugiyama F, Yamamura K, Serikawa T (2010) Survey of live laboratory animals reared in Japan. Exp Anim 59(4):531–535. https://doi.org/10.1538/expanim59.531

Epigenetics and the Evolution of Human Social Cognition

Gillian Ragsdale and Robert A. Foley

Abstract Human social cognitive abilities share derived attributes with other primates, but humans excel uniquely with regard to skills such as theory of mind, perspective-taking, sharing intentions and using language. Even these apparently unique abilities, however, must be supported by neurochemistry that is in turn developmentally dependent on genes derived from our primate lineage. In the last decade, epigenetic processes have emerged as an influence on differences both within and between mammalian species, and the brain is a major target for epigenetic processes regulating gene expression. This short review looks at how epigenetic processes may have contributed to human social cognitive evolution, how this relates to differences between human and non-human primate social cognition and to what extent this is consistent with the social brain hypothesis, i.e. increasing human sociality as a driver rather than a consequence of human brain evolution. The evidence ranges from general trends in maternally and paternally expressed genes influencing different parts of the brain, quantitative differences in mechanisms such as DNA methylation and gene expression in the brains of humans and other primates and examples of species-specific epigenetic regulation of genes associated with social cognition and language.

Keywords Epigenetics · Imprinted genes · Parent of origin · Language · Methylation · GABA · Serotonin · Human

---

G. Ragsdale (✉)
The Ronin Institute for Independent Scholarship, Montclair, NJ, USA
e-mail: gill.ragsdale@roninstitute.org

R. A. Foley
The Leverhulme Centre for Human Evolutionary Studies, University of Cambridge, Cambridge, UK
e-mail: raf10@cam.ac.uk

L. D. Di Paolo et al. (eds.), *Evolution of Primate Social Cognition*, Interdisciplinary Evolution Research 5, https://doi.org/10.1007/978-3-319-93776-2_11

1 Introduction

While the second half of the twentieth century saw a rise in evolutionary approaches to cognition and behaviour, it also led to a detachment of the field from genetics. For primates, including humans, social learning and its behavioural outcomes became the core adaptation, the distinctive element of being a primate, and this emphasised the flexibility and adaptability of behaviour. Primates do not depend on the genetically determined stereotypical behaviour of less cognitively endowed creatures. Primate social cognition and behaviour seemed at the nurture end of the spectrum, not the nature.

However, this in turn led to a disjunction between a primate as a phenotype produced by some form of genetic blueprint and a creature of learned behaviour. That gap is a problem, but one that is increasingly being filled by epigenetics. Epigenetics enables environmental factors to regulate long-term gene expression without altering the genetic code itself. This process can enable behaviours to be both flexible and variable and still rooted in the biology of the organism. This field has begun to flourish, but very much in terms of the regulation of individual differences in behaviour. Thus far very little attention has been paid to the potential of epigenetics to throw light on differences between species and so on broader macroevolutionary patterns. In this paper, we will explore the potential for epigenetics to provide insights into the evolution of primate (and human) social cognition. In the first section, we will introduce the field of epigenetics and why it is relevant to social cognition, considering also the evidence for its influence on social behaviour. From there we will consider how an epigenetic approach can also inform us about differences between species. Finally, we will explore some aspects of human social cognition, in a comparative context, where epigenetics might be a particularly important element.

2 Epigenetics and Social Cognition

The stability of genes across generations has been at the heart of the mystery concerning how the environment could make long-term changes to gene expression without changing the DNA sequence itself (such as promoter regions and other regulatory sequences that control how the gene is switched on and off). Epigenetic changes, however, do not alter the DNA code itself but include a number of processes that can increase or decrease the level of gene expression in a way which is both long term (even heritable) and potentially reversible. For example, the most researched epigenetic mechanism is the addition (or removal) of a small methyl (CH_3) group to a cytosine base in the DNA (methylation). The presence of methyl groups at the promotor affects the binding of transcription factors and tends to suppress gene expression. These small, potentially reversible chemical changes can switch genes on or off, partially or completely. And the process that leads to placing or erasing these epigenetic marks can be triggered by the environment (Sadakierska-Chudy et al. 2015a, b).

The first environmentally induced epigenetic processes uncovered in humans were responses to poor nutrition and famine (Heijmans et al. 2008; Tobi et al. 2009; Hajj et al. 2014). Although the process was surprising, the context was not—famine is a recurring event in human history and rapid adaptation could be the difference between survival and extinction. Many genes have been found to be involved in this response, for example, the genes influencing the insulin response to blood sugar can be programmed to respond to famine. This is consistent with the Barker hypothesis proposing a link between foetal programming in response to poor nutrition and restricted growth (resulting in a thrifty phenotype) and later health problems in adult life if nutrition improved, such as obesity and heart disease (Bateson et al. 2004; Hales and Barker 2001; Shultz 2010). This has developed into a general theory, developmental origins of health and disease (DOHaD) and broad area of research, including foetal programming via epigenetic mechanisms (Lillycrop and Burdge 2015).

Initially, it seemed that all epigenetic processes took place during foetal development—but as research expanded, so did the range of epigenetic processes. The environment triggering epigenetic changes can range from the physical and social environment to the internal environment of the body including the sex of the cell. Although epigenetic responses to the environment appear to be strongest during both foetal and child development, they can be triggered over the adult lifespan and by paternal lifestyle as well as maternal lifestyle (Hsieh and Zhou 2016; Gershon and High 2015; Stuppia et al. 2015). In some cases, epigenetic influences can be inherited across generations: a radical departure from classical genetics implying that some acquired characteristics can be inherited. Epigenetic effects can be maintained over the lifetime and in some instances transmitted to future generations as a kind of biological memory of environmental stress (Nagy and Turecki 2015).

The full extent of possibilities is still unfolding, but they range from the regulation of small, time-limited, tissue-specific phenotypes to those that are enduring and species-defining (Chadwick 2015; Jirtle and Skinner 2016). Epigenetic mechanisms effectively multiply the phenotypic possibilities of the genome. The mammalian brain is a hotspot for epigenetic regulation, and this is hardly surprising as epigenetics is all about adapting quickly to a changing environment—and the most powerful method humans possess in the face of adversity is our highly flexible behaviour. Since the human mind is possibly our single most species-defining characteristic, it is reasonable to propose that changes brought about by epigenetic mechanisms may have played a role in the evolution of primate cognition overall as well as the transition to more specifically human cognitive skills.

3 Epigenetics and Individual Variation in Social Cognition

Human social cognition is characterised by a range of skills including theory of mind, face recognition, joint attention, social learning and, not least, language. The neural basis for these traits is influenced by many genes, some of which are known to be epigenetically regulated.

In the 1960s, psychologist Harry Harlow's sad, motherless monkeys grew up to be anxious and unable to behave normally with others of their kind. It seems astonishing now that it was ever necessary to prove how important attachment and care are to infant well-being. In the last 10 years, biological research has begun to join the dots between early care and lifelong brain chemistry. One major link between nature and nurture turns out to be epigenetics, enabling patterns of behaviour over a lifetime to be influenced by our earliest experiences. The stress response itself be epigenetically programmed via demethylation of the glucocorticoid receptor gene in response to early maternal care (citations). Maternal stress prenatally and the quality of care in early life can act as a signal about the world and whether it is safe to trust others or not. In a stressful, unsafe world, trust can be dangerous. For adults conditioned to mistrust others, aspects of attachment, such as affection and intimacy, can trigger stress rather than pleasure and lead to a cycle of troubled attachments throughout adult life and even across generations (Gershon and High 2015; Monk et al. 2012; Strathearn 2011).

There are several neurochemical processes associated with social cognition and behaviour that are epigenetically regulated. Oxytocin, for example, has been called 'the love chemical'. It has a major influence on both parental and romantic love alongside others such as vasopressin, dopamine and the opioids—all of which can be environmentally influenced via epigenetics (Auger et al. 2011; Jack et al. 2012; Moriam and Sobhani 2013; Munoa et al. 2015; Puglia et al. 2015). (Oxytocin spray is marketed as increasing personal appeal by triggering increased trust and can be purchased online.) Oxytocin and vasopressin can even induce monogamy—at least, in prairie voles (Keebaugh and Young 2011). Prairie voles are monogamous, while meadow voles are not. Receptors for oxytocin in the reward system of the brain are greatly increased when monogamous voles mate. This is brought about by increasing expression of the oxytocin receptor gene. This is an epigenetic process triggered in prairie voles but not meadow voles. When prairie voles mate, histone acetylation leads to increased oxytocin receptor gene expression linking the sexual reward with the odour of that particular partner (if they stay together for at least 6 h) (Wang et al. 2013). Clearly epigenetics has the potential to strongly influence a broad spectrum of mammalian social behaviours.

4 Epigenetics and Differences Between Species

The term 'species' has a complex and controversial history beyond the scope of this paper. For our purposes, we assume that species as mentioned below exhibit phenotypes resulting at least in part from species-specific genotypes. It is increasingly clear, however, that just as the number of genes carried by individual genomes is less than expected, the genetic differences between species, in terms of DNA coding, are also narrower than might be expected (Chimpanzee Sequencing Analysis Consortium 2005).

In the case of individual variation, epigenetic regulation of gene expression is emerging as a key factor in explaining how so few genes (some with few, if any common alleles) can produce such a range of phenotypic variation. There are two processes by which this could be extended to differences between species. Firstly, and most controversially, it now appears that epigenetic modifications themselves can be inherited in some instances (Jablonka and Raz 2009; Nagy and Turecki 2015; Prokopuk et al. 2015). The potential stability of epigenetic modifications over multiple generations is currently uncertain in mammals although it has been observed in plants and is theoretically possible (citations). Although it would remain at least potentially reversible, it could result in permanent genetic adaptation by influencing the evolutionary process itself as discussed in a later section below.

Secondly (and possibly facilitated by the process described above), there may be changes to the regulatory elements that direct the processes of epigenetic modification. To give a theoretical example as illustration, a gene may be regulated by methylation at a non-coding CpG site. If there is an otherwise neutral mutation at that site, from CpG to TpG (a relatively common mutation), the regulation of that gene will be altered because it has lost a methylation site. If methylation was downregulating that gene, that is, supressing expression, this gene will henceforth be upregulated—more of the gene product, that specific protein, will be made. Whether this produces individual variation within a species or leads to divergence between species will depend on what that protein does and how it influences reproductive fitness in the current environment.

5 Imprinted Genes, Brain Size and the Social Brain

One of the ongoing debates in the study of human social cognition is the extent to which it comprises traits which are continuous with our primate lineage—and the extent to which it is qualitatively unique. This is really part of the larger debate concerning the evolution of human cognition in general: to what extent has primate cognitive evolution been driven by selection for increased social skills. The leading theory in this context is the social brain hypothesis proposing that selection for maintaining co-operation and communication within increasing group sizes drove the expansion of the neocortex—culminating in the evolution of language in humans (Dunbar 1998, 2009).

One particular kind of epigenetic process, genomic imprinting, has been found to influence brain growth. The expression of imprinted genes, contrary to classical genetic theory, depends on their parent of origin. A mother's and father's genes are not always equally expressed in their child. Sometimes the mother's gene is preferred and sometimes the father's. The pattern also varies depending on the developmental stage of the child and the body tissue. The preference can be complete, where one parent's allele is totally silenced, or more subtle, with an unequal split in the influence of the two parental alleles. When one parent's genes are expressed more than the other, this is termed 'genomic imprinting'. To set the preference, one allele of a pair is 'imprinted', that is, epigenetically marked. The main mechanism for this is DNA methylation such that the methylated—imprinted—copy of a gene is not expressed (or its expression is reduced). However, in some cases, it is the methylated copy which is expressed rendering the term 'imprinted' somewhat problematic in interpretation so that the term 'preferentially expressed' is increasingly preferred. The methylation pattern is reset appropriately during early embryonic development so that, for example, a maternally expressed gene passed on from a mother and expressed in her son will not be expressed when that son becomes a father.

The overall influence of maternal and paternal genes is illustrated by studies using chimeric mice, that is, mice having cell lines with different genotypes in the same individual. When chimeric mice were produced having half their cells containing only chromosomes from one parent, maternal and paternal genes were found to have very different influences on brain growth (Keverne et al. 1996). Mice lacking maternal genes had smaller brains and larger bodies, while those lacking paternal genes had larger brains but smaller bodies. In particular, maternal genes drive growth of the neocortex, striatum and hippocampus, while paternal genes are more active in the hypothalamus. Within these overall trends in brain growth, there are more specific parental influences such as maternal expression of *UBE3A* and *GABRB3* in the Purkinje cells and paternal expression of *SNRPN* and *MAGEL2* in the hypothalamus. *UBE3A* codes for E6-AP ubiquitin ligase which influences context-dependent learning via long-term potentiation in the hippocampus (Jiang et al. 1998), and the role of GABAergic activity in social cognition is discussed further below.

Imprinted genes come at a cost and their selection despite such a cost requires explanation. Having two alleles at any locus may act as an insurance policy against mutation in one of the alleles so that the deleterious effects can be overridden either entirely or partially. Imprinted genes are therefore at a disadvantage and render individuals more susceptible to genetic diseases associated with that locus. Having only one allele expressed at a locus also increases the likelihood of the organism exhibiting extreme phenotypes. However, although these extreme phenotypes may be less adaptive, there may also be advantages in specific cases. The greater the variance in the phenotype, the stronger the effect of natural selection will be and consequently, the faster the associated trait can evolve. This logic suggests that genes influencing traits whose evolution (rather than conservation) provides significant improvement in fitness are likely candidates for imprinting. This also further supports a role for imprinted genes in speciation and maintaining species boundaries. Given that imprinted genes are over-represented in the brain and that higher cognitive skills, including those supporting complex social behaviour and language, are defining traits for *Homo sapiens*, it is reasonable to expect that imprinted genes may influence species-defining cognitive skills.

Overall mammalian brain growth, especially of the neocortex, is driven by maternally expressed genes. Since the human neocortex is approximately three times larger than that of a chimpanzee, this suggests a role for changes in maternal gene expression that has further increased brain growth in humans. This requires an extension of the conflict theory of imprinting that proposes a tug-of-war between maternal and paternal genes such that paternal gene expression tends to draw food and care from the mother, while maternal gene expression tends to limit these demands on the mother. This tug-of-war may have arisen in mammals due to greater investment of resources by the mother compared to the father compounded by paternity uncertainty. This conflict may extend into adulthood for species with sex-biased dispersal—a common occurrence in mammals (Haig 2002, 2014). The predominant tendency in mammals is for males to disperse so that group members are more related matrilineally than patrilineally. This matrilineal bias could support selection for maternally expressed genes promoting social bonding and both maternal and paternal preferential gene expression facilitating incest avoidance as appropriate. Rodent studies have provided some support for this, such as male dispersal triggered by the odour of maternally related females (Isles et al. 2002).

Although male dispersal predominates among catarrhine primates, the dispersal patterns of many other primates are more variable, both within and between species (Jack and Isbell 2009), and the case for aligning parent-of-origin effects with affected kinship groups must be, if not weaker, then certainly more complex. Dispersal patterns vary among the great apes, for example, and it is difficult to establish the dispersal patterns of extinct hominins with any certainty: modern humans tend to exhibit female rather than male dispersal (Koenig and Borries 2012).

It is also possible for imprinted genes to be further epigenetically regulated in response to environmental cues such as chemical, physical, social, or maternal stress (Kappil et al. 2015; Vangeel et al. 2015). Alternatively, other factors may have a stronger influence on the fitness benefits of social bonding in some primates, such as the transition from a nocturnal to diurnal lifestyle (Schultz et al. 2011).

The differential influence of parental genes in the brain has been interpreted as support for the social brain hypothesis, i.e. that selection for increasingly complex social skills to maintain increasing group size was the principal reason for increasing brain-body ratios in primates generally and especially in humans (citation). This view interprets the preferential maternal expression in the neocortex as reflecting the strong social influence of matrilineal kin in Old World monkeys. Although the pattern of maternal and paternal gene expression in the brain is consistent with the social brain hypothesis, the necessary extension of this work from rodents to primates is largely lacking. Indeed most of the research on epigenetics has been carried out using rodents with a tendency to extrapolate findings as being generally 'mammalian'. There is, however, evidence for the influence of imprinted genes on social cognition in humans which supports a role for these imprinted genes in the evolution of human social cognition whether or not that directly supports the social brain hypothesis (Curley 2011; Ragsdale and Foley 2011, 2012).

6 Species Differences in Epigenetic Regulation of Genes Underpinning Social Cognition

Most homologous gene sequences between humans and chimpanzees are highly conserved, but there are parts of the human genome, particularly in non-coding regions, that have been subject to human-specific accelerated evolution (Webster 2013). While the functions of non-coding DNA are, as yet, poorly characterised, it is likely that their role is regulatory and that altered regulation of otherwise conserved genes is important in phenotypic variation both within and between species. Changes to gene expression, rather than sequences, are likely to have driven the evolution of the human brain, and differences in levels of gene expression between the brains of humans and other primates are well-documented (Cáceres et al. 2003; Nowick et al. 2009). Such changes can be brought about by a combination of mechanisms, involving changes to DNA-encoded regulatory elements, some of which may regulate epigenetic processes such as the methylation of cytosine bases and the methylation, acetylation or deacetylation of histones. In general, DNA methylation in humans has been found to be less than that seen in macaques, African green monkeys and squirrel monkeys, and differences in promoter methylation alone may be responsible for 12–18% of quantitative differences between human and chimpanzee gene expression (Pai et al. 2011).

Zeng et al. (2012) identified 1055 genes in the human cortex that were significantly less methylated than in the chimpanzee cortex. Many of these human hypomethylated genes showed increased expression compared to their chimpanzee homologues. Furthermore, this set of hypomethylated genes is associated with human-specific disease vulnerabilities and includes a number of genes associated with autism such as *GABRA2*, *GABRG1* (GABA receptor genes) and *SLC6A4* (the serotonin transporter gene). In humans, a gene's influence on social cognition can be implied when gene dysfunction is associated with communication disorders or autism spectrum traits such as lack of empathy or social skills. Further research on these particular genes as described below illustrates how epigenetic processes may facilitate both individual and species-level variation in social cognition.

Levels of the neurotransmitter serotonin are regulated by a number of genes, one of the most influential being the serotonin transporter gene. This gene has been extensively studied, and its level of expression is associated with the development of depression and anxiety as well as influencing social learning and social vs. antisocial behaviour (Crişan et al. 2009; Hariri et al. 2002; Karg et al. 2011; Petito et al. 2016).

A polymorphism producing short and long alleles illustrates the possible range of influence: carriers of the short allele are more likely to become anxious and depressed as adults and more likely still if they have experienced childhood abuse. The effect is also influenced by sex, so that females homozygous for the short allele who also experience childhood abuse have a greater risk of developing depression (Uher et al. 2011). Neuroimaging studies show differential activation in brain areas involved in social cognition and communication depending on the combination of short vs. long alleles suggesting that psychosocial stress may interact with serotonin transporter gene alleles to influence social behaviour (Canli and Lesch 2007). Polymorphic variation in the promotor region of the serotonin transporter gene is associated with variation in the emotional response to threatening facial expression, especially as mediated by the eyes—employing the characteristic enlarged whites of the human eye (Skuse 2006). This strongly suggests that altering the level of expression of the serotonin transporter gene can influence social behaviour via social cognitive responses to social cues such as facial expression.

Differences in serotonin levels have been associated with interspecies differences in social behaviour such as despotic vs. tolerant primate species—for example, the more tolerant, less aggressive social behaviour typical of bonobos may reflect greater serotonin levels in the amygdala due to greater serotonin transporter gene expression as well as a greater density of enervating serotonergic neurons (Stimpson et al. 2016).

A serotonin transporter gene polymorphism similar to that identified in humans is associated with aggression and social organisation in macaque species. Different macaque species with distinct levels of tolerance and conciliatory behaviours vs. dominance and hierarchy are associated with species-specific allele variation (Wendland et al. 2006). Until recently this polymorphic variation in gene sequence was thought to be the major pathway by which serotonin levels influenced stress reactivity and depression via the serotonin transporter gene.

Independently of genotype, however, methylation of the serotonin transporter gene promotor also predicts amygdala reactivity and consequently the response to emotionally salient signals such as anger or anxiety in response to the threatening facial expression of a conspecific (Nikolova et al. 2014). More methylation is associated with greater reactivity which could lead to a stronger stress response and vulnerability to anxiety and depression. Moreover 'the amount of methylation was a better predictor of amygdala activity than DNA sequence variation'. There are several epigenetic influences on the serotonin signalling pathway making it potentially highly responsive to conditions such as maternal stress during foetal development (Holloway and Gonzalez-Maeso 2015), but in this instance serotonin transporter methylation patterns were consistent across tissue sources suggesting it was heritable rather than acquired though gene-environment interaction.

Decreased GABAergic neuron activity has been associated with poorer social cognition among both typically developing individuals and those with autism spectrum disorder (Cochran et al. 2015). Furthermore, treatment with a GABAergic agonist reversed social deficits in mouse models of autism (Silverman et al. 2015). GABAergic neurons synchronise and integrate throughout the neocortex, and the development of GABAergic neurons is epigenetically regulated in several ways—by microRNAs, retrotransposons and imprinted genes (Keverne 2014).

Taken together, the evidence supports a role for species-specific changes in the epigenetic regulation of serotonin and GABAergic activity in adapting the social cognitive phenotype of primates. In the case of humans, relative hypomethylation and consequent increased gene expression have resulted in increased GABAergic and serotonergic activity supporting increased social cognitive skills and prosocial behaviour.

7 Epigenetics and the Evolution of Language

The search for specific genes influencing cognition which peaked in optimism in the 1980s has floundered among thousands of candidate genes with tiny effect sizes. Against this largely uninformative backdrop, however, a few genes have been identified as having a relatively strong influence on social cognition and language—including *FOXP2* and *CNTNAP2*. Human *FOXP2*, the first gene specifically associated with language development, differs from its chimpanzee homologue by only two amino acids. FOXP2 is a transcription factor that appears to bring about its major influence by regulating many other genes in the brain that influence neural growth and connectivity. It is possible that major changes in the influence of FOXP2 could be brought about by changes in the downstream target genes' expression as well as changes to the timing of *FOXP2* expression itself. (*FOXP2*, on chromosome 7q31, is adjacent to the *CNTNAP2* gene, which it downregulates.) Although Neanderthals shared the modern human FOXP2 amino acid sequence, there is evidence for several selective sweeps during the last 200,000 years for *FOXP2* non-coding regulatory sequence changes (Ayub et al. 2013; Maricic et al. 2013; Mozzi et al. 2016).

Disruption of *CNTNAP2* expression is associated with several forms of specific language impairment, autism and social problems (Vernes et al. 2008; Dennis et al. 2011). Schneider et al. (2014) compared the DNA methylation pattern of human and chimpanzee *CNTNAP2* in the cortex and found significant differences. One site showing reduced methylation in humans is only 300 base pairs upstream from a site associated with autism and parent-of-origin effects. One of the protein-coding sequences is expressed from this gene in 1.6 times greater quantity in the human cortex compared to the chimpanzee. *CNTNAP2* is one of the largest human genes with very large introns (non-coding regions), and 87% of methylation sites were located in these non-coding regions. Schneider et al. speculate that methylation differences between humans and chimpanzees may regulate exon splicing, thereby altering the possible gene products resulting from transcription of the original gene. Preliminary evidence for sex-specific methylation differences was also found.

Evidence for the influence of genomic imprinting on *FOXP2* is both controversial and contradictory with preferred paternal expression (Feuk et al. 2006), no imprinting (Thomas et al. 2012) and random mono-allelic expression (Adegbola et al. 2015) all documented. The possibility that genes associated with human language might be imprinted would support theories of human language evolution derived from conflict theory where, for example, paternal gene expression increases communication skills in the infant in order to extract attention and resources from the mother and in competition with its siblings who may not have the same father (Crespi 2007; Brown 2011).

8 Epigenetics Influencing the Evolutionary Process

An example of this process can be seen in the evolution of new, adaptive genes following gene duplication. The development of gene families via gene duplication is an important source of phenotypic variability, the substrate for selection. The expression of duplicate genes is regulated by several epigenetic processes such as both random and non-random methylation (Keller and Yi 2014). For example, there appears to have been human-specific gene duplication of the *SRGAP2* gene, a gene strongly associated with brain function. While the original *SRGAP2* gene product acts to slow down migration of neurons in the cortex, the function of one of these duplicates in particular, *SRGAP2C*, has the opposite influence. This later duplicate, arising 2.4 mya, may therefore have played a role in increasing neocortex size and neural plasticity in the transition from *Australopithecus* to *Homo* (Charrier and Polleux 2012; Dennis et al. 2012).

Epigenetic processes and their consequences in terms of a near-Lamarckian transmission of acquired traits across generations have major implications for evolution generally. Epigenetic variation, especially under stressful conditions, is more rapid and produces more variation than genetic mutation. This enables an organism to develop adaptive phenotypes without genetic changes. Yet any epigenetic information ultimately depends on the associated DNA sequences—with the potential for epistasis between the two. The epigenetically adapted phenotype, having increased fitness, increases the time over which genetic changes may occur and accumulate to support that phenotype. Thus epigenetic changes can accelerate and facilitate the adaptive process (Klironomos et al. 2013).

Since epigenetic changes allow the adaptive phenotype to precede genetic changes, this allows the organism to undergo niche construction which in turn acts to select both supporting epigenetic and genetic changes. This may be especially important in the evolution of highly responsive and flexible phenotypes such as animal behaviour (Jablonka 2013). This is a major departure from previous neo-Darwinian thinking which views the organism as a passive vehicle for the true substrate of selection—the genes. Epigenetic inheritance systems give the organism 'the ability to generate relevant selectable phenotypic variation' (Jablonka and Lamb 2005, p. 345). Primate social cognition and behaviour is known to be epigenetically regulated both within and between species, and this does make possible a range of behaviours which has allowed both species and individuals to be responsive and flexible on the one hand—while also enabling some permanent, species-defining adaptive responses.

9 Epigenetics and Current Human Adaptiveness

There has been much debate about the 'true nature' of our species. Are we naturally aggressive and war-faring like our more common chimpanzee cousins—or are we more peaceable and affiliative, like bonobos? In fact human behaviour appears to range naturally to both extremes. While humans show some continuity in their social cognitive abilities and behaviours with other primates and mammals generally, humans typically show an extraordinary range of behaviours for a single species. Our capacity for both co-operation and conflict is complex and unusual.

Within an individual's lifetime, epigenetic regulation of genes underpinning social cognition and behaviour prepares that individual to respond to current physical and social conditions in the most adaptive way. In this way, stress, including the social stresses of neglect or violence, predisposes people towards less affiliative, more defensive, hostile behaviour, increasing both the tendency towards anxiety and aggression. In the animal kingdom generally, this is an adaptive behavioural strategy when there is aggressive competition for resources. In better circumstances, epigenetic regulation predisposes to more prosocial behaviour and increases empathy and trust between people.

Some of these processes can act across generations, being reset as conditions improve or deteriorate. But as discussed above, epigenetics can have wider implications and can influence the evolutionary process itself. Where, for example, intense social stress persists across many generations, the likelihood of the concomitant adaptive strategy becoming the permanent default increases. This has the potential to become a vicious cycle. Children growing up in traumatic, violent, unpredictable environments, such as are all too common around the world, will tend to grow into adults predisposed towards aggression and hostility as a defence. In this way, humans may create their own environmental niche of perpetuated social stress. Epigenetic processes provide a feedback mechanism for selection driven by social stress—potentially magnifying and prolonging the selective pressure. While some individuals remain resilient (a phenomena only recently coming under scrutiny), whole populations may be influenced by severe stress resulting in population differences in the epigenetic status of genes. For example, the degree of methylation of the glucocorticoid receptor gene, and hence the sensitivity of the stress response, is altered in children of Rwandan holocaust survivors with posttraumatic stress disorder compared to those of Rwandans not exposed to the holocaust (Perroud et al. 2014; Yehuda et al. 2014).

Such population differences may have arisen among primate species in the past, associated with different environmental niches, and in some cases, this may have contributed to speciation events. Humans, however, are unique in the degree to which they are the architects of their own environment and potentially the drivers of their current and future adaptiveness. Their diverse environmental contexts may promote the conditions under which epigenetics are the most effective way for selection and adaptation to occur.

10 Conclusions

There are interspecies differences in gene coding, polymorphic variation and gene expression, including the epigenetic regulation of genes implicated in species differences in social cognition and behaviour. These differences are both quantitative and qualitative, influencing such species-defining traits as theory of mind (and hence self-awareness) and possibly language. Interspecies differences in epigenetic regulation greatly increase the potential for interspecies differences that do not depend on differences in the protein-encoding regions of homologous genes. Such population differences may have arisen among primate species in the past, associated with different environmental niches, and in some cases, this may have contributed to speciation events.

Research on epigenetics is proceeding rapidly, changing ideas on how selection can fine-tune organisms, and it is likely that this will increase as genomics and molecular biology become more functionally intertwined. Do these developments have implications for how we see human evolution having occurred? We can perhaps suggest some areas where this may be the case, each operating at a different evolutionary scale. The first is the transition to becoming human—or more precisely, the transition from ape-like state to human-like state. Current evidence would suggest that this was not a single, major shift but cumulative ones over millions of years (Foley 2016). At this large scale, the increased reliance on epigenetic mechanisms may have been a slow process, because there are major costs as well as benefits. Stable, gene-based adaptations have many advantages, and so the evolutionary advantages of a more flexible process of adaptation may have only occurred under particular conditions. Discovering those conditions is an important evolutionary question.

The second scale is the evolution of human diversity. Given that living humans are descended from a small population relatively recently, human diversity is also a relatively recent phenomenon, and the extensive evidence for gene flow across populations suggests that our diversity has a very complex pattern. The flexibility of epigenetics is one means by which we can resolve the paradox of high levels of human behavioural flexibility, in an evolutionary context, and the relatively low level of genetic diversity.

Finally, at the third scale lies the question of interindividual differences. Humans occupy not only many different natural habitats but also an ever-increasing range of socioculturally constructed niches. Within human societies, our complex relationships and hierarchies produce micro-niches—so that individuals living within the same area may experience radically different circumstances with regard, for example, to resources (in terms of access and competition), status and various forms of stress. The most adaptive social behaviour to maximise reproductive fitness will vary depending on the micro-niche inhabited by the individual. The unusual complexity of human social cognition and the range and variation in human social behaviours could provide a kind of mental immune system whereby behaviours are triggered that will be optimal for different physical and social environments. Gene-environment interaction facilitated by epigenetic processes could enable such a system to adapt more rapidly.

For all organisms, there is a tension between the need for stability of form and the potential for adaptive change. For some traits, the need for stability is greater than the need for change, and the trait becomes a stable species characteristic. But where the advantages of rapid adaptation are stronger, mechanisms facilitating gene-environment interaction, within the lifetime or across generations, may be under selection. Mammalian social behaviour may be one such trait, and the need to adapt rapidly to changes in our natural and social environments is especially pronounced in humans. This need may have driven selection for changes in the epigenetic regulation of genes influencing social cognition and behaviour to maximise the adaptive responses of humans as a species and as individuals.

References

Adegbola AA et al (2015) Monoallelic expression of the human FOXP2 speech gene. Proc Natl Acad Sci USA 112(22):6848–6854

Auger CJ et al (2011) Epigenetic control of vasopressin expression is maintained by steroid hormones in the adult male rat brain. Proc Natl Acad Sci USA 108(10):4242–4247

Ayub Q et al (2013) FOXP2 targets show evidence of positive election in European populations. Am J Hum Genet 92(5):696–706

Bateson P et al (2004) Developmental plasticity and human health. Nature 430:419–421

Brown WM (2011) The parental antagonism theory of language evolution: preliminary evidence for the proposal. Hum Biol 83(2):213–245

Cáceres M et al (2003) Elevated gene expression levels distinguish human from non-human primate brains. Proc Natl Acad Sci USA 100(22):13030–13035

Canli T, Lesh K-P (2007) Long story short: the serotonin transporter in emotion regulation and social cognition. Nat Neurosci 10(9):1103–1109

Chadwick BP (2015) Epigenetics: current research and emerging trends. Caister Academic, Norfolk

Charrier C, Polleux F (2012) Role of partial duplication of the SRGAP2 gene in the evolution and development of the human brain. Med Sci 28(11):911–914

Chimpanzee Sequencing Analysis Consortium (2005) Initial sequence of the chimpanzee genome and comparison with the human genome. Nature 437(7055):69–87

Cochran DM et al (2015) Relationship among glutamine, gamma-aminobutyric acid, and social cognition in autism spectrum disorders. J Child Adolesc Psychopharmacol 25(4):314–322

Crespi BJ (2007) Sly FOXP2: genomic conflict in the evolution of language. Trends Ecol Evol 22(4):174–175

Crişan LG et al (2009) Genetic contributions of the serotonin transporter to social learning of fear and economic decision making. Soc Cogn Affect Neurosci 4(4):399–408

Curley JP (2011) Is there a genomically imprinted social brain? Bioessays 33(9):662–668

Dennis EL et al (2011) Altered structural brain connectivity in healthy carriers of the autism risk gene, CNTNAP2. Brain Connect 1(6):447–459

Dennis MY et al (2012) Evolution of human-specific neural SRGAP2 genes by incomplete segmental duplication. Cell 149(4):912–922

Dunbar RIM (1998) The social brain hypothesis. Evol Anthropol 6(5):178–190

Dunbar RI (2009) The Social Brain Hypothesis and its implications for social evolution. Ann Hum Biol 36(5):562–572

Feuk L et al (2006) Absence of a paternally inherited FOXP2 gene in developmental verbal dyspraxia. Am J Hum Genet 79(5):956–972

Foley RA (2016) Mosaic evolution and the pattern of transitions in the hominin lineage. Philos Trans R Soc Lond B Biol Sci 371:20150244

Gershon NB, High PC (2015) Epigenetics and child abuse: modern-day Darwinism -the miraculous ability of the human genome to adapt, and then adapt again. Am J Med Genet C Semin Med Genet 169(4):353–360

Haig D (2002) Genomic imprinting and kinship. Rutgers University, New Brunswick

Haig D (2014) Coadaptation and conflict, misconception and muddle, in the evolution of genomic imprinting. Heredity 113(2):96–103

Hajj EI et al (2014) Epigenetics and life-long consequences of an adverse nutritional and diabetic intrauterine environment. Reproduction 148(6):R111–R120

Hales CN, Barker DJ (2001) The thrifty phenotype hypothesis. Br Med Bull 60:5–20

Hariri AR et al (2002) Serotonin transporter genetic variation and the response of the human amygdala. Science 297:400–403

Heijmans BT et al (2008) Persistent epigenetic differences associated with prenatal exposure to famine in humans. Proc Natl Acad Sci USA 105:17046–17049

Holloway T, Gonzalez-Maeso J (2015) Epigenetic mechanisms of serotonin signalling. ACS Chem Neurosci 6(7):1099–1199

Hsieh J, Zhao X (2016) Genetics and epigenetics in adult neurogenesis. Cold Spring Harb Perspect Biol 8(6):a018911

Isles AR et al (2002) A possible role for imprinted genes in inbreeding avoidance and dispersal from the natal area in mice. Proc Biol Sci 269(1492):665–670

Jablonka E (2013) Epigenetic inheritance and plasticity: the responsive germline. Prog Biophys Mol Biol 111:99–107

Jablonka E, Lamb MJ (2005) Evolution in four dimensions: genetic, epigenetic, behavioural and symbolic variation in the history of life. MIT, Cambridge

Jablonka E, Raz G (2009) Transgenerational epigenetic inheritance: prevalence, mecahnisms and implications for the study of heredity and evolution. Q Rev Biol 84(2):131–176

Jack KM, Isbell LA (2009) Dispersal in primates: advancing an individualized approach. Behaviour 146:429–436

Jack A et al (2012) DNA methylation of the oxytocin receptor gene predicts neural response to ambiguous social stimuli. Front Hum Neurosci 6:a280

Jiang YH et al (1998) Mutation of the angelman ubiquitin ligase in mice causes increased cytoplasmic p53 and deficits of contextual learning and long-term potentiation. Neuron 21(4): 799–911

Jirtle RL, Skinner MK (2016) Environmental epigenomics and disease susceptibility. Nat Rev Genet 8:253–262

Kappil M et al (2015) Environmental influences on genomic imprinting. Curr Environ Health Rep 2:155. https://doi.org/10.1007/s40572-015-0046-z

Karg K et al (2011) The serotonin transporter promoter variant (5-HTTLPR), stress, and depression meta-analysis revisited: evidence of genetic moderation. Arch Gen Psychiatry 68(5):444–454

Keebaugh AC, Young LJ (2011) Increasing oxytocin receptor expression in the nucleus accumbens of pre-pubertal female prairie voles enhances alloparental responsiveness and partner preference formation as adults. Horm Behav 60(5):498–504

Keller TE, Yi SV (2014) DNA methylation and evolution of duplicate genes. Proc Natl Acad Sci USA 111(16):5932–5937

Keverne EB (2014) Significance of epigenetics for understanding brain development, brain evolution and behaviour. Neuroscience 264:207–217
Keverne EB et al (1996) Genomic imprinting and the differential roles of parental genomes in brain development. Dev Brain Res 92(1):91–200
Klironomos FD, Berg J, Collins S (2013) How epigenetic mutations can affect genetic evolution: model and mechanism. Bioessays 35(6):571–578
Koenig A, Borries C (2012) Hominoid dispersal and human evolution. Evol Anthropol 21(3): 108–112
Lillycrop KA, Burdge GC (2015) Maternal diet as a modifier of offspring epigenetics. J Dev Orig Health Dis 6(2):88–95
Maricic T et al (2013) A recent evolutionary change affects a regulatory element in human FOXP2 gene. Mol Biol Evol 30(4):844–852
Monk C et al (2012) Linking prenatal maternal adversity to developmental outcomes in infants: the role of epigenetic pathways. Dev Psychopathol 24(4):1361–1376
Moriam S, Sobhani M (2013) Epigenetic effect of chronic stress on dopamine signaling and depression. Genet Epigenet 5:11–16
Mozzi A et al (2016) The evolutionary history of genes involved in spoken and written language: beyond FOXP2. Sci Rep 6:a22157
Munoa I et al (2015) The epigenetic regulation of the opioid system: new individualized prompt prevention and treatment strategies. J Cell Biochem 116(11):2419–2426
Nagy C, Turecki G (2015) Transgenerational epigenetic inheritance: an open discussion. Epigenomics 7(5):781–790
Nikolova YS et al (2014) Beyond genotype: serotonin transporter epigenetic modification predicts human brain function. Nat Neurosci 17(9):1153–1155
Nowick K et al (2009) Differences in human and chimpanzee gene expression patterns define an evolving network of transcription factors in brain. Proc Natl Acad Sci USA 106(52): 22358–22363
Pai AA et al (2011) A genome-wide study of DNA methylation patterns and gene expression in multiple human and chimpanzee tissues. PLoS Genet 7(2). https://doi.org/10.1371/journal.pgen.1001316
Perroud N et al (2014) The Tutsi genocide and transgenerational transmission of maternal stress: the epigenetics and biology of the HPA axis. World J Biol Psychiatry 15(4):334–345
Petito A et al (2016) The relationship between personality traits, the 5HTT polymorphisms, and the occurrence of anxiety and depressive symptoms in elite athletes. PLoS One 11(6):e0156601. https://doi.org/10.1371/journal.pone.0156601
Prokopuk L et al (2015) Transgenerational epigenetic inheritance: adaptation through the germline epigenome? Epigenomics 7(5):829–846
Puglia MH et al (2015) Epigenetic modification of the oxytocin receptor gene influences the perception of anger and fear in the human brain. Proc Natl Acad Sci USA 112(11):3308–3313
Ragsdale G, Foley RA (2011) A maternal influence on theory of mind mediated by executive function: differential parental influences on full and half-siblings. PLoS One 6(8). https://doi.org/10.1371/journal.pone.0023236)
Ragsdale G, Foley RA (2012) Testing the imprinted brain: parent-of-origin effects on empathy and systemising. Evol Hum Behav 33(4):402–410
Sadakierska-Chudy A et al (2015a) A comprehensive view of the epigenetic landscape part I: DNA methylation, passive and active DNA demethylation pathways and histone variants. Neurotox Res 27(1):84–97
Sadakierska-Chudy A et al (2015b) A comprehensive view of the epigenetic landscape part II: histone post-translational modification, nucleosome level, and chromatin regulation by ncRNAs. Neurotox Res 27(2):172–197
Schneider E et al (2014) Widespread differences in cortex DNA methylation of the "language gene" CNTNAP2 between humans and chimpanzees. Epigenetics 9(4):533–545

Schulz LC (2010) The Dutch Hunger Winter and the developmental origins of health and disease. Proc Natl Acad Sci USA 107(39):16757–16758

Shultz S, Opie C, Atkinson QD (2011) Stepwise evolution of stable sociality in primates. Nature 479(7372):219–U96

Silverman JL et al (2015) GABA(B) receptor agonist R-Baclofen reverses social deficits and reduces repetitive behaviour in two mouse models of autism. Neuropsychopharmacology 40(9):2228–2239

Skuse D (2006) Genetic influences on the neural basis of social cognition. Philos Trans R Soc Lond Ser B Biol Sci 361(1476):2129–2141

Stimpson CD et al (2016) Differential serotonergic innervation of the amygdala in bonobos and chimpanzees. Soc Cogn Affect Neurosci 11(3):413422. https://doi.org/10.1093/scan/nsv128

Strathearn L (2011) Maternal neglect: oxytocin, dopamine and the neurobiology of attachment. J Neuroendocrinol 23(11):1054–1065

Stuppia L et al (2015) Epigenetics and male reproduction: the consequences of paternal lifestyle on fertility, embryo development, and children lifetime health. Clin Epigenetics 7:a120

Thomas AC et al (2012) The speech gene FOXP2 is not imprinted. J Med Genet 49(11):669–670

Tobi EW et al (2009) DNA methylation differences after exposure to prenatal famine are common and timing and sex-specific. Hum Mol Genet 18:4046–4053

Uher R et al (2011) Serotonin transporter gene moderates childhood maltreatment's effects on persistent but not single-episode depression: replications and implications for resolving inconsistent results. J Affect Disord 135(1–3):56–65

Vangeel EB et al (2015) DNA methylation in imprinted genes IGF2 and GNASXL is associated with prenatal maternal stress. Genes Brain Behav 14(8):573–582. https://doi.org/10.1111/gbb.12249

Vernes SC et al (2008) A functional genetic link between distinct developmental language disorders. N Engl J Med 359(22):2337–2345

Wang H et al (2013) Histone deacetylase inhibitors facilitate partner preference formation in female prairie voles. Nat Neurosci 16(7):919–924

Webster MT (2013) Human-specific accelerated evolution of noncoding sequences. eLS. https://doi.org/10.1002/9780470015902.a0020849.pub2

Wendland JR et al (2006) Differential functional variability of serotonin transporter and monoamine oxidase a genes in macaque species displaying contrasting levels of aggression-related behaviour. Behav Genet 36(2):163–172

Yehuda NP et al (2014) Influences of maternal and paternal PTSD on epigenetic regulation of the glucocorticoid receptor gene in holocaust survivor offspring. Am J Psychiatry 171(8):872–880. https://doi.org/10.1176/appi.ajp.2014.13121571

Zeng J, Konopka G, Hunt BG, Preuss TM, Geschwind D, Soojin VY (2012) Divergent whole-genome methylation maps of human and chimpanzee brains reveal epigenetic basis of human regulatory evolution. Am J Hum Genet 91(3):455–465

Neanderthals and *Homo sapiens*: Cognitively Different Kinds of Human?

Eiluned Pearce

Abstract Membership of an extensive social network is imperative for human survival. However, maintaining network cohesion is particularly challenging for hunter–gatherers because they are dispersed over large home ranges and need to keep track of absent social partners for extended periods. The archaeological record suggests that compared to Neanderthals, contemporary modern humans maintained social ties between greater numbers of individuals over greater distances. I argue that such differences would have influenced neural development, driving differences in brain structure and the degree of social complexity that each taxon could sustain cognitively. Following recent suggestions that modern humans' larger parietals might suggest an enhanced ability to create a 'virtual inner world', I hypothesise that this capacity allowed them to monitor larger numbers of absent social partners and thus maintain larger dispersed social networks than their Neanderthal counterparts. Larger social networks would have boosted the ability of modern humans to insure against local resource failure, sustain demographic stability and conserve cultural innovations.

Keywords Social networks · Sociocognition · Neural plasticity · Hunter–gatherers · Social archaeology

1 The Importance and Challenges of Maintaining Social Networks

Social relationships are crucial for human survival. Indeed, the number and quality of such ties are positively related to life expectancy, and the protective effect on health is greater than well-known factors such as giving up smoking and doing exercise (e.g. Holt-Lunstad et al. 2010). Furthermore, being part of a cohesive social

E. Pearce (✉)
Social and Evolutionary Neuroscience Research Group, Department of Experimental Psychology, University of Oxford, Oxford, UK
e-mail: eiluned.pearce@psy.ox.ac.uk

L. D. Di Paolo et al. (eds.), *Evolution of Primate Social Cognition*, Interdisciplinary Evolution Research 5, https://doi.org/10.1007/978-3-319-93776-2_12

network improves not only an individual's health but also that of their children, both in humans (Adams et al. 2002; Iaupuni et al. 2005) and in baboons (Silk et al. 2003). Since social behaviour impacts on individuals' genetic fitness, it is likely to be under strong selection and is therefore fundamental to understanding the evolutionary history of primates such as modern humans and Neanderthals (Pearce et al. 2014).

For mobile hunter–gatherers, maintaining friendly links with other social units minimises the risk of conflict with other groups as they move around the landscape. Moreover, maintaining connections with individuals who habitually reside in distant regions provides insurance against local resource failure. Such long-distance contacts provide access to both up-to-date information on the distribution and yields of alternative resources in different areas and the resources themselves (e.g. Pearce and Moutsiou 2014). Furthermore, a large network is better able to conserve knowledge and transmit innovations (e.g. Pearce 2014). Given such advantages, those with more extensive social connections will generally do better in a changing environment, especially if those changes are unpredictable: they will hear news more quickly and have more numerous independent sources of help as a safety net (Whallon 2006). Consequently, the size of social network that Neanderthals and contemporary modern humans were able to maintain would have had major implications for their survival.

Ensuring that social networks remain connected is particularly challenging for hunter–gatherers given their fission–fusion social system, whereby units periodically disperse and reaggregate according to changing resource distributions (Grove et al. 2012). When residential bands are dispersed, individuals need to factor in their absent social partners' perspectives in decision-making and keep tabs on them over extended periods, so that they know who they can currently count on and where they are. These abilities relate to the capacity for creating a 'virtual inner reality' (Bruner 2010b; Devlin 2000). Such 'offline thinking' is required in order to imagine alternative solutions to challenges and has to happen in the face of interference from online events by switching routine tasks to autopilot. In order to have the cognitive capacity for offline thinking, individuals require sufficient neural architecture in the associated brain areas, which, as described below, are predominately located in the upper parietal lobe.

In this chapter I propose that positive feedback between their social environment, neural architecture and sociocognitive capacities allowed modern humans to gradually enlarge their social networks relative to those of the Neanderthals (Fig. 1). Such feedback may have occurred in a similar way to Iriki and Taoka's (2012) 'triadic niche construction' (ecological, neural and cognitive) model, although they proposed this specifically in relation to tool use. Although the exact sequence and direction of causality remains unclear, the more conservative hypothesis proposed here is that differences in social environment led to altered neural architecture, but it is possible that neurocognitive differences initially drove these social disparities (for a discussion of different possible mechanisms of neural change, see Bruner and Iriki 2016). I first review the archaeological evidence indicating that compared to Neanderthals, contemporary modern humans had larger social networks. I then use studies on neural plasticity in living humans and macaques to propose that the consequently

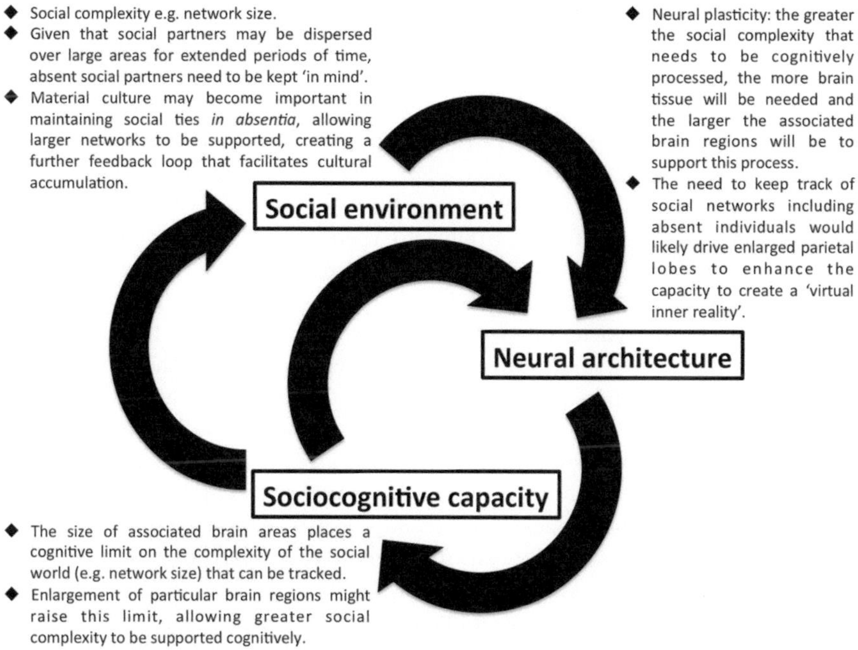

Fig. 1 The proposed interaction between social environment, neural architecture and sociocognitive capacity

more complex social environments in which modern humans lived became reflected in their brain structure and their associated cognitive capacities. In particular, I propose that the enlarged upper parietals demonstrated by the fossil record allowed modern humans to maintain cohesion in larger dispersed tribes by enabling them to keep track of more social partners in their 'mind's eye'. Finally, I suggest possible ways of starting to test the proposed hypothesis and discuss the implications of social network size for hominin survival.

2 The Archaeological Record

Several lines of evidence suggest that although Neanderthals had large social networks compared to other primates, they nonetheless had smaller ones than contemporaneous modern humans (e.g. see Pearce and Moutsiou 2014). Although social behaviour can only be studied indirectly in past populations, archaeological indices of social group or network size have been proposed, and I examine these below. Although there would have been variation in network size, here I address the maximum network size that each taxon was able to maintain: what might be considered the 'best case scenario' for each.

The size of residential bands can be approximated from archaeological sites using the area of occupation and the number of 'sleeping hearths' (Hayden 2012). For Neanderthals, these proxies suggest that residential groups most often comprised 12–14 individuals but could number up to ~30 (Hayden 2012). Tentative support for the former value comes from the 12 El Sidron Neanderthals, who are purported to represent a single social group (Lalueza-Fox et al. 2011). In contrast, the butchery evidence for Neanderthals at Mauran suggests a minimum of 30 individuals and perhaps an aggregation of several hundred (Farizy 1994). Unfortunately, these estimates suffer from the palimpsest nature of the archaeological record: even single archaeological strata do not necessarily signify occupation by a single group. Despite this, these data seem to match recent modern human hunter–gatherer group sizes fairly well: the median size of a recent hunter–gatherer residential band size is 14 during the most dispersed phase of the seasonal round and 37 during the most aggregated phase, whereas the median size of periodic aggregations of a number of bands is 150 (data from Binford 2001). However, despite the archaeological group estimates suggesting a Neanderthal band size of up to ~30 individuals, ethnographic data for recent hunter–gatherers suggests that the average band comprises 37 individuals. Modern humans may thus have had larger bands than Neanderthals (Hayden 2012). Although this difference may seem minimal, even an incremental difference at the level of the band might actually have had substantial implications in terms of survival. This is due to the organisation of human social networks, which comprise a series of cumulative nested layers (the residential band, the personal network, the mating pool and the tribe) in which each layer contains approximately three times as many members as the previous one (e.g. Pearce and Moutsiou 2014; Zhou et al. 2005). Due to this scaling, even a small difference in band size at the inner layers may have translated into marked differences at the level of ethnolinguistic tribe, the outermost layer comprising the largest identifiable social unit in hunter–gatherers (Pearce and Moutsiou 2014). For instance, a conservative difference of 7 individuals at the level of the band would have translated into a difference of 21 members of an individual's personal network, a difference of 63 individuals at the level of the mating pool (i.e. 31 fewer potential mates) and a difference of 189 at the level of the tribe. In this conservative case, modern humans would have been able to count on help from ~5 more bands than Neanderthals if their local resources failed or for support in competition for resources. Although the exact impact of having 189 more, or fewer, people who could be called on in times of need and act as reservoirs of cultural knowledge remains to be modelled and tested, the effect could have been substantial, particularly if such differences were sustained over time.

This suggested difference in the size of Neanderthal and modern human tribes appears to be borne out by evidence that Neanderthals exploited smaller geographic regions than contemporary modern humans living in similar habitats (e.g. Burke 2012; Pearce and Moutsiou 2014). The maximum distance over which raw materials move from their geological source to an archaeological site can give an indication of the maximum area covered by a social network, either through directly representing face-to-face exchanges or by representing potential interactions between individuals or subgroups exploiting the same geographic area (Pearce and Moutsiou 2014). Whereas Neanderthals transferred obsidian up to 300 km, European modern humans from similar dates transferred obsidian more than 400 km, a distance that accords well with the geographic areas exploited by recent hunter–gatherer tribes living in analogous habitats (Pearce and Moutsiou 2014). This difference indicates that Neanderthal tribes probably exploited smaller areas than modern humans. Moreover, Neanderthals seem to have lived at lower population densities (Mellars and French 2011). Given larger home ranges and higher population densities, it is likely that modern human tribes comprised greater numbers of individuals than the tribes of contemporary Neanderthals (Pearce and Moutsiou 2014).

The higher frequency of 'symbolic' artefacts in the modern human record, such as ornaments and figurines, also supports their networks being larger than those of Neanderthals. Such artefacts can stand as a mnemonic for social ties as well as creating obligations to reciprocate and maintain that tie, so face-to-face interactions can be less frequent, and more ties can be maintained (e.g. Gamble 1999). Moreover, material culture representing group membership can create connections between personal networks with no overlapping members, through ties to a shared group such as a tribe. Consequently, if modern humans made more use of 'symbolic' artefacts as a mechanism of social bonding than the Neanderthals did, they would have been able to sustain more extensive social networks.

Although there are purported examples of nonutilitarian artefacts in the Neanderthal record, apparently demonstrating that they did have the cognitive capacity for such creative behaviour, these are far less prevalent and diverse than those attributed to modern humans. For example, Neanderthals have been linked with possible examples of ornamentation (e.g. Finlayson et al. 2012; Zilhão et al. 2010), ochre use (e.g. Roebroeks et al. 2012) and some examples of cave art date to the overlap between the last Neanderthals and the earliest European modern humans (Pike et al. 2012). Nonetheless, the Neanderthal symbolic record appears very modest compared to the many shell beads from modern human sites (e.g. d'Errico et al. 2005; Vanhaeren and D'Errico 2006), the distinctive regional tool styles of modern humans (e.g. McBrearty and Brooks 2000) and the figurative art and evidence for music found in Europe from the earliest modern human occupations onwards (Conard 2009). Collective music making is a powerful way of bonding large groups of individuals simultaneously, and so far musical instruments have only been found associated with modern humans (Morley 2013; Pearce et al. 2015; Weinstein et al. 2015). Though it remains possible that we simply cannot recognise objects used by Neanderthals to support large-scale community cohesion, the contrast in the archaeological records between the two taxa suggests that although Neanderthals may have been technically capable of creating 'symbolic' objects, they did not do so to the same extent as contemporary modern humans. Given the Neanderthals' consequent need to meet social partners face-to-face rather than being able to rely on material culture to support their social ties, this may have limited them to smaller networks.

Taken together, the archaeological data suggest that modern human bands and tribes were larger than those of contemporary Neanderthals living in similar habitats. This implies that Neanderthals, though by no means asocial or isolated, would have interacted with fewer individuals both daily and across their lifespans. Consequently, a modern human would have had to function in more complicated social contexts than the average Neanderthal, and this likely had implications for their brain structure.

3 Neural Plasticity

The fact that the relatively recent cultural innovations of reading and arithmetic are processed in well-defined cerebral areas suggests that neural circuitry that evolved for a particular function can be co-opted to support a different, though probably related, function associated with novel behaviours, at least within certain limits (the neuronal recycling hypothesis: Dehaene and Cohen 2007). Along similar lines, engaging in maintaining more extensive, and more complex, social worlds may have had repercussions for how neural circuits were allocated in the modern human brain. Support for the impact of behavioural and environmental variation on brain structure comes from the findings that extensive use of complicated cognitive maps leads to enlargement of the posterior hippocampus in London taxi drivers (Maguire et al. 2000) and that the longer a taxi driver has been working, the more their brain changes (Maguire et al. 2003). Thus, even during adulthood routine tasks can have a marked impact on brain structure. For instance, when adults learn to juggle, there is selective structural change in brain areas underpinning the complex processing of visual motion (Draganski et al. 2004). That neural plasticity happens across primates is illustrated by the finding that macaques trained to use tools show increased grey matter in specific regions of their parietal and temporal lobes (Quallo et al. 2009). If neural plasticity occurs in both living humans and macaques as a result of training, it may also have occurred in Neanderthals and their modern human contemporaries, particularly since cortical organisation seems to be more plastic in humans compared to other primates (Gómez-Robles et al. 2015).

Since the cognitive processes which individuals engage in habitually appear to influence their brain structure, it seems likely that the more often someone processes complicated social stimuli, the larger and more interconnected their associated brain regions would become (Fig. 1). This is indeed the case for macaques: monkeys housed with a larger number of cage mates show increased grey matter volume in the specific regions of their temporal and prefrontal cortices that underlie social cognition, as well as more connectivity between them (Sallet et al. 2011). Both personal social network size and the capacity to understand the mental states of others are associated with the size of analogous areas in humans (Powell et al. 2012; Van Overwalle 2009). In humans these associations between social environment and brain structure are only correlations, so the direction of causality remains unclear. However, the macaques were not free to choose how many conspecifics they lived with, and this means that the size of the macaques' social networks *caused* the development of more neural tissue dedicated to processing social information, rather than vice versa (Sallet et al. 2011). Accordingly, given the archaeological indicators that modern humans maintained links with more individuals, neural networks associated with processing social complexity may have been enlarged in modern humans relative to Neanderthals.

Since neural plasticity seems to continue into adulthood, a Neanderthal living in a modern human group and exposed to large dispersed social networks might have developed the requisite neural architecture and sociocognitive abilities to keep track of them. However, it seems likely that the majority of Neanderthals would have remained in Neanderthal groups and thus would have developed the social brain areas that went along with these less complicated social milieus. In living humans, the fine-tuning of the neural and cognitive processes for handling complex social information continues into early adulthood (Deeley et al. 2008; Dumontheil et al. 2010). Consequently, if Neanderthals matured more quickly than modern humans (Smith et al. 2010), this might have limited the refinements they could make to cognitive processes dealing with social complexity. In contrast, modern humans were interacting within the context of more extensive social networks throughout their longer adolescences and would have developed larger and more interconnected 'social brain' structures to cope with this added social complexity. Although the issue of causality requires much further debate (cf. Bruner and Iriki 2016), even if neurological differences were not the initial reason for differences in social complexity between these taxa, they would have arisen as a consequence.

4 Virtual Worlds

Although Neanderthals and their modern human contemporaries had similarly large brains in terms of absolute volume, this does not necessarily mean that they were organised identically. For instance, Pearce et al. (2013) proposed that modern humans had smaller visual brain areas compared to contemporary Neanderthals and therefore that more of the modern human brain could be taken up with processing of social information, allowing modern humans to deal with greater social complexity, including larger cohesive groups. The suggestion of differential brain organisation is supported by external brain shape differences between Neanderthals and modern humans that are present from early postnatal development (e.g. Gunz et al. 2010). I have suggested that differences in brain structures underlying social cognition might arise through training, but such early ontogenetic cranial differences between modern humans and Neanderthals seem to suggest that these brain differences were inherited traits rather than purely being the result of varying (social) environments. It is possible that as modern humans started to disperse out of Africa and across the globe, adaptive learning (and neural changes associated with such learning) allowed them to increase their fitness by enlarging their social networks (illustrating the Baldwin effect: Sznajder et al. 2012). This enhanced social capacity might then have become inherited through genetic (selection for gene variants that either coded for enlargement of 'social brain' regions *per se* or for genes that regulated phenotypic plasticity and sensitivity to the social environment) or epigenetic (changes of gene expression driven by environmental factors) processes (see Bruner and Iriki 2016). Whatever the exact mechanism, interactions between social environment, neural architecture and sociocognitive capacity (Fig. 1) are likely to have occurred throughout hominin evolution.

If, as the archaeology suggests, modern humans had larger social networks and more complex social environments compared to Neanderthals, we might expect differences in 'social brain' areas in the frontal and temporo-parietal regions. Although both Neanderthals and modern humans display wider frontal lobes than other hominins (Bruner and Holloway 2010), there is some suggestion that modern humans had relatively wider orbitofrontal regions (Bastir et al. 2011) as well as differences in their temporal lobes that might reflect differing mentalising and social memory capacities between these taxa (Bastir et al. 2008, 2011; cf. Olson et al. 2013). However, frontal and temporal morphology is influenced by other regions of the skull, so these apparent differences may be associated with secondary structural variations rather than neural function (Bastir et al. 2004; Bruner and Ripani 2008; Lieberman et al. 2000).

Despite these controversies, what is generally accepted is that modern humans have larger upper parietal lobes than Neanderthals (Bruner 2010a; Bruner et al. 2003). The upper parietals have strong connections to the frontal lobe and integrate multimodal information to represent an individual's body and the world (Bruner and Iriki 2016; see Bruner et al. 2018). The superior parietal lobule is involved in working memory tasks requiring information manipulation, visual attention and spatial orientation (Culham and Valyear 2006; Koenigs et al. 2009), and the intraparietal sulcus dividing the superior and inferior parietal lobules is thought to play a role in directed action and interpreting the goals of others, visuospatial working memory and processing symbolic numerical information (Cantlon et al. 2006; de Hamilton and Grafton 2006). In addition, the retrosplenial cortex (RSC) is linked to navigation, episodic memory of recent autobiographical information and planning (Vann et al. 2009). Moreover, the medial precuneus is involved in reflective self-awareness as part of the 'default mode network', which is active during daydreaming, thinking about the past, or planning, and overlaps 'social brain' areas associated with understanding others (e.g. Cavanna 2006; Land 2014; Mars et al. 2012). Peer et al. (2015) found that both differentiated and overlapping areas in the precuneus and inferior parietal lobes are activated when someone is orientating to a person, to space and to time, forming 'a specific brain system with a highly ordered internal organization, closely related to the default-mode network'. Although the default mode network was active for all three orientation domains, it particularly overlapped with brain areas linked to person-orientation in terms of thinking about how emotionally close another individual is to oneself (Peer et al. 2015). In sum, the upper parietal cortex integrates memory with the manipulation of sensory information to allow navigation, mental orientation towards other people and interpretation of their goals, as well as replaying of past events and simulation of possible future ones. As I discuss further below, these spatial integration, memory and imaginative functions are likely to be imperative to cognitively representing a dynamic and extensive social network distributed across space, whose membership needs to bear each other in mind in order to maintain social connection despite only infrequently meeting face-to-face (see also Bruner et al. 2018). Parietal differences between hominin species might point to differences in the ability to do this.

However, a note of caution is required in relation to interpretations of the fossil record with regard to parietal anatomy and function: since both the precuneus and the RSC are located deep within the brain, it is impossible to study their anatomy directly from fossil endocasts. Nonetheless, recent work indicates that precuneal morphology has a strong influence on midsagittal geometry and that much of the difference in midbrain morphology between humans and chimpanzees is attributable to differences in the size of the precuneus, which is enlarged in humans (Bruner et al. 2014, 2016). These findings suggest that internal differences in parietal organisation do affect surface shape. Consequently, external differences between the parietals of Neanderthals and modern humans are likely to reflect internal differences as well.

Given their associated functions, the larger superior parietal lobules of modern humans are likely associated with an enhanced capacity to create a 'virtual inner reality' compared to other hominins (though it should be noted that Bruner et al. (2015) did not find associations between precuneal size and some cognitive abilities *within* humans). Such an ability as this would support individuals to take account of absent social partners in decision-making, as well as in the spatial tracking of network members across a home range (i.e. knowing where they are), and the navigation required to travel to their location to meet them. For instance, Burke (2012) argues that modern humans translated egocentric navigation based on detailed knowledge of the local landscape into a cognitive representation resembling an aerial map (interestingly, the RSC has been associated with switching between these egocentric and allocentric frames of reference: Vann et al. 2009). If correct, such a shift in spatial perspective may have allowed modern humans to mentally map their social partners across the landscape and be able to find them more easily if the need arose. The development of such search and navigational abilities would have required abstract spatial representation and the capacity to follow ordered sequences of navigational cues, processes involving various regions of the parietal lobe. In addition, a distributed cognitive system may have developed through pooling information on the changing spatial distribution of the collective network, for example, through different residential bands sharing gossip on encountering each other (Hills et al. 2015). Repeated engagement in these visuospatial cognitive processes is likely to have been accompanied by upper parietal enlargement.

Spatial cognition is not only relevant to social network maintenance in terms of locating members but also seems to be the key to cognitive processes underlying face-to-face social engagement, such as empathy: mentally 'putting yourself in someone else's place'. For example, the way in which another's body is represented in relation to the self seems to drive associated attitudes, and increasing the sharing of bodily representations has been shown to increase positivity to members of an out-group (Maister et al. 2015). Differences in spatial cognition between Neanderthals and modern humans may thus also hint at differences in the degree of self-other distinction and associated interpersonal traits such as empathy.

As well as tracking the spatial distribution of their network, hunter–gatherers need to factor in absent partners during their current interactions. This would engage at least the precuneus/RSC (incorporating social memories and planning) and the intraparietal sulcus (visuospatial integration to understand the goals of self and others) (Bruner et al. 2014). The more social partners an individual needs to keep in their mental picture, the more complicated that picture and the larger the precuneus/RSC and intraparietal sulcus required. In addition, individuals need to visualise social partners despite interference from their current surroundings, requiring enhanced working memory and focused attention (e.g. Coolidge and Wynn 2005). Again, this extra cognitive burden on working memory would probably require enlarged superior parietals. Overall, the larger parietal lobes exhibited by modern humans, and in particular the expansion of the precuneus, may be linked to their more complex social environments, which would have required the development of enhanced capacities for imagination, including complex processing of spatial and goal-related information (Bruner 2004; Bruner et al. 2016; Coolidge and Overmann 2012).

The lateral widening of the Neanderthals' superior parietal lobules (Bruner et al. 2003) implies that Neanderthals may have found offline thinking easier than their *Homo heidelbergensis* ancestors. However, Neanderthals do not show upper parietal enlargement to the same extent as modern humans and thus may have been less specialised for spatial and social processing (in terms of keeping track of the spatial distribution of their social network and taking account of absent social partners in their decisions). As mentioned above, although this parietal difference could be purely developmental, arising each generation in response to social environments, over time the globularisation of the modern human brain associated with enlargement of the precuneus (Bruner et al. 2014), seems to have become an evolved trait that arises soon after birth and deviates from the general allometric hominin trajectory (Bruner et al. 2003; Gunz et al. 2010). Either way, it is likely that the complexity of the social environments inhabited by Neanderthals and modern humans would have been manifested in their respective brain structures, and this in turn may have fed back into differential cognitive limits on social network size (Fig. 1).

Although testing the hypothesis that differences in social environment drove brain differences between Neanderthals and modern humans would be challenging to say the least, some of the underlying claims could be investigated. Firstly, future work could explore whether better visuospatial capacities, particularly in relation to keeping track of conspecifics across space and time, in extant primates is associated with greater social complexity: the expectation would be that primate species with larger groups and/or fission–fusion social systems would have greater abilities of visuospatial integration, particularly in relation to social stimuli (for instance, in knowing the relative locations of conspecifics). However, solving how to collect comparable data on visuospatial capacities across species might prove challenging. Secondly, the hypothesis that better performance on visuospatial tasks is linked to having (1) larger networks and (2) more spatially distributed social networks could also be tested within humans. For instance, one might expect mobile hunter–gatherers to outperform individuals whose social networks are more clustered and sedentary: individuals who see the members of their social network regularly face-to-face (e.g. because they all reside in the same local area) might not develop their visuospatial abilities to the same extent as individuals who need to keep track of a spatially dispersed network. Thirdly, whether larger parietals are linked to larger social groups or networks, and greater visuospatial abilities, could be tested both across extant primate species and in humans (though the work of Bruner et al. (2015) suggests that this might not be the case in the latter). Fourthly, longitudinal studies could examine whether childhood social environments impact on adult visuospatial abilities and brain morphology.

5 Cognitively Different, but Not Stupid

Although the initial portrayal of Neanderthals as stupid proved incorrect, this does not necessarily exclude the possibility of some cognitive differences existing between Neanderthals and modern humans (Langbroek 2012; Wynn et al. 2016). Since Neanderthals and contemporary modern humans exploited similar resources using similar subsistence strategies, regarding 'technical' cognition there may have been little difference (Villa and Roebroeks 2014). However, Neanderthals and modern humans still may have differed in their abilities to process complex social information. That social and technical intelligence relate to different cognitive domains is supported by the finding that chimpanzees do as well on mechanistic tasks as 2.5-year-old humans but perform significantly worse on social tasks (Herrmann et al. 2007). Moreover, individuals with autistic traits are often good at understanding abstract concepts and mechanisms ('technical' cognition) but either fail or show marked developmental delays in passing, tasks requiring understanding others' mental states or identifying the emotional expression on human faces ('social' cognition) (Baron-Cohen 2009). Given their different social environments, Neanderthals and modern humans likely differed in their capacities to cognitively process social complexity.

Having the cognitive capacity to sustain more extensive social networks would have had a number of implications for modern humans. Even if differences with Neanderthals were initially relatively small, the impact of the associated effects would have built up over time. Firstly, larger modern humans social networks would have meant that contact could be kept with a greater number of bands inhabiting distant areas, providing more independent sources of help if local resources failed. Although Neanderthal bands could take refuge in new areas as well, without social ties to resident groups the incomers would have lacked local knowledge and up-to-date information about resources in those unfamiliar areas. Even if they could find available resources, they would risk violent altercations with other groups in order to access them, and in such intergroup conflicts they would suffer through having fewer supporters to rely on to provide reinforcement. In contrast, if modern humans encountered competition for resources, they could either call on more coalitionary partners for support in order to win through brute force, or they could access resources in other areas where they knew friendly groups.

Secondly, larger modern human social networks would have been less vulnerable to demographic instability, for example, by providing a larger pool of mates. In addition, any deaths would have had a proportionally smaller impact on the availability of individuals to forage, reproduce and provide specialist knowledge. A large social network thus decreases the risk of local population extinction and, ultimately, species extinction.

Similarly, larger social networks are associated with a decreased risk of losing cultural innovations and information due to a larger pool of experts and innovators, boosting the development of cumulative culture (Henrich 2004; Powell et al. 2009). In contrast, it may be that cultural innovations made by some local Neanderthal groups never spread or persevered due to their smaller social networks. Moreover, constraints on transmission of cultural innovations may have further limited Neanderthal network size, since exchange of 'symbolic' artefacts is a well-established means of sustaining social ties (e.g. Pearce 2014). Taken together, these implications suggest that the ability to sustain larger social networks, particularly through 'offline' thinking, was a crucial factor in why modern humans could survive both unpredictable environments and intergroup competition for resources.

Acknowledgments I thank Emiliano Bruner, Thomas Wynn, Kit Opie, Robin Dunbar, Prajñaketu Holden and Iain Morley for comments on various versions of this paper. This work was funded by the European Research Council (295663). The author declares no conflict of interest.

References

Adams AM, Madhavan S, Simon D (2002) Women's social networks and child survival in Mali. Soc Sci Med 54(2):165–178. https://doi.org/10.1016/s0277-9536(01)00017-x

Baron-Cohen S (2009) Autism: the empathizing–systemizing (E-S) theory. Ann N Y Acad Sci 1156(1):68–80. https://doi.org/10.1111/j.1749-6632.2009.04467.x

Bastir M, Rosas A, Kuroe K (2004) Petrosal orientation and mandibular ramus breadth: evidence for an integrated petroso-mandibular developmental unit. Am J Phys Anthropol 123(4): 340–350. https://doi.org/10.1002/ajpa.10313

Bastir M, Rosas A, Lieberman DE, O'Higgins P (2008) Middle cranial fossa anatomy and the origin of modern humans. Anat Rec Adv Integr Anat Evol Biol 291(2):130–140. https://doi.org/10.1002/ar.20636

Bastir M, Rosas A, Gunz P, Pena-Melian A, Manzi G, Harvati K et al (2011) Evolution of the base of the brain in highly encephalized human species. Nat Commun 2:588

Binford L (2001) Constructing frames of reference: an analytical method for archaeological theory building using ethnographic and environmental data sets. University of California Press, Berkeley

Bruner E (2004) Geometric morphometrics and paleoneurology: brain shape evolution in the genus *Homo*. J Hum Evol 47(5):279–303. https://doi.org/10.1016/j.jhevol.2004.03.009

Bruner E (2010a) Morphological differences in the parietal lobes within the human genus: a neurofunctional perspective. Curr Anthropol 51(s1):S77–S88. https://doi.org/10.1086/650729

Bruner E (2010b) The evolution of the parietal cortical areas in the human genus: between structure and cognition. In: Yuan M, Schick K, Toth N (eds) The human brain evolving: paleoneurological studies in honor of Ralph L. Holloway. Stone Age Institute, Gosport, IN, pp 83–96

Bruner E, Holloway RL (2010) A bivariate approach to the widening of the frontal lobes in the genus *Homo*. J Hum Evol 58(2):138–146

Bruner E, Iriki A (2016) Extending mind, visuospatial integration, and the evolution of the parietal lobes in the human genus. Quat Int 405(A):98–110. https://doi.org/10.1016/j.quaint.2015.05.019

Bruner E, Ripani M (2008) A quantitative and descriptive approach to morphological variation of the endocranial base in modern humans. Am J Phys Anthropol 137(1):30–40. https://doi.org/10.1002/ajpa.20837

Bruner E, Manzi G, Arsuaga JL (2003) Encephalization and allometric trajectories in the genus *Homo*: evidence from the Neanderthal and modern lineages. Proc Natl Acad Sci USA 100(26):15335–15340

Bruner E, Rangel de Lázaro G, De la Cuétara JM, Martín-Loeches M, Colom R, Jacobs HIL (2014) Midsagittal brain variation and MRI shape analysis of the precuneus in adult individuals. J Anat 224(4):367–376. https://doi.org/10.1111/joa.12155

Bruner E, Román FJ, de la Cuétara JM, Martin-Loeches M, Colom R (2015) Cortical surface area and cortical thickness in the precuneus of adult humans. Neuroscience 286:345–352. https://doi.org/10.1016/j.neuroscience.2014.11.063

Bruner E, Preuss TM, Chen X, Rilling JK (2016) Evidence for expansion of the precuneus in human evolution. Brain Struct Funct. https://doi.org/10.1007/s00429-015-1172-y

Bruner E, Spinapolice E, Burke A, Overmann K (2018) Visuospatial integration: paleoanthropological and archaeological perspectives. In: di Paolo LD, di Vincenzo F, Almeida A (eds) Evolution of primate social cognition. Springer, Cham

Burke A (2012) Spatial abilities, cognition and the pattern of Neanderthal and modern human dispersals. Quat Int 247:230–235. https://doi.org/10.1016/j.quaint.2010.10.029

Cantlon JF, Brannon EM, Carter EJ, Pelphrey KA (2006) Functional imaging of numerical processing in adults and 4-y-old children. PLoS Biol 4(5):e125. https://doi.org/10.1371/journal.pbio.0040125

Cavanna AE (2006) The precuneus: a review of its functional anatomy and behavioural correlates. Brain 129(3):564–583. https://doi.org/10.1093/brain/awl004

Conard NJ (2009) A female figurine from the basal Aurignacian of Hohle Fels Cave in southwestern Germany. Nature 459(7244):248–252

Coolidge FL, Overmann KA (2012) Numerosity, abstraction, and the emergence of symbolic thinking. Curr Anthropol 53(2):204–225. https://doi.org/10.1086/664818

Coolidge FL, Wynn T (2005) Working memory, its executive functions, and the emergence of modern thinking. Camb Archaeol J 15(1):5–26. https://doi.org/10.1017/S0959774305000016

Culham JC, Valyear KF (2006) Human parietal cortex in action. Curr Opin Neurobiol 16(2):205–212. https://doi.org/10.1016/j.conb.2006.03.005

d'Errico F, Henshilwood C, Vanhaeren M, van Niekerk K (2005) Nassarius kraussianus shell beads from Blombos Cave: evidence for symbolic behaviour in the Middle Stone Age. J Hum Evol 48(1):3–24

de Hamilton AFC, Grafton ST (2006) Goal representation in human anterior intraparietal sulcus. J Neurosci 26(4):1133–1137

Deeley Q, Daly EM, Azuma R, Surguladze S, Giampietro V, Brammer MJ et al (2008) Changes in male brain responses to emotional faces from adolescence to middle age. NeuroImage 40(1):389–397. https://doi.org/10.1016/j.neuroimage.2007.11.023

Dehaene S, Cohen L (2007) Cultural recycling of cortical maps. Neuron 56(2):384–398. https://doi.org/10.1016/j.neuron.2007.10.004

Devlin K (2000) The role of conceptual structure in human evolution. Brain 1867:1–12

Draganski B, Gaser C, Busch V, Schuierer G, Bogdahn U, May A (2004) Neuroplasticity: changes in grey matter induced by training. Nature 427(6972):311–312

Dumontheil I, Apperly IA, Blakemore S-J (2010) Online usage of theory of mind continues to develop in late adolescence. Dev Sci 13(2):331–338. https://doi.org/10.1111/j.1467-7687.2009.00888.x

Farizy C (1994) Spatial patterning of Middle Paleolithic sites. J Anthropol Archaeol 13(2):153–160

Finlayson C, Brown K, Blasco R, Rosell J, Negro JJ, Bortolotti GR et al (2012) Birds of a feather: Neanderthal exploitation of raptors and corvids. PLoS One 7(9):e45927. https://doi.org/10.1371/journal.pone.0045927

Gamble C (1999) The Palaeolithic societies of Europe. Cambridge world archaeology. Cambridge University Press, Cambridge

Gómez-Robles A, Hopkins WD, Schapiro SJ, Sherwood CC (2015) Relaxed genetic control of cortical organization in human brains compared with chimpanzees. Proc Natl Acad Sci USA 112(48):14799–14804. https://doi.org/10.1073/pnas.1512646112

Grove M, Pearce E, Dunbar RIM (2012) Fission-fusion and the evolution of hominin social systems. J Hum Evol 62(2):191–200

Gunz P, Neubauer S, Maureille B, Hublin J-J (2010) Brain development after birth differs between Neanderthals and modern humans. Curr Biol 20(21):R921–R922

Hayden B (2012) Neanderthal social structure? Oxf J Archaeol 31(1):1–26. https://doi.org/10.1111/j.1468-0092.2011.00376.x

Henrich J (2004) Demography and cultural evolution: how adaptive cultural processes can produce maladaptive losses: the Tasmanian case. Am Antiq 69(2):197–214

Herrmann E, Call J, Hernàndez-Lloreda MV, Hare B, Tomasello M (2007) Humans have evolved specialized skills of social cognition: the cultural intelligence hypothesis. Science 317(5843):1360–1366. https://doi.org/10.1126/science.1146282

Hills TT, Todd PM, Lazer D, Redish AD, Couzin ID, Cognitive Search Research Group (2015) Exploration versus exploitation in space, mind, and society. Trends Cogn Sci 19(1):46–54. https://doi.org/10.1016/j.tics.2014.10.004.Exploration

Holt-Lunstad J, Smith TB, Layton JB (2010) Social relationships and mortality risk: a meta-analytic review. PLoS Med 7(7):e1000316. https://doi.org/10.1371/journal.pmed.1000316

Iaupuni SMK, Donato KM, Thompson-Colón T, Stainback M (2005) Counting on kin: social networks, social support, and child health status. Soc Forces 83(3):1137–1164

Iriki A, Taoka M (2012) Triadic (ecological, neural, cognitive) niche construction: a scenario of human brain evolution extrapolating tool use and language from the control of reaching actions. Philos Trans R Soc Lond B Biol Sci 367(1585):10–23. https://doi.org/10.1098/rstb.2011.0190

Koenigs M, Barbey AK, Postle BR, Grafman J (2009) Superior parietal cortex is critical for the manipulation of information in working memory. J Neurosci 29(47):14980–14986. https://doi.org/10.1523/JNEUROSCI.3706-09.2009

Lalueza-Fox C, Rosas A, Estalrrich A, Gigli E, Campos PF, GarcÃa-Tabernero A et al (2011) Genetic evidence for patrilocal mating behavior among Neanderthal groups. Proc Natl Acad Sci USA 108(1):250–253

Land MF (2014) Do we have an internal model of the outside world? Philos Trans R Soc Lond B Biol Sci 369:20130045. https://doi.org/10.1098/rstb.2013.0045

Langbroek M (2012) Trees and ladders: a critique of the theory of human cognitive and behavioural evolution in Palaeolithic archaeology. Quat Int 270:4–14. https://doi.org/10.1016/j.quaint.2011.03.007

Lieberman DE, Ross CF, Ravosa MJ (2000) The primate cranial base: ontogeny, function, and integration. Am J Phys Anthropol Suppl 31:117–169

Maguire EA, Gadian DG, Johnsrude IS, Good CD, Ashburner J, Frackowiak RSJ, Frith CD (2000) Navigation-related structural change in the hippocampi of taxi drivers. Proc Natl Acad Sci USA 97(8):4398–4403

Maguire EA, Spiers HJ, Good CD, Hartley T, Frackowiak RSJ, Burgess N (2003) Navigation expertise and the human hippocampus: a structural brain imaging analysis. Hippocampus 13(2): 250–259. https://doi.org/10.1002/hipo.10087

Maister L, Slater M, Sanchez-Vives MV, Tsakiris M (2015) Changing bodies changes minds: owning another body affects social cognition. Trends Cogn Sci 19(1):6–12. https://doi.org/10.1016/j.tics.2014.11.001

Mars RB, Neubert F-X, Noonan MP, Sallet J, Toni I, Rushworth MFS (2012) On the relationship between the "default mode network" and the "social brain". Front Hum Neurosci 6:1–9. https://doi.org/10.3389/fnhum.2012.00189

McBrearty S, Brooks AS (2000) The revolution that wasn't: a new interpretation of the origin of modern human behavior. J Hum Evol 39:453–563

Mellars P, French JC (2011) Tenfold population increase in Western Europe at the Neanderthal to modern human transition. Science 333(6042):623–627. https://doi.org/10.1126/science.1206930

Morley I (2013) The prehistory of music: human evolution, archaeology, and the origins of musicality. Oxford University Press, Oxford

Olson IR, McCoy D, Klobusicky E, Ross LA (2013) Social cognition and the anterior temporal lobes: a review and theoretical framework. Soc Cogn Affect Neurosci 8(2):123–133. https://doi.org/10.1093/scan/nss119

Pearce E (2014) Modelling mechanisms of social network maintenance in hunter–gatherers. J Archaeol Sci 50:403–413. https://doi.org/10.1016/j.jas.2014.08.004

Pearce E, Moutsiou T (2014) Using obsidian transfer distances to explore social network maintenance in late Pleistocene hunter–gatherers. J Anthropol Archaeol 36:12–20. https://doi.org/10.1016/j.jaa.2014.07.002

Pearce E, Stringer C, Dunbar RIM (2013) New insights into differences in brain organization between Neanderthals and anatomically modern humans. Proc R Soc B Biol Sci 280(1758). https://doi.org/10.1098/rspb.2013.0168

Pearce E, Shuttleworth A, Grove M, Layton R (2014) The costs of being a high latitude hominin. In: Dunbar RIM, Gamble C, Gowlett J (eds) The Lucy project: benchmark papers. Oxford University Press, Oxford, pp 356–379

Pearce E, Launay J, Dunbar RIM (2015) The ice-breaker effect: singing mediates fast social bonding. R Soc Open Sci 2:150221. https://doi.org/10.1098/rsos.150221

Peer M, Salomon R, Goldberg I, Blanke O, Arzy S (2015) Brain system for mental orientation in space, time, and person. Proc Natl Acad Sci U S A 112(35):11072–11077

Pike AWG, Hoffmann DL, García-Diez M, Pettitt PB, Alcolea J, De Balbín R et al (2012) U-series dating of Paleolithic art in 11 caves in Spain. Science 336(6087):1409–1413. https://doi.org/10.1126/science.1219957

Powell A, Shennan S, Thomas MG (2009) Late Pleistocene demography and the appearance of modern human behavior. Science 324(5932):1298–1301. https://doi.org/10.1126/science.1170165

Powell J, Lewis PA, Roberts N, Garcia-Finana M, Dunbar RIM (2012) Orbital prefrontal cortex volume predicts social network size: an imaging study of individual differences in humans. Proc R Soc B Biol Sci 279(1736):2157–2162. https://doi.org/10.1098/rspb.2011.2574

Quallo MM, Price CJ, Ueno K, Asamizuya T, Cheng K, Lemon RN, Iriki A (2009) Gray and white matter changes associated with tool-use learning in macaque monkeys. Proc Natl Acad Sci U S A 106(43):18379–18384

Roebroeks W, Sier MJ, Nielsen TK, De Loecker D, Parés JM, Arps CES, Mücher HJ (2012) Use of red ochre by early Neanderthals. Proc Natl Acad Sci USA 109(6):1889–1894

Sallet J, Mars RB, Noonan MP, Andersson JL, O'Reilly JX, Jbabdi S et al (2011) Social network size affects neural circuits in macaques. Science 334(6056):697–700. https://doi.org/10.1126/science.1210027

Silk JB, Alberts SC, Altmann J (2003) Social bonds of female baboons enhance infant survival. Science 302(5648):1231–1234. https://doi.org/10.1126/science.1088580

Smith TM, Tafforeau P, Reid DJ, Pouech J, Lazzari V, Zermeno JP et al (2010) Dental evidence for ontogenetic differences between modern humans and Neanderthals. Proc Natl Acad Sci USA 107(49):20923–20928. https://doi.org/10.1073/pnas.1010906107

Sznajder B, Sabelis MW, Egas M (2012) How adaptive learning affects evolution: reviewing theory on the baldwin effect. Evol Biol 39(3):301–310. https://doi.org/10.1007/s11692-011-9155-2

Van Overwalle F (2009) Social cognition and the brain: a meta-analysis. Hum Brain Mapp 30(3): 829–858. https://doi.org/10.1002/hbm.20547

Vanhaeren M, D'Errico F (2006) Aurignacian ethno-linguistic geography of Europe revealed by personal ornaments. J Archaeol Sci 33(8):1105–1128. https://doi.org/10.1016/j.jas.2005.11.017

Vann SD, Aggleton JP, Maguire EA (2009) What does the retrosplenial cortex do? Nat Rev Neurosci 10(11):792–802. https://doi.org/10.1038/nrn2733

Villa P, Roebroeks W (2014) Neanderthal demise: an archaeological analysis of the modern human superiority complex. PLoS One 9(4):e96424. https://doi.org/10.1371/journal.pone.0096424

Weinstein D, Launay J, Pearce E, Dunbar RIM, Stewart L (2015) Singing and social bonding: changes in connectivity and pain threshold as a function of group size. Evol Hum Behav. https://doi.org/10.1016/j.evolhumbehav.2015.10.002

Whallon R (2006) Social networks and information: non-"utilitarian" mobility among hunter-gatherers. J Anthropol Archaeol 25(2):259–270

Wynn T, Overmann KA, Coolidge FL (2016) The false dichotomy: a refutation of the Neanderthal indistinguishability claim. J Anthropol Sci 94:1–21. https://doi.org/10.4436/JASS.94022

Zhou WX, Sornette D, Hill RA, Dunbar RIM (2005) Discrete hierarchical organization of social group sizes. Proc R Soc B Biol Sci 272(1561):439–444

Zilhão J, Angelucci DE, Badal-García E, D'Errico F, Daniel F, Dayet L et al (2010) Symbolic use of marine shells and mineral pigments by Iberian Neanderthals. Proc Natl Acad Sci USA 107(3): 1023–1028

Part III
Cultural Artefacts and Transmission in Primates

Recognizing Culture in Wild Primate Tool Use

Michael Haslam, Tiago Falótico, and Lydia Luncz

Abstract Cultural differences between animal groups offer a means of tracing social relationships and cognition through time and across space. Where behaviours include tool use, we can observe the influence of available materials and role models on the development of tool-based activities. Here, we discuss the ways that we can study the social influence of tool-use behaviour in wild primates, focusing on two species that use durable stone tools: bearded capuchin monkeys (*Sapajus libidinosus*) and Western chimpanzees (*Pan troglodytes verus*). We concentrate on durable tools, as these provide an archaeologically recoverable record of activities. However, we also consider the influence of less durable tools when examining behavioural patterns in capuchins and chimpanzees. In order to study abstract concepts like culture and cognition, we identify socially learned behavioural diversity that is not influenced by environmental circumstances. This diversity, when compared among social units, allows us to detect cultural differences. Our bottom-up approach identifies some of the opportunities and challenges in studying social cognition through tool use in wild-ranging primates.

Keywords Social learning · Stone tools · Chimpanzee · Capuchin · Primate archaeology

M. Haslam (✉)
School of Archaeology, University of Oxford, Oxford, UK

T. Falótico
Institute of Psychology, University of São Paulo, São Paulo, Brazil

L. Luncz
Institute of Cognitive and Evolutionary Anthropology, University of Oxford, Oxford, UK

L. D. Di Paolo et al. (eds.), *Evolution of Primate Social Cognition*, Interdisciplinary Evolution Research 5, https://doi.org/10.1007/978-3-319-93776-2_13

1 Introduction

This chapter discusses the notion of culture, with a focus on tool use in wild primate groups. Specifically, we look at methods that help us detect and understand the development and transmission of socially learned tool-use patterns. We concentrate on durable objects, namely, stones, used by wild primates, as these objects form a salient and long-lasting marker of tool use both for the animals involved and for us as researchers (Fragaszy et al. 2013; Haslam 2012; Haslam et al. 2009, 2016a, b, 2017; Luncz et al. 2015). This focus naturally restricts our discussion primarily to three primate species that are currently known to use stone tools in the wild: bearded capuchins (*Sapajus libidinosus*) in Brazil, Western chimpanzees (*Pan troglodytes verus*) in west Africa and long-tailed macaques (*Macaca fascicularis aurea*) in SE Asia (Boesch and Boesch-Achermann 2000; Gumert et al. 2009; Haslam et al. 2016c; Luncz et al. 2017; Matsuzawa et al. 2011; Ottoni and Izar 2008; Visalberghi and Fragaszy 2013). In the remainder of this chapter, we concentrate on the first two of these taxa, as they are longer studied in the wild at multiple field sites.

To begin with, we note that we work with wild animals and with the patchy archaeological record created by both wild animals and human ancestors (Falótico and Ottoni 2013; Haslam 2014; Haslam et al. 2014, 2013; Luncz and Boesch 2015; Luncz et al. 2015, b; Proffitt et al. 2016). Under natural conditions, behavioural observations are often difficult, and fully controlled experiments are impossible. Because of this, our data are always incomplete and our conclusions typically inferential. Further, because of our emphasis on recording natural behaviour under natural conditions, we must always begin with empirical observations and build upwards from those to more abstract concepts such as culture.

In the following, therefore, we base our discussion on the concrete notion of observed, socially influenced and shared behaviours. The inclusion of a social component inevitably implies regularly interacting individuals. We identify socially influenced behaviour by detecting and analysing behavioural diversity that is not solely influenced by environmental circumstances. We discount entirely a second criterion often applied to studies of animal culture, namely, ruling out genetic causation for a given behaviour. The reason for ignoring genetics is practical: there are currently no studies that reliably identify a specific genetic pathway for any tool-use activity, in humans or other animals. However, to minimize the potential effect of genetically underlying predispositions for certain behaviours, we focus in our work on animals of the same subspecies.

Beyond directly observable phenomena, *culture* is for us simply a label that can be applied to a set of widespread, socially influenced behaviours exhibited by interacting members of a particular group or community of individuals (McGrew 2004; Morin 2016; Whitehead and Rendell 2015). One individual can be part of many different and potentially overlapping cultures; in humans, for example, the same person may be labelled an academic, a conservative, a fan of death metal and a football supporter. Each of these labels defines a set of behaviours that such an individual is likely to exhibit. This perspective also emphasizes the two main benefits of cultural labelling: (1) it is descriptive, it delineates for analytical purposes who are inside or outside of a particular cultural group as defined by a researcher, and (2) it is informative; a given individual part of a given cultural group informs on which behaviour they are likely to exhibit (and perhaps suggests that other as-yet unobserved but typically linked behaviours may also be followed).

The culture label may also inform on where and when an individual lived, although linking behaviour to geographic origin becomes more difficult the less is known about the behavioural repertoire of individuals. Where reliable observations have been collected during long-term studies of wild primates, we can use sets of co-occurring behaviours to distinguish location from behaviour alone. For example, if the behaviour of a given capuchin monkey includes use of stick probes, digging stones and stone throwing for sexual display, we know that it belongs to one of the extant *Sapajus libidinosus* groups in Serra da Capivara National Park, Northeast Brazil (Falótico and Ottoni 2013, 2014; Haslam and Falótico 2015; Ottoni and Izar 2008). If they include hand-sniffing and eye-poking, then they are almost certainly part of a *Cebus capucinus* group in Lomas Barbudal, Costa Rica (Perry 2011). The information in the cultural primatology dataset assembled by Whiten et al. (2001) can be used in a similar fashion for chimpanzees, and that of van Schaik et al. (2003) for orangutans.

In our empirical formulation, therefore, there is no such thing as a cultural process, only social transmission and learning processes. Documentation of behavioural variants, and assessment of social input to the learning and spread of those variants, comprises our fundamental research focus among wild primates. Everything else involves analytical distinctions made away from the field. In the remainder of this chapter, we will outline what we know of technological traditions within wild stone-tool-using capuchins and chimpanzees. We concentrate on tool-use behaviours and the factors that suggest that these behaviours are socially learned. We then give examples of how those behaviours may be assembled under cultural labels at different scales of analysis and for different purposes.

2 Chimpanzees

Observing wild chimpanzees in their natural habitat is a difficult and time-consuming challenge. Impenetrable tropical rain forests have many unpredictable obstacles and visibly can be very low. Nevertheless the rich behavioural repertoire of wild chimpanzees offers unique opportunities to answer lingering questions of culture and cognition of our highly sociable sister clade. Long-term field projects and detailed observations provide insightful information of complex social structures of our closest living relatives. Historically, identifying culture in wild chimpanzees has involved detecting diversity in behaviour where underlying ecological driving factors can be excluded (Whiten et al. 1999). The unique behavioural repertoire of each participating chimpanzee community was described as the culture of that community. However, the chimpanzee groups under study ranged over thousands of kilometres, and therefore underlying ecological or genetic factors for behavioural diversity could not be entirely excluded.

A long-term research project in the Taï National Park in Ivory Coast has been ongoing for more than 30 years (Boesch and Boesch-Achermann 2000). Over time and with patience, three neighbouring communities were gradually habituated to the presence of humans. All three communities live in one continuous stretch of tropical rainforest with bordering territories, and ecological circumstances are similar throughout the forest. This setting enabled the search for socially influenced behaviours, and our recent research has shown that many aspects of chimpanzees' behaviour are influenced by their social group (Luncz and Boesch 2014; Luncz et al. 2012, 2018).

One of the most striking differences between neighbours involves their nut-cracking behaviour. Chimpanzees in the Taï forest are known to crack five different nut species. Some of them are only eaten opportunistically and observations are rare. However, from November to March the group spends hours each day cracking the highly nutritious *Coula edulis* nut. Those nuts resemble walnuts, and at the beginning of each new nut season, when they first ripen in the tree, they are protected by a thick green layer of outer skin. During this time chimpanzees often transport a tool up into the trees where they pick some nuts and use a branch as an anvil to crack them. This requires good coordination of positioning the tool, nut and their own body, which often leaves juvenile learners frustrated by fallen tools or nuts. When nuts start to ripen and fall to the ground chimpanzees collect them underneath a tree, using their hands, feet and mouth to carry multiple nuts at a time to a suitable anvil. The anvils in most cases are *Coula* roots that spread out underneath the nut tree. The chimpanzees place one nut at a time on the root and hit them with their hammer carefully, so as to not damage the inner kernel of the nut but still crack the outer hard shell.

Coula nut-cracking is a very social activity, where usually the entire group is foraging together in at least audio distance to one another. Family groups forage alongside each other with infants close by their mothers' side begging for bits and pieces of the nuts. The tasty nut gets a lot of attention already by very young individuals. Only much later, at the age of 3 or 4 years, will young chimpanzees start their own first nut-cracking attempts. It then takes many more years before they are skilled and proficient nut-crackers. The foraging circumstances of *Coula* nuts offer grounds for excellent social learning opportunities in young chimpanzee offspring.

Taï forest chimpanzees use wooden and stone hammers to crack the relatively soft *Coula* nut. Selected tools can vary in their size, shape and material. *Coula* nut-cracking is an excellent candidate to search for behavioural diversity in wild chimpanzees, as experiments in captivity have suggested that nut-cracking is potentially influenced by social learning and prone to variation. The tool selection of each individual is the only information we have available when identifying behaviour that is potentially influenced by social learning. We first investigate if tool selection is random or if we are able to pick up repeated pattern of preferred tool choice made by different individuals. From the individual we can move on to investigating the similarity of specific subgroups in a community, for example, families or individuals with high association index (i.e. friends).

The fact that we can observe the behaviour and the tool choice of three neighbouring groups in the Taï forest provides the unique opportunity to take the comparison of tool selection pattern to a group level. After years of observations, comparisons of tool patterns show that neighbouring chimpanzee communities exhibit distinct tool selection which differ between groups (Luncz et al. 2012). The variety of tool patterns within the same community has been shown to be very low. When female chimpanzees reach sexual maturity, they leave their native community and immigrate into a different group. Our research has shown that at the onset of immigration, they quickly adopt the tool selection pattern of their new community which leads to long-lived, stable nut-cracking cultures in the Taï chimpanzee communities (Luncz and Boesch 2014, 2015; Luncz et al. 2018).

We therefore were able to observe cultural transmission in action. The underlying mechanisms that drive new females to conform to the nut-cracking norm of their new community remain to be identified. Natural observations do not allow for controlled tests, which leave multiple possibilities for the observed transmission. New females either adopt the tool selection or other prominent behavioural patterns to reinforce group belonging and minimize behavioural diversity between themselves and their new group. A more socially passive explanation could be that in order to stay with the group, they will have to adopt the foraging speed of their new group. Therefore they need to use the material the group is using at this given foraging location without regarding personal preference.

The transmission of cultural repertoire in chimpanzee groups can lead to very long-lasting and stable behavioural traits. The durability of stone tools has allowed an archaeological approach in recovering chimpanzee nut-cracking tools. The recovery of ancient chimpanzee nut-cracking sites revealed that this behaviour is at least about 4000 years old in the Taï National Park (Mercader et al. 2002, 2007).

Chimpanzees in a number of West African sites (although not all) are known to use pounding tools to crack open nuts. The next step in identifying different chimpanzee cultures therefore should include the comparisons of other nut-cracking communities. For example, extensive research has been done at the long-term field site of Bossou in Guinea (Carvalho et al. 2009), a site where chimpanzee nut-cracking culture does not involve the use of wooden clubs as hammers. The extended geographic distance between two nut-cracking populations that live further apart from each other can therefore be described as a greater cultural distance as well. This is one example of how we can investigate cultural behaviour starting from direct observation of individuals to populations across long distances.

3 Capuchins

Capuchin monkeys' behavioural studies in wild populations are more recent than those of chimpanzees. Several studies have been done regarding ecology and behaviour since the 1970s, most in rainforest populations, including the Amazon and Atlantic forests (Izawa 1978, 1979, 1980; Janson 1988). Although those studies comprised several hundred hours of observations, only a few cases of tool use were observed. This was unexpected, because capuchins are very manipulative towards objects (Fernandes 1991; Izawa and Mizuno 1977), and in captivity they have long been known to use objects as tools (Visalberghi and Trinca 1989; Westergaard and Fragaszy 1987).

Only when capuchin studies began in semiarid environments in Brazil (Cerrado and Caatinga) were regular stone tool-use behaviours reported in wild groups (Fragaszy et al. 2004; Mannu and Ottoni 2009; Moura and Lee 2004). To date, two long-term field sites in Piaui, Brazil, have been established for bearded capuchin monkey research, and although these cover similar environments and the same species (*Sapajus libidinosus*), the tool-use behaviour presented by those populations are different (Table 1). The Fazenda Boa Vista (FBV) site has been the home of the Etho*Cebus* project since 2003 (Visalberghi and Fragaszy 2013), and two groups have been closely studied in this site. The Serra da Capivara National Park (SCNP) capuchin population has been studied sporadically since 2000 (Moura and Lee 2004) and constantly since 2004 (Falótico and Ottoni 2013, 2014, 2016; Haslam and Falótico 2015; Haslam et al. 2016b; Mannu and Ottoni 2009; Proffitt et al. 2016). Four groups have been studied at SCNP to date, two of them long term.

Several other populations in Brazilian Cerrado and Caatinga environments have been surveyed in the last few years, and some of them present direct and indirect evidence of stone tool use by capuchin monkeys of two species, *S. libidinosus* and *S. xanthosternos* (Canale et al. 2009; De Moraes et al. 2014; Ferreira et al. 2010; Mendes et al. 2015). The reported pounding stone tools are primarily used to process

Table 1 Known tool-use behaviour in the two capuchin monkey long-term research sites, Fazenda Boa Vista and Serra da Capivara National Park, PI, Brazil

Behaviour	Fazenda Boa Vista	Serra da Capivara National Park
Pounding stone tools		
Palm nuts	Yes	No (ecology)
Cashew nuts (*Anacardium* sp.)	Yes (dry nuts)	Yes (fresh and dry nuts)
Grão-de-galo (*Cordia rufescens*)	No (ecology?)	Yes
Manihot seed	Yes	Yes
Stone-on-stone pulverization	No	Yes
Digging stone tools		
Underground storage organs	No	Yes
Trap-door spiders	No	Yes
Stone throwing (sexual display)	No	Yes (one group)
Probe stick tool	No	Yes (males)

palm nuts, when this resource is available, although in some areas the capuchins also process gastropods, other seeds and even cactuses. We can identify some similarities between those populations, such as the use of stones to process locally available, encased resources.

Some of the observed differences in tool use are clearly due to ecological variance, including the absence of palm nut processing in SCNP, at which the nut species opened by FBV capuchins are absent. An example of differences unrelated to ecology at the two sites involves cashew nut processing. FBV capuchins use two techniques to extract the cashew nut kernel: they rub the fresh nut on an abrasive surface to open a hole on the husk and use stone tools to crack the dry nuts (Sirianni and Visalberghi 2013; Visalberghi et al. 2016). On the other hand, capuchins at SCNP use stone tools to process all stages of cashew nut development and rarely use the rub technique (Falótico and Ottoni 2016; Falótico et al. 2016; Luncz et al. 2016a). These differences are not currently attributable to ecological variances but instead likely represent cultural differences. The same appears to be the explanation for ubiquitous digging with stones and probe tool use in SCNP (Falótico and Ottoni 2014; Falótico et al. 2017), behaviours never seen at FBV, as similar resources are present in both environments.

Tool-use variance can even be found within a well-studied population, such as stone throwing by females in one group in SCNP (Falótico and Ottoni 2013). This behaviour has not been registered in neighbouring groups in the area, which means that it could be a useful marker of cultural distinction. Females are the philopatric sex in capuchins, which may prevent the transmission of this variation to other groups at SCNP. Only long-term monitoring of these groups, spanning multiple generations, can determine whether or not the observed behaviours spread or disappear.

As we noted, studies on social learning in wild primates are challenging. The lack of control and the large number of variables make strong results difficult to obtain. Laboratory experiments and some field experiments can help in some instances to understand the learning involved, once field observations have found an ecologically relevant instance of behaviour. For example, from experiments we do know that socially biased learning is involved in the acquiring of stone tool-use behaviour by capuchins (Coelho et al. 2015; De Resende et al. 2008; Eshchar et al. 2016) and the same is likely true for other manipulative behaviours.

4 Conclusion

The results gained from years of patient field observations of wild chimpanzees and capuchins have demonstrated a high probability that cultural differences exist not just between widely separated groups of the one species but between neighbouring groups of the same species. This conclusion is one that cannot be reached merely by subjecting small groups of captive animals to a series of tests, no matter how rigorously controlled those tests may be. To some extent, this situation occurs because of factors inherent in keeping captive animals far from their natural environment, such as humans providing food and medical care (Haslam 2013). But more importantly, it occurs because wild animals are encountering real problems, and devising novel solutions, as part of their everyday activities.

The costs of discovering and maintaining technological solutions may be much higher in the wild than would be the case in the sheltered world of captivity, with the result that cultural differences may be slow to develop or may appear independently in different groups faced with similar problems. However, it is these differences, when maintained through social learning, which allow us to identify how tool -using primates succeed or fail over evolutionary timescales. The continued development of primate archaeology as a means to record and interpret the behaviour of past primates (Luncz et al. 2015; Haslam et al. 2016a, b, 2017) will allow us to assess whether the small variations in technique or tool choice that we have observed over the past few decades reflect modern innovations or entrenched cultural differences, and it is the only way to study tool use and technology across many generations. That approach is complementary to current experimental approaches, but it allows us to go beyond the recent past in understanding how primate cultures emerge and change.

Acknowledgements This work was supported by the European Research Council grant #283959 (Primate Archaeology) and São Paulo Research Foundation (FAPESP) grant #2014/18364-1.

References

Boesch C, Boesch-Achermann H (2000) The chimpanzees of the Tai forest: behavioural ecology and evolution. Oxford University Press, Oxford

Canale GR, Guidorizzi CE, Kierulff MCM, Gatto CAFR (2009) First record of tool use by wild populations of the yellow-breasted capuchin monkey (*Cebus xanthosternos*) and new records for the bearded capuchin (*Cebus libidinosus*). Am J Primatol 71:366–372

Carvalho S, Biro D, McGrew WC, Matsuzawa T (2009) Tool-composite reuse in wild chimpanzees (*Pan troglodytes*): archaeologically invisible steps in the technological evolution of early hominins? Anim Cogn 12:S103–S114

Coelho C, Falotico T, Izar P, Mannu M, Resende B, Siqueira J, Ottoni E (2015) Social learning strategies for nut-cracking by tufted capuchin monkeys (*Sapajus* spp.). Anim Cogn 18:911–919

De Moraes BLC, Da Silva Souto A, Schiel N (2014) Adaptability in stone tool use by wild capuchin monkeys (*Sapajus libidinosus*). Am J Primatol 76:967–977

De Resende BD, Ottoni EB, Fragaszy DM (2008) Ontogeny of manipulative behavior and nut-cracking in young tufted capuchin monkeys (*Cebus apella*): a perception–action perspective. Dev Sci 11:828–840

Eshchar Y, Izar P, Visalberghi E, Resende B, Fragaszy D (2016) When and where to practice: social influences on the development of nutcracking in bearded capuchins (*Sapajus libidinosus*). Anim Cogn 19:605–618

Falótico T, Ottoni E (2013) Stone throwing as a sexual display in wild female bearded capuchin monkeys, *Sapajus libidinosus*. PLoS One 8:e79535

Falótico T, Ottoni E (2014) Sexual bias in probe tool manufacture and use by wild bearded capuchin monkeys. Behav Process 108:117–122

Falótico T, Ottoni E (2016) The manifold use of pounding stone tools by wild capuchin monkeys of Serra da Capivara National Park, Brazil. Behaviour 153:421–442

Falótico T, Siqueira JO, Ottoni EB (2017) Digging up food: excavation stone tool use by wild capuchin monkeys. Sci Rep 7:6278

Falótico T, Luncz L, Svensson MS, Haslam M (2016) Cashew nut positioning during stone tool use by wild bearded capuchin monkeys (*Sapajus libidinosus*). Folia Primatol 87:392–397
Fernandes ME (1991) Tool use and predation of oysters (*Crassostrea rhizophorae*) by the tufted capuchin, *Cebus apella appella*, in brackish water mangrove swamp. Primates 32:529–531
Ferreira R, Emidio R, Jerusalinsky L (2010) Three stones for three seeds: natural occurrence of selective tool use by capuchins (*Cebus libidinosus*) based on an analysis of the weight of stones found at nutting sites. Am J Primatol 72:270–275
Fragaszy D, Izar P, Visalberghi E, Ottoni E, de Oliveira M (2004) Wild capuchin monkeys (*Cebus libidinosus*) use anvils and stone pounding tools. Am J Primatol 64:359–366
Fragaszy D, Biro D, Eshchar Y, Humle T, Izar P, Resende B, Visalberghi E (2013) The fourth dimension of tool use: temporally enduring artefacts aid primates learning to use tools. Philos Trans R Soc Lond B Biol Sci 368:20120410
Gumert M, Kluck M, Malaivijitnond S (2009) The physical characteristics and usage patterns of stone axe and pounding hammers used by long-tailed macaques in the Andaman Sea region of Thailand. Am J Primatol 71:594–608
Haslam M (2012) Towards a prehistory of primates. Antiquity 86:299–315
Haslam M (2013) ‘Captivity bias’ in animal tool use and its implications for the evolution of hominin technology. Philos Trans R Soc Lond B Biol Sci 368:20120421
Haslam M (2014) Primate archaeobotany: the potential for revealing nonhuman primate plant-use in the African archaeological record. In: Nixon S, Murray MA, Fuller D (eds) The archaeology of African plant use. Left Coast Press, Walnut Creek, CA, pp 25–35
Haslam M, Falótico T (2015) Nasal probe and toothpick tool use by a wild female bearded capuchin (*Sapajus libidinosus*). Primates 56:211–214
Haslam M, Hernandez-Aguilar A, Ling V, Carvalho S, de la Torre I, DeStefano A, Du A, Hardy BL, Harris J, Marchant L, Matsuzawa T, McGrew W, Mercader J, Mora R, Petraglia M, Roche H, Visalberghi E, Warren R (2009) Primate archaeology. Nature 460:339–344
Haslam M, Gumert M, Biro D, Carvalho S, Malaivijitnond S (2013) Use-wear patterns on wild macaque stone tools reveal their behavioural history. PLoS One 8:e72872
Haslam M, Cardoso RM, Visalberghi E, Fragaszy D (2014) Stone anvil damage by wild bearded capuchins (*Sapajus libidinosus*) during pounding tool use: a field experiment. PLoS One 9: e111273
Haslam M, Luncz L, Pascual-Garrido A, Falótico T, Malaivijitnond S, Gumert M (2016a) Archaeological excavation of wild macaque stone tools. J Hum Evol 96:134–138
Haslam M, Luncz LV, Staff RA, Bradshaw F, Ottoni EB, Falótico T (2016b) Pre-Columbian monkey tools. Curr Biol 26:R521–R522
Haslam M, Pascual-Garrido A, Malaivijitnond S, Gumert M (2016c) Stone tool transport by wild Burmese long-tailed macaques (*Macaca fascicularis aurea*). J Archaeol Sci Rep 7:408–413
Haslam M, Hernandez-Aguilar RA, Proffitt T, Arroyo A, Falótico T, Fragaszy D, Gumert M, Harris JWK, Huffman MA, Kalan AK, Malaivijitnond S, Matsuzawa T, McGrew W, Ottoni EB, Pascual-Garrido A, Piel A, Pruetz J, Schuppli C, Stewart F, Tan A, Visalberghi E, Luncz LV (2017) Primate archaeology evolves. Nat Ecol Evol 1:1431–1437
Izawa K (1978) Frog-eating behavior of wild black-capped capuchin (*Cebus apella*). Primates 19:633–642
Izawa K (1979) Foods and feeding behavior of wild black-capped capuchin (*Cebus apella*). Primates 20:57–76
Izawa K (1980) Social behavior of the wild black-capped capuchin (*Cebus apella*). Primates 21:443–467
Izawa K, Mizuno A (1977) Palm-fruit cracking behavior of wild black-capped capuchin (*Cebus apella*). Primates 18:773–792
Janson CH (1988) Food competition in brown capuchin monkeys (*Cebus apella*): quantitative effects of group size and tree productivity. Behaviour 105:53–76

Luncz L, Boesch C (2014) Tradition over trend: neighboring chimpanzee communities maintain differences in cultural behavior despite frequent immigration of adult females. Am J Primatol 76:649–657

Luncz L, Boesch C (2015) The extent of cultural variation between adjacent chimpanzee (*Pan troglodytes verus*) communities; a microecological approach. Am J Phys Anthropol 156:67–75

Luncz L, Mundry R, Boesch C (2012) Evidence for cultural differences between neighboring chimpanzee communities. Curr Biol 22:922–926

Luncz L, Wittig R, Boesch C (2015) Primate archaeology reveals cultural transmission in wild chimpanzees (*Pan troglodytes verus*). Philos Trans R Soc Lond B Biol Sci 370:20140348

Luncz L, Falótico T, Pascual-Garrido A, Corat C, Mosley H, Haslam M (2016a) Wild capuchin monkeys adjust stone tools according to changing nut properties. Sci Rep 6:33089

Luncz L, Proffitt T, Kulik L, Haslam M, Wittig RM (2016b) Distance-decay effect in stone tool transport by wild chimpanzees. Proc R Soc B Biol Sci 283:20161607

Luncz L, Tan A, Haslam M, Kulik L, Proffitt T, Malaivijitnond S, Gumert M (2017) Resource depletion through primate stone technology. Elife 6:1–16

Luncz L, Sirianni G, Mundry R, Boesch C (2018) Costly culture: differences in nut-cracking efficiency between wild chimpanzee groups. Anim Behav 137:63–73

Mannu M, Ottoni E (2009) The enhanced tool-kit of two groups of wild bearded capuchin monkeys in the caatinga: tool making, associative use, and secondary tools. Am J Primatol 71:242–251

Matsuzawa T, Humle T, Sugiyama Y (2011) The chimpanzees of Bossou and Nimba. Springer, Dordrecht

McGrew WC (2004) The cultured chimpanzee: reflections on cultural primatology. Cambridge University Press, Cambridge

Mendes FDC, Cardoso RM, Ottoni E, Izar P, Villar DNA, Marquezan RF (2015) Diversity of nutcracking tool sites used by *Sapajus libidinosus* in Brazilian *cerrado*. Am J Primatol 77:535–546

Mercader J, Panger M, Boesch C (2002) Excavation of a chimpanzee stone tool site in the African rainforest. Science 296:1452–1455

Mercader J, Barton H, Gillespie J, Harris J, Kuhn S, Tyler RT, Boesch C (2007) 4,300-year-old chimpanzee sites and the origins of percussive stone technology. Proc Natl Acad Sci U S A 104:3043–3048

Morin O (2016) How traditions live and die. Oxford University Press, Oxford

Moura A, Lee P (2004) Capuchin stone tool use in caatinga dry forest. Science 306:1909

Ottoni E, Izar P (2008) Capuchin monkey tool use: overview and implications. Evol Anthropol 17:171–178

Perry S (2011) Social traditions and social learning in capuchin monkeys (*Cebus*). Philos Trans R Soc Lond B Biol Sci 366:988–996

Proffitt T, Luncz LV, Falótico T, Ottoni EB, de la Torre I, Haslam M (2016) Wild monkeys flake stone tools. Nature 539:85–88

Sirianni G, Visalberghi E (2013) Wild bearded capuchins process cashew nuts without contacting caustic compounds. Am J Primatol 75:387–393

van Schaik C, Ancrenaz M, Borgen G, Galdikas B, Knott C, Singleton I, Suzuki A, Utami SS, Merrill M (2003) Orangutan cultures and the evolution of material culture. Science 299:102–105

Visalberghi E, Fragaszy D (2013) The Etho-*Cebus* Project: stone-tool use by wild capuchin monkeys. In: Sanz C, Call J, Boesch C (eds) Tool use in animals: cognition and ecology. Cambridge University Press, Cambridge, pp 203–223

Visalberghi E, Trinca L (1989) Tool use in capuchin monkeys: distinguishing between performing and understanding. Primates 30:511–521

Visalberghi E, Albani A, Ventricelli M, Izar P, Schino G, Fragazsy D (2016) Factors affecting cashew processing by wild bearded capuchin monkeys (*Sapajus libidinosus*, Kerr 1792). Am J Primatol 7:799–815

Westergaard G, Fragaszy D (1987) The manufacture and use of tools by capuchin monkeys (*Cebus apella*). J Comp Psychol 101:159–168
Whitehead H, Rendell L (2015) The cultural lives of whales and dolphins. University of Chicago Press, Chicago
Whiten A, Goodall J, McGrew WC, Nishida T, Reynolds V, Sugiyama Y, Tutin C, Wrangham R, Boesch C (1999) Cultures in chimpanzees. Nature 399:682–685
Whiten A, Goodall J, McGrew WC, Nishida T, Reynolds V, Sugiyama Y, Tutin C, Wrangham R, Boesch C (2001) Charting cultural variation in chimpanzees. Behaviour 138:1481–1516

Culture and Selective Social Learning in Wild and Captive Primates

Stuart K. Watson, Jennifer Botting, Andrew Whiten, and Erica van de Waal

Abstract Once thought to be a unique human trait, the presence of culture in non-human primates has been confirmed and studied by researchers for several decades. What has been discovered is evidence for between-group traditions in a wide range of primate taxa, including all of the great apes, macaques, capuchins and spider monkeys, as well as many non-primate species. The capacity to learn from others is a powerful means by which animals can acquire adaptive ways of interacting with their environment and each other without engaging in time-consuming and potentially risky trial-and-error learning. However, much remains to be understood about the exact mechanisms and processes that underpin social learning and how these lead to the cultures identified in wild populations of primates, including humans. In the current chapter, we review what is known about

The authors "Stuart K. Watson" and "Jennifer Botting" contributed equally. Order of authorship was determined by coin-toss.

S. K. Watson
Centre for Social Learning and Cognitive Evolution, and Scottish Primate Research Group, School of Psychology and Neuroscience, University of St Andrews, St Andrews, UK

J. Botting
Centre for Social Learning and Cognitive Evolution, and Scottish Primate Research Group, School of Psychology and Neuroscience, University of St Andrews, St Andrews, UK

Inkawu Vervet Project, Mawana Game Reserve, Swart Mfolozi, KwaZulu Natal, South Africa

Smithsonian National Zoological Park, Washington, DC, USA

A. Whiten
Centre for Social Learning and Cognitive Evolution, and Scottish Primate Research Group, School of Psychology and Neuroscience, University of St Andrews, St Andrews, UK

Inkawu Vervet Project, Mawana Game Reserve, Swart Mfolozi, KwaZulu Natal, South Africa
e-mail: aw2@st-andrews.ac.uk

E. van de Waal (✉)
Inkawu Vervet Project, Mawana Game Reserve, Swart Mfolozi, KwaZulu Natal, South Africa

Anthropological Institute and Museum, University of Zurich, Zurich, Switzerland
e-mail: erica.vandewaal@unil.ch

L. D. Di Paolo et al. (eds.), *Evolution of Primate Social Cognition*, Interdisciplinary Evolution Research 5, https://doi.org/10.1007/978-3-319-93776-2_14

non-human primate culture with a particular emphasis on the emerging field of social learning biases. Theoreticians and field researchers alike have suggested that animals may exhibit biases in whom they obtain information from, either as by-products of social dynamics or as adaptive strategies that allow animals to selectively acquire the most useful information. Here, we review the theoretical arguments and current empirical evidence for proposed biases in social learning, including majority-based biases and model-based biases. We draw from field observations and experiments in both captive and wild populations to examine how information may be transferred between individuals and how this may affect the emergence of cultural behaviours across primate species.

Keywords Culture · Primates · Social learning biases · Social transmission mechanisms · Conformity · Model-based biases

1 Introduction

Culture pervades every aspect of human life, from the way we communicate and the values we hold to the way we think and shape our environment. The extent and complexity of human culture has created such a palpable gulf between us and other members of the animal kingdom that the capacity for culture has historically been argued to be unique to humans (e.g. Galef 1992). However, this depends on how culture is defined, operationalised and tested for. Here we adopt the definition offered by Reader and Laland (2003) that cultures are 'group-typical behaviour patterns shared by members of a community that rely on socially learned and transmitted information' (p. 151). This seems sufficiently broad to account for all instances which one might consider 'cultural' in humans, but it means that theoretically, culture might emerge in any species with sufficient propensity for social learning. We humans are pre-eminent (Dean et al. 2014) in our capacity for 'cumulative culture'—the ability to modify learned behaviours to become more complex and/or efficient, which can be transmitted and further improved by others—but an increasingly large body of research shows that our fellow animals are also capable of a functionally significant degree of cultural inheritance and diversity.

The adaptive benefits of social learning appear clear; the alternative, individual innovation, is potentially costly as it may require a significant time or energy investment, produce only minor or non-existent rewards or be physically dangerous. Instead, the capacity to learn from others allows individuals to reap the benefits of others' useful innovations whilst minimising the costs. Furthermore, this may benefit not only the learner but also any offspring or other kin to whom the behaviour subsequently spreads. In this sense, culture acts as nature's 'second inheritance system' (Whiten 2005). However, social learning may not always be adaptive. An indiscriminate social learner runs the risk of copying costly or other suboptimal behaviours (Laland and Williams 1998; Kendal et al. 2005). To avoid these, the evolution of social learning 'biases' (or 'strategies' in some literature) has been predicted, to guide when to use social information and whom best to acquire it from (Giraldeau et al. 2002; Laland 2004).

This chapter reviews what is known about social learning in non-human primates, with particular emphasis on the cognitive biases that underpin it, and the important light it sheds on our understanding of the emergence, propagation and maintenance of culture. Studies of social learning in non-primate species will also be noted when pertinent, in order to situate the primate literature within a broader biological context.

2 Observations of Culture in the Wild

Early whispers of non-human primate culture were heard when, in 1953, a young Japanese macaque, Imo, carried pieces of sand-covered sweet potato to a stream and washed off the sand before eating them (Kawai 1965). In the months and years that followed, other macaques began using this technique, and it eventually became a common behaviour in Imo's group, leading the researchers studying the macaques to label this behaviour as 'pre-culture'. Whilst later scrutiny of the spread of sweet potato washing behaviour suggested it did not provide strong evidence for cultural transmission (e.g. due to factors such as human provisioning; Galef 1992), the case of the Japanese macaques established the study of social transmission and culture in wild primates, which led to a plethora of exciting discoveries in the animal kingdom.

A further influential finding came when, following observations of group differences in chimpanzee behaviours across different sites (Goodall 1986; Boesch et al. 1994), Whiten et al. (1999, 2001) systematically collated data from seven long-term chimpanzee field sites across Africa and found evidence of multiple variations in behaviours between communities, inferred to be cultural. The researchers listed 39 behavioural traditions which were common (either customary or habitual) in some groups, yet absent in others, without obvious ecological explanation, ranging from handclasp grooming to different termite fishing and nut-cracking behaviours (see Fig. 1). The discovery of this large number of putative cultural variations in our closest relatives challenged the assumption of complex culture being uniquely human.

On the heels of these findings, researchers collated data from other great ape species. Van Schaik et al. (2003) conducted a similar analysis of behavioural traditions in orangutans across six different field sites and identified 24 cultural variants that were habitual or customary at some sites and absent at others. The authors concluded that orangutans also possess multiple-tradition cultures. A more recent analysis supports these earlier conclusions, finding that ecological and genetic differences accounted for only a small proportion of the variation seen between groups (Krützen et al. 2011). Emerging evidence has also been found for a number of traditions in wild gorillas (Robbins et al. 2016, but see Neadle et al. (2017) for a discussion on the ontogeny of food cleaning behaviours) and bonobos (Hohmann

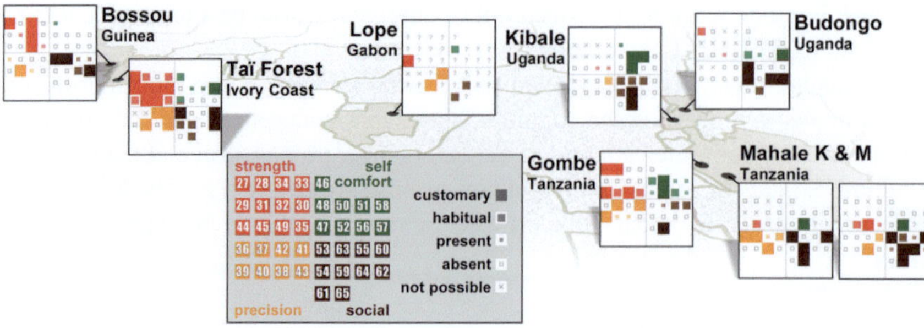

Fig. 1 The putative cultures of wild chimpanzees (after Whiten 2005). 'Customary' acts are those typical in a community; 'habitual' are less frequent yet consistent with social learning. Each community displays its own profile of such local behavioural variants, providing evidence of a unique culture for each locality. Numbers identify behaviour patterns in the catalogue attached to Whiten et al. (1999). For a more recent and detailed version focused on nut-cracking variations, see Carvalho and McGrew (2012)

and Fruth 2003), although we have less information on bonobos than for chimpanzees or orangutans.

Whilst the 'exclusion' method of inferring social learning where ecological or genetic explanations appear untenable can never be conclusively watertight (Laland and Janik 2006; Langergraber et al. 2011: Robbins et al. 2016, explicitly refer to the traditions they identify as 'putative'), additional lines of evidence have more recently supported the role of social learning in the maintenance of the above behaviours. First, a series of diffusion experiments in which different foraging techniques are 'seeded' in different groups has shown that these will spread in ways that confirm a capacity to transmit and sustain multiple traditions (Whiten 2011; Whiten et al. 2016). Second, recent advances in statistical techniques have allowed researchers to assess the role of social learning in the spontaneous spread of a novel behaviour in the wild. For example, Hobaiter et al. (2014) used a dynamic form of network-based diffusion analysis (NBDA, Franz and Nunn 2009) to systematically chart the spread of a novel leaf-sponging behaviour in a group of wild chimpanzees via social transmission, providing the first direct evidence that at least some of the observed behavioural variation in wild chimpanzees is likely due to cultural learning. This conclusion is further supported by Lonsdorf (2006), who found a correlation between the amount of time infant chimpanzees spend with their mother and the rate at which they become proficient at termite fishing, as well as Luncz and Boesch (2014, see below) comparing neighbouring communities where neither genetic nor ecological appear able to explain behavioural variations. It is also worth noting that the exclusion method of identifying animal cultures has been criticised for being too conservative, as it precludes examination of ecologically driven cultural differences (Koops et al. 2014; Sanz and Morgan 2013).

Whilst the great apes, in particular chimpanzees, display the most diverse repertoire of cultural behaviours documented thus far, the first candidates for cultural behaviours came from monkeys. Evidence has since accrued that several monkey species exhibit their own forms of culture. Notably, Perry et al. (2003) described a number of 'social customs' found in some, but not all, groups of capuchins studied in Costa Rica. These included a number of social 'games' and other, seemingly bizarre, conventions such as poking fingers in each other's nostrils and eyes. Whilst the functions of these conventions are difficult to pin down, the authors suggest that they may serve to enhance social bonds between participants and perhaps highlight them to others. Whatever the function, the distributions of conventions across groups are strongly suggestive of cultural transmission. Additionally, Santorelli et al. (2011) identified a number of behavioural traditions, many in the social domain, in wild spider monkeys, and peculiar stone-handling behaviours seen in numerous provisioned groups of Japanese macaques also appear to display the pattern of multiple, if narrowly constrained, cultural traditions (Leca et al. 2007).

Whilst it might be tempting for us to think of primates as distinctive in their capacity for cultural transmission, researchers have uncovered convincing evidence for culture in other animal taxa, notably in cetaceans and in birds. Whitehead and Rendell (2014) discuss evidence for a range of putative traditions in cetaceans, most notably in the domain of song transmission in the baleen whales (also well documented in birds; Slater 1986), but also including a number of foraging traditions (Krützen et al. 2005; Allen et al. 2013). Implementation of the aforementioned technique, NBDA, has provided some of the strongest evidence for social learning in the wild in any species, documenting the gradual spread of a particular foraging method, lob-tail feeding, in a population of over 600 humpback whales (Allen et al. 2013).

In addition to vocal culture, there may also be tentative evidence for tool-making traditions in birds. Hunt and Gray (2003) described evidence for sophisticated tool-making in wild New Caledonian crows and, finding that the distribution could not be linked to ecological correlates, suggested a role for social learning. They argued that the distribution of tool types across New Caledonia indicates cumulative technological culture. However, the extent to which the tool manufacture is necessarily socially learned remains in doubt, as hand-reared crows in captivity also display tool-making skills (Kenward et al. 2005, but see also Holzhaider et al. 2010), and experiments have failed to support social learning of alternative techniques (Logan et al. 2016). Recently, researchers also found that the experimental removal and replacement of individuals from a group of homing pigeons improved the efficiency of homing routes over successive generations through social learning and refinement, satisfying the main criteria for cumulative culture (Sasaki and Biro 2017). This finding is important not only for identifying a putative case of cumulative culture in a non-human species, a capacity many argue to be uniquely human, but also for emphasising that culture extends beyond foraging behaviour which is the primary focus of a majority of social learning studies.

3 Mechanisms of Information Transmission

The study of the social learning processes that underlie culture can broadly be split into two main categories: mechanisms and biases. Mechanisms, or psychological processes, refer to the *how* of information transmission. Whereas biases (or strategies) refer to *when* and *whom* to copy (Laland 2004; Hoppitt and Laland 2013, but see Heyes (2016) for discussion of limitations in the identification of underlying cognitive mechanisms). Whilst this chapter focuses primarily upon the biases of social learning, here we provide a brief introduction to transmission mechanisms. Some researchers argued early on that only the (supposedly) more cognitively complex processes of imitation and teaching would allow for transmission of sufficient fidelity to create stable, between-group traditions (Galef 1992, but see Claidiere and Sperber 2010). Thus, it was the study of mechanisms and particularly the exploration of the process of imitation which formed the primary research focus in experiments on social learning that until recently were largely restricted to captive primates and other animals, where rigorous experimental and contrasting control conditions can be engineered.

Social learning mechanisms range from the simplest processes of local and stimulus enhancement (increased attention respectively towards locations and objects one observes others acting on) to more complex mechanisms such as teaching and copying (Heyes 1994). Experimenters have distinguished between two principal types of copying mechanisms: imitation, which refers to copying the *actions* (often conceived of a *bodily* actions) of another individual (Whiten et al. 2004), and emulation, which refers to learning focused only on desirable environmental *results* of other's actions (Tomasello et al. 1987). Whilst initial tests with chimpanzees suggested they are capable of only emulation (Tomasello et al. 1987), later studies provided a diversity of evidence for imitation, including recognisable successes in 'do-as-I-do' games and other tests that require matching of bodily actions [Custance et al. (1995), Buttelmann et al. (2007), although it should be noted that some of these studies were conducted with hand-reared chimpanzees; see also Fuhrmann et al. (2014), but see Tomasello et al. (1997), Tennie et al. (2012) for evidence of marked limitations in copying arbitrary or novel gestures in untrained chimpanzees]. Other studies with captive chimpanzees (and children) found evidence for flexible use of imitation and emulation in chimpanzees (Hopper et al. 2008), with replication of a whole sequence of actions being preferred when a complex task was relatively opaque ("program-level imitation", Byrne and Russon 1998) and a more emulative response made when it was sufficiently transparent that some actions could be seen to be redundant and were not copied (Horner and Whiten 2005). Other studies revealed some contexts where emulation does not enable chimpanzees to solve a complex task, whereas seeing another chimpanzee complete it allows success by copying (Hopper et al. 2007, 2015) and other contexts where chimpanzees are flexible enough to succeed by emulation when direct imitation is made impossible (Tennie et al. 2010).

Limited evidence for bodily imitation in monkeys has also emerged, from species including marmosets (Voelkl and Huber 2000), vervet monkeys (van de Waal and Whiten 2012) and capuchins (Fragaszy et al. 2011, although see Dean et al. 2012). There is thus limited evidence in monkeys and apes for one of the two features that Galef emphasised as crucial to human culture: imitation. However, perhaps more important is that a range of cultural diffusion experiments have demonstrated that species of monkeys and apes can transmit and sustain traditions, whether or not these are driven by imitative or emulative copying processes (reviewed in Whiten et al. 2016, but see Bandini and Tennie 2017). By contrast, there is little evidence for Galef's other factor, teaching, in non-human primates (although see Musgrave et al. 2016 for a recent example of tool transfer interpreted as teaching in chimpanzees). There is, however, evidence for teaching (defined in functional rather than intentional terms) in non-primate animals such as meerkats (Thornton and McAuliffe 2006), ants (Franks and Richardson 2006) and pied babblers (Raihani and Ridley 2008). It should be noted, however, that whilst we include teaching here to illustrate its importance in the study of cultural transmission, it is not a mechanism in the learner, but rather in the teacher, coupled with complementary social learning processes in the learner. For example, the teaching process of *demonstration* couples well with a capacity for *imitation* in the learner (Hoppitt and Laland 2008). For discussions on the evidence for teaching and its significance in non-human animals, see Thornton and Raihani (2010) and Hoppitt and Laland (2008).

4 Social Learning Biases

More recently, research has begun to tackle the potential biases or 'strategies' which influence when and from whom animals learn socially. Given that learning from others indiscriminately may result in the transmission of maladaptive behaviours (Laland and Williams 1998; Pongracz et al. 2003), Laland (2004) suggested that individuals should be selective in when and from whom they engage in social learning, highlighting a number of potential social learning strategies (or biases as we refer to them here). Indeed Coussi-Korbel and Fragaszy (1995) had earlier suggested that the social dynamics of groups would likely lead to biases in social learning. An adaptive bias may allow individuals to select the most productive behaviour between multiple options and overwrite pre-existing behaviours when innovations or a changing environment renders them inefficient. Whilst a number of such biases have been suggested by researchers, empirical investigation of many remains lacking, and Heyes (2016) points out that the underlying mechanisms remain largely unspecified. Here we limit our discussion to the social learning biases that have received the most research attention to date (Table 1), ranging from frequency-based copying (e.g. copying common behaviours) to model-based biases (e.g. copying high-ranked individuals). One of the most studied—and contentious—biases addressed in recent years, in both humans and animals, is that of conformity.

Table 1 Definitions and examples of selected social learning biases

Category	Term	Definition	Selected examples
Frequency-based biases	Majority bias	A tendency to copy the behaviour of the majority when learning a task	Chimpanzees, Haun et al. (2012) Cf. van Leeuwen and Haun (2013)
	Conformity	A tendency to forgo one's own behavioural preference in favour of that used by the majority	Fish, Pike and Laland (2010) Vervet monkeys, van de Waal et al. (2013) Chimpanzees, Luncz and Boesch (2014) Birds, Aplin et al. (2015a, b)
Model-based biases	Rank bias	A tendency to copy individuals of high social rank	Chimpanzees, Horner et al. (2010); Kendal et al. (2015)
	Sex bias	A tendency to copy individuals of one sex over the other	Vervet monkeys, van de Waal et al. (2010)
	Kin bias	A tendency to copy one's own kin	Chimpanzees, Matsuzawa et al. (2001) Vervet monkeys, van de Waal et al. (2012, 2014)

For a more exhaustive review, particularly of 'when' biases, see Laland (2004) and Rendell et al. (2011)

4.1 Conformity and Majority Bias

In the 1950s, Solomon Asch (Asch 1951) showed that when people were faced with a unanimous majority giving incorrect answers on a line judgement task, roughly one third of individuals conformed to the group decision even though it was visibly wrong. These findings were replicated across cultures (Bond and Smith 1996) and more recently in children (Haun and Tomasello 2011), suggesting a powerful predisposition for humans to conform to the judgement of the majority.

Circumstantial evidence for conformist behaviour in primates arose in open diffusion studies such as that by Whiten et al. (2005) which utilised a puzzle box (the 'panpipes') that could be opened using either of two tool-based techniques ('poke' or 'lift'). A high-ranking female from each of the two groups was trained in one of these methods and demonstrated it in front of the rest of their group, who later had opportunities to interact with the panpipes. After 2 months, it was found that although some individuals did open with the alternative method, there was a significant tendency for individuals to converge on the option most common in their group. Whiten et al. (2005) suggested this to be indicative of a conformity effect. A similar effect has since been recorded in capuchin monkeys (Dindo et al. 2008, 2009). Critics have pointed out that chimpanzees are often reluctant to give up a first-learned behaviour (Hopper et al. 2011; Marshall-Pescini and Whiten 2008; Hrubesch et al. 2009; Harrison and Whiten 2018), so it may be that individuals were simply returning to their original method after a period of exploration (van Leeuwen and Haun 2013). However, there are also several documented examples of flexible, non-conservative problem-solving by chimpanzees (Hopper et al. 2015; Yamamoto et al. 2013)

Subsequent evidence for conformity in chimpanzees has been mixed. No study has yet found an example of conformity in which an individual gives up a first-learned behaviour in favour of that used by the majority. However, Luncz and Boesch (2014), having earlier identified consistent differences between neighbouring groups in aspects of nut-cracking tool selection, found that females who transfer between these groups are as likely as males to match their local group preferences, suggesting they must have conformed to them since their arrival. A case study of one wild female chimpanzee that had recently migrated to a new group showed that, over time, her behavioural repertoire became progressively more similar to that of the group. However, without systematic testing we cannot be sure this was driven by majority influence, rather than other biases. Working with captive chimpanzees on a token exchange task, van Leeuwen and Haun (2013) found that individuals were motivated to switch methods by greater rewards, but not by the behaviour of the majority. This suggests other factors, such as maximising payoffs, may be more critical in motivating apparently conformist behaviour than majority influence. Moreover, Vale et al. (2017) found that captive chimpanzees who were trained to prefer one colour of food and then moved to a group with a strong preference for the alternative colour did not conform to the foraging preferences of their new 'host' group, although they did feed from this food more frequently than before. Watson et al. (2018) found that lone minority individuals trained on a method of opening a puzzle box, whilst the rest of their group were trained on an alternative method, rapidly converged on the behavioural preference of the majority. However, this occurred after observing just one or two groupmates, meaning that they did not have first-hand knowledge of the majority preference. In contrast, dyads of chimpanzees trained on alternative methods never switched to using their partner's preferred method. The authors argue that it is possible that, when in a group context, chimpanzees may make inferences about group-level behaviour preferences based on a limited sample and act accordingly, in a potentially conformist manner.

Arguably, the strongest evidence for conformist transmission in non-human animals has been found in non-primate species. In Pike and Laland (2010), nine-spined stickleback fish independently learned that one of the two feeders was richer in food than the other. The positions of the feeders were then reversed, and the fish could observe others feeding at the previously less rewarding location. When these demonstrator fish were removed and the observer was once again allowed to choose a feeding location, they preferentially used the one favoured by the majority. This effect increased disproportionately according to the relative size of the majority, thereby fulfilling the most stringent definitions of conformist bias. In another convincing example of conformity in a non-primate species, Aplin and colleagues (Aplin et al. 2015a, b) found that not only did experimentally induced innovations (alternative methods of operating a foraging box) spread throughout groups of great tits in a manner consistent with majority-biased transmission but they expressed an exaggerated tendency to do so. In addition, when individuals migrated between groups, if the local method was different from their own, then a majority of birds adopted the behaviour most common in their new community [but see van Leeuwen et al. (2015) and Aplin et al. (2015b) for further debate]. This echoes the results from an experiment that seeded alternative options in wild vervet monkeys (van de Waal et al. 2013). Four groups of wild vervet monkeys were repeatedly provided with two boxes of coloured corn, one pink and one blue. For each group, one of these colours was initially made to taste bitter and unpalatable. Then, after monkeys had learned to avoid one colour, no more bitter material was added to either colour of corn. When males later migrated between groups with different preferences, and once they were not outranked by residents at the boxes, nine of ten switched to eat the colour that was locally preferred, expressing an apparently striking degree of conformity. In a follow-up study, the preferences of low-ranking females who had permanently splintered from their natal group have been monitored. The low rank of these females meant that in their original group, they ate more of the previously bad-tasting food than others yet, after the split, they ate exclusively the colour preferred in the group from which they split, even though both colours were available and neither tasted bad, emphasising the potency and durability of socially learned behaviours (van de Waal et al. 2017). However, whether the individuals were influenced by the behaviour of the majority of individuals or by certain classes of individual within that group (as discussed below) remains unclear. The lack of clear experimental evidence for conformity in chimpanzees may suggest that other biases are more central to the transmission of information in this species or that it is only expressed by certain individuals. It is very possible that primates may display conformist transmission in only certain contexts, such as after a recent immigration to a new group. Further experimental research into these questions is required in order to better understand these distinctions (Fig. 2).

To date, only one primate study has directly and experimentally tested for majority-biased transmission. Haun et al. (2012) carried out a study in which human children, chimpanzees and orangutans were exposed to conspecific models depositing tokens in a receptacle in order to receive a reward. Three receptacles were on display, one of which was used by a single individual three times, another by three individuals a single time and another by none. Observers had no previous exposure to the task. When given a token to deposit for themselves, both children and chimpanzees (but not orangutans) showed a tendency to choose the receptacle used by the majority of demonstrators. Related findings have been reported in such distantly related species as rats (Chou and Richerson 1992), pigeons (Lefebvre and Giraldeau 1994), dogs (Kundey et al. 2012) and fruit flies (Battesti et al. 2012).

Most studies of majority influence have focussed on foraging behaviour, as this is relatively straightforward to model in experimental designs. A disadvantage of this approach is that cognition is being examined within a single behavioural context, when our understanding could be enriched by taking a broader approach (Watson and Caldwell 2009). For example, there is a wealth of evidence for the social learning of vocalisations in a wide range of species (e.g. bats, birds, bears, cetaceans and primates), which Janik and Slater (1997) argue serves (amongst other functions) the purpose of identifying oneself with the group, just as has been argued to motivate

Fig. 2 An illustration from the follow-up experiments of van de Waal et al. (2013) showing a group of vervets crowding around their preferred colour of corn (pink, left) and avoiding the other (blue, right)

human conformist behaviour. Collecting data before and after wild group migrations to test for vocal conformity is inherently difficult, but this can be achieved in captivity. Watson et al. (2015a) examined the referential 'food grunts' of chimpanzees moved from the Netherlands and integrated with a resident group at Edinburgh Zoo in Scotland. Over the period of time that it took for the two groups to socially integrate, so too did the referential food calls of the 'Dutch' chimps converge on that of the 'local' group, despite the arousal levels associated with the referent foods remaining constant. This draws an interesting parallel with examples of conformity in migrating animals (Luncz and Boesch 2014; van de Waal et al. 2013), although in this case the immigrant subgroup was equal in size to that which they converged on. This is suggestive of conformist behaviour in the vocal domain, although it is worth noting that there is debate over the extent of and likely motivations for this convergence (Fischer et al. 2015; Watson et al. 2015b).

The few studies purporting to have identified conformity or majority bias transmission in non-human animals have come under criticism in part for not adequately ruling out other types of social influence that might explain their findings, including biases towards certain individuals within the majority (van Leeuwen and Haun 2013; Haun et al. 2013). Rather than being attentive to the behaviour displayed by the majority of conspecifics, it is argued that animals may instead direct social learning towards specific individuals, perhaps an individual that is regarded as more successful or more knowledgeable or an individual who presents the learner with more opportunities for social learning.

4.2 Model-Based Biases in Primates

Also termed 'who strategies' (Laland 2004), model-based biases (Wood et al. 2013) focus on characteristics of the model, for instance, the age, sex, rank or previous success of the model. Whilst there is a growing body of research with humans identifying a number of biases causing directed social learning [Mesoudi and Whiten 2008; Molleman et al. 2014; for reviews in children, see Wood et al. (2013) and Price et al. (2017)], research with non-human primates has only recently begun to gain momentum. Here we discuss the evidence gathered for these biases in non-human primates, both in captivity and the field, and discuss the implications of these biases for the social transmission of behaviours.

The initial models selected by experimenters in open diffusion paradigms with non-human primates are often dominant members in the group (Hopper et al. 2011; Whiten et al. 2005). This is often because a dominant individual has first access to new food/stimuli and can thus act as a model for the rest of the group without interference, particularly in captivity. But do non-human primates have a natural bias towards copying a dominant individual? If rank serves as a proxy for success, then animals may benefit from copying the behaviour of a high-ranking individual. However, whilst a high-ranking individual may be seen as successful in some regards, there is no guarantee that copying their behaviours in all domains would be an adaptive strategy. Such a bias could certainly have important implications for the spread of innovations and emergence of traditions. In chimpanzees, there is a relatively high level of innovation compared to the number of traditions observed (Nishida et al. 2009). A review of the primate literature found that most innovations are performed by low-ranking individuals (Reader and Laland 2001), and it has been suggested that a bias towards attending to and copying higher-ranking individuals may explain the discrepancy between innovation rates and the relatively low number of traditions that become established in chimpanzee groups (Kendal et al. 2015).

Whilst there remains a lack of data from the field to corroborate this assertion, two recent studies have found biases towards copying dominant and knowledgeable individuals in captive chimpanzees. Horner et al. (2010) investigated model biases in chimpanzees by training two females in each of two groups of captive chimpanzees to deposit tokens in different receptacles in exchange for a reward. One of the models (Model A) was older than the other, higher-ranking and had previously introduced novel tasks to the group; a collection of traits labelled by the authors as 'prestige' (a term used by Henrich and Gil-White 2001 in discussing such phenomena in humans). In both groups, after the models had demonstrated their methods, naive observers dropped significantly more tokens into the receptacle demonstrated by Model A, thereby suggesting that the chimpanzees were biased towards copying the behaviour of this 'prestigious' female, over the non-prestigious female. This study, however, provides only limited support for a bias based upon dominance rank, since the models differed in age and previous success rates, either or both of which may have influenced the observer's actions. Following this, Kendal et al. (2015) used an open diffusion method and sophisticated statistical techniques to investigate transmission biases in captive chimpanzees. Alternative sliding movements of the door on an artificial fruit (left or right) were first seeded into two groups by mid-ranking females. The diffusion of behaviours and patterns of observation revealed biases towards copying dominant and knowledgeable individuals. Any existing rank bias is unlikely to be totally rigid, however, as both Watson et al. (2017) and Hopper et al. (2013) report diffusion of social information from low-ranking chimpanzees and squirrel monkeys, respectively. In a field experiment with wild vervet monkeys, Botting, Whiten, Grampp and van de Waal (2018) found that monkeys showed no consistent preference for a foraging behaviour demonstrated by a high-ranking, rather than that displayed by a low-ranking, female. Similarly, examining model-based biases in captive capuchin monkeys, Dindo et al. (2011) found no effect of watching a high-ranking or low-ranking female model on an individual's choice of action in a foraging task. However, in a subset of participants, there was a significant bias towards a related model (i.e. kin), which leads us to another bias which has been examined in non-human primates.

Observations from the field suggest that chimpanzee mothers act as the primary model in an infant's early life and are highly tolerant of their offspring. It is suggested that this may allow young chimpanzees to learn such behaviours as nut-cracking through a 'master-apprentice' style relationship (Matsuzawa et al. 2001). Field experiments have shown that infants are significantly more likely than juveniles or adults to observe a related model nut-cracking (Biro et al. 2003), perhaps reflecting a 'critical age' (Matsusawa and Yamakoshi 1996) at which this bias may operate to allow infants to learn these skills. Similarly, wild orangutan infants engage in increased 'peering' towards their mother whilst she is performing difficult extractive foraging techniques (Jaeggi et al. 2010; Schuppli et al. 2016). Field experiments with wild vervet monkeys have provided further evidence of a bias towards copying the mother in a primate's early life. Van de Waal et al. (2012) presented wild groups of vervet monkeys with grapes covered in sand and recorded the different techniques used by individuals to clean them prior to ingestion. It was found that individuals within the same matriline used similar techniques. This experiment was later extended to reveal that infants showed significant fidelity to their mother's favoured technique (van de Waal et al. 2014), indicating a strong bias towards the mother as a model for social learning in infancy, as has also been reported for humans (Henrich and Broesch 2011). Preferring kin as models likely stems from the increased tolerance, and thus social learning opportunities, that they provide (Coussi-Korbel and Fragaszy 1995), although it has also been suggested that learning from kin may prove adaptive since they are more likely to have similar behaviours and reactions to the learner (Hoppitt and Laland 2013). With respect to the emergence of traditions, a kinship bias may promote between-group differences (Wrangham et al. 2016), but this will depend upon the dispersion patterns of the species and might also be affected by sex-based biases in social learning (e.g. Lonsdorf et al. 2004).

In addition to this kinship-based bias, a sex-based bias has also been found in wild vervet monkeys, highlighting the fact that different social learning biases may operate concurrently in a community. van de Waal et al. (2010) found that monkeys who witnessed a dominant female model showed higher rates of participation and acted on the same part of an artificial fruit when compared with individuals who witnessed a dominant male model. The authors also found this to be associated with selective attention to female models, rather than a greater tolerance for observers from these models. Given this selective attention to females in a species with female philopatry, we might expect to see localised traditions between groups, as the migrating males would be far less likely to spread socially learned information. Lonsdorf et al. (2004) also found a bias in the acquisition of tool-use skills in juvenile chimpanzees, with females watching others termite fishing at an earlier age and becoming proficient at an earlier age than males. This highlights that the sex of the learner may also bias social learning processes (although it should be noted that the number of juveniles analysed here was relatively low).

Looking briefly beyond the primates, a handful of studies have examined model-based biases in other species. A bias towards copying dominant individuals was found in domestic hens (Nicol and Pope 1999), and biases towards copying older and larger individuals have been found in both fish (Duffy et al. 2009) and mice (Choleris et al. 1997). Biases towards copying from kin (ravens, Schwab et al. 2008) and a certain sex (finches, Katz and Lachlan 2003; hens, Nicol and Pope 1999) have also been reported. Thus, model biases appear to be a widespread adaptation in the animal kingdom, which suggests that they are an efficient way of obtaining information. Much more research is needed to discover when each bias is utilised, as the presence of multiple biases in a species suggests contextual implementation. Indeed, in humans, there is evidence that different individuals rely on different social learning strategies in the same contexts (Molleman et al. 2014). This highlights the individual variation seen in social learning and thus makes the task of elucidating which biases are utilised by animals and their role in cultural transmission all the more complex.

5 Conclusions

The last few decades have seen a huge leap forward in our understanding of non-human culture, with a particular intensity of research on primates. Phenomena once thought to be uniquely human have been found again and again amongst primates and, indeed, across the animal world. It seems that the more we examine social learning in non-human animals including our fellow primates, the less we find it to be limited to humans. This allows us not only to understand more of how and why cultural behaviours are present in modern humans but also to appreciate the richness of the cultural lives of these species and expand our conceptions of the role such second inheritance systems play in evolutionary processes.

There are, of course, many questions which remain unanswered. In particular, whilst research into social learning biases has yielded deeper understanding, it has also revealed the complexity of learning biases; how there may be several biases operating within a single species, or even within a single individual; and how a multitude of factors (context, sex, age, environment, etc.) can affect their subsequent expression. Only when we better understand how the range of social learning biases and mechanisms operate in different contexts can we fully understand how they function to contribute to the emergence and maintenance of traditions and culture. Our current knowledge is the result of an amalgamation of the complementary techniques of field observations, controlled captive experimentation and more recently, bringing these together, the more challenging achievement of field experiments. These methods all provide us with valuable complementary insights, for example, by using captive experimentation to test predictions based upon wild observation e.g. as in the case of wild chimpanzee traditions (Whiten et al. 1999) and the underlying role of social learning within these (Whiten et al. 2005). By continuing to utilise all of the methods that are available to us and in particular maintaining the inclusion of studies of free-ranging animals, which have provided some of the most exciting findings in recent years, we can expect to gain a clearer understanding of non-human and, indeed, human culture.

Acknowledgements JB, SKW and AW are grateful for the support of grant ID40128, 'Exploring the evolutionary foundations of cultural complexity, creativity and trust', from the John Templeton Foundation, and JB is grateful for the support of a grant from the David Bohnett Foundation during the writing of this paper. EW was supported by the Swiss National Science Foundation (P300P3_151187 and 31003A_159587) and Society in Science—Branco Weiss Fellowship. We thank two anonymous referees for discussion and comments on the manuscript.

References

Allen J, Weinrich M, Hoppitt W, Rendell L (2013) Network-based diffusion analysis reveals cultural transmission of lobtail feeding in humpback whales. Science 340(6131):485–488

Aplin LM, Farine DR, Morand-Ferron J, Cockburn A, Thornton A, Sheldon BC (2015a) Experimentally induced innovations lead to persistent culture via conformity in wild birds. Nature 518 (7540):538–541

Aplin LM, Farine DR, Morand-Ferron J, Cockburn A, Thornton A, Sheldon BC (2015b) Counting conformity: evaluating the units of information in frequency-dependent social learning. Anim Behav 110:e5–e8

Asch SE (1951) Effects of group pressure upon the modification and distortion of judgments. In: Guetzkow H (ed) Groups, leadership, and men. Carnegie, Pittsburgh, pp 222–236

Bandini E, Tennie C (2017) Spontaneous reoccurrence of "scooping", a wild tool-use behaviour, in naïve chimpanzees. PeerJ 5:e3814

Battesti M, Moreno C, Joly D, Mery F (2012) Spread of social information and dynamics of social transmission within Drosophila groups. Curr Biol 22(4):309–313

Biro D, Inoue-Nakamura N, Tonooka R, Yamakoshi G, Sousa C, Matsuzawa T (2003) Cultural innovation and transmission of tool use in wild chimpanzees: evidence from field experiments. Anim Cogn 6(4):213–223

Boesch C, Marchesi P, Marchesi N, Fruth B, Joulian F (1994) Is nut cracking in wild chimpanzees a cultural behaviour? J Hum Evol 26(4):325–338

Bond R, Smith PB (1996) Culture and conformity: a meta-analysis of studies using Asch's (1952b, 1956) line judgment task. Psychol Bull 119(1):111

Botting J, Whiten A, Grampp M, van de Waal E (2018) Field experiments with wild primates reveal no consistent dominance-based bias in social learning. Anim Behav 136:1–2

Buttelmann D, Carpenter M, Call J, Tomasello M (2007) Enculturated chimpanzees imitate rationally. Dev Sci 10(4):F31–F38

Byrne RW, Russon AE (1998) Learning by imitation: a hierarchical approach. Behav Brain Sci 21 (05):667–684

Carvalho S, McGrew W (2012) The origins of the Oldowan: why chimpanzees (Pan troglodytes) are still good models for technological evolution in Africa. In: Dominguez-Rodrigo M (ed) Stone tools and fossil bones. Cambridge University Press, Cambridge

Choleris E, Guo C, Liu H, Mainardi M, Valsecchi P (1997) The effect of demonstrator age and number on duration of socially-induced food preferences in house mouse (Mus domesticus). Behav Process 41(1):69–77

Chou L-S, Richerson PJ (1992) Multiple models in social transmission of food selection by Norway rats, *Rattus norvegicus*. Anim Behav 44:337–343

Claidiere N, Sperber D (2010) Imitation explains the propagation, not the stability of animal culture. Proc R Soc Lond B 277:651–659

Coussi-Korbel S, Fragaszy DM (1995) On the relation between social dynamics and social learning. Anim Behav 50(6):1441–1453

Custance DM, Whiten A, Bard KA (1995) Can young chimpanzees (Pan troglodytes) imitate arbitrary actions? Hayes & Hayes (1952) revisited. Behaviour 132(11):837–859

Dean LG, Kendal RL, Schapiro SJ, Thierry B, Laland KN (2012) Identification of the social and cognitive processes underlying human cumulative culture. Science 335:1114–1118

Dean LG, Vale GL, Laland KN, Flynn E, Kendal RL (2014) Human cumulative culture: a comparative perspective. Biol Rev 89(2):284–301

Dindo M, Thierry B, Whiten A (2008) Social diffusion of novel foraging methods in brown capuchin monkeys (Cebus apella). Proc R Soc Lond B Biol Sci 275(1631):187–193

Dindo M, Whiten A, de Waal FB (2009) In-group conformity sustains different foraging traditions in capuchin monkeys (Cebus apella). PLoS One 4(11):e7858–e7858

Dindo M, Leimgruber K, Ahmed R, Whiten A, de Waal F (2011) Observer choices during experimental foraging tasks in brown capuchin monkeys (Cebus apella). Am J Primatol 73 (9):920–927

Duffy GA, Pike TW, Laland KN (2009) Size-dependent directed social learning in nine-spined sticklebacks. Anim Behav 78(2):371–375

Fischer J, Wheeler BC, Higham JP (2015) Is there any evidence for vocal learning in chimpanzee food calls? Curr Biol 25:R1019–R1031

Fragaszy DM, Deputte B, Cooper EJ, Colber-White EN, Hémery C (2011) When and how well can human-socialized capuchins match actions demonstrated by a familiar human? Am J Primatol 73(7):643–654

Franks NR, Richardson T (2006) Teaching in tandem-running ants. Nature 439(7073):153–153

Franz M, Nunn CL (2009) Network-based diffusion analysis: a new method for detecting social learning. Proc R Soc Lond B Biol Sci 276(1663):1829–1836

Fuhrmann D, Ravignani A, Marshall-Pescini S, Whiten A (2014) Synchrony and motor mimicking in chimpanzee observational learning. Sci Rep 4:5283

Galef BG (1992) The question of animal culture. Hum Nat 3(2):157–178

Giraldeau LA, Valone TJ, Templeton JJ (2002) Potential disadvantages of using socially acquired information. Philos Trans R Soc B Biol Sci 357(1427):1559–1566

Goodall J (1986) The chimpanzees of Gombe: patterns of behavior. Belknap Press of Harvard University Press, Cambridge

Harrison RA, Whiten A (2018) Chimpanzees (Pan troglodytes) display limited behavioural flexibility when faced with a changing foraging task requiring tool use. PeerJ 6:e4366

Haun D, Tomasello M (2011) Conformity to peer pressure in preschool children. Child Dev 82 (6):1759–1767

Haun DB, Rekers Y, Tomasello M (2012) Majority-biased transmission in chimpanzees and human children, but not orangutans. Curr Biol 22(8):727–731

Haun DB, Van Leeuwen EJ, Edelson MG (2013) Majority influence in children and other animals. Dev Cogn Neurosci 3:61–71

Henrich J, Broesch J (2011) On the nature of cultural transmission networks: evidence from Fijian villages for adaptive learning biases. Philos Trans R Soc Lond B Biol Sci 366(1567):1139–1148

Henrich J, Gil-White FJ (2001) The evolution of prestige: freely conferred deference as a mechanism for enhancing the benefits of cultural transmission. Evol Hum Behav 22(3):165–196

Heyes CM (1994) Social learning in animals: categories and mechanisms. Biol Rev 69(2):207–231

Heyes C (2016) Blackboxing: social learning strategies and cultural evolution. Philos Trans R Soc B 371(1693):2015.0369

Hobaiter C, Poisot T, Zuberbühler K, Hoppitt W, Gruber T (2014) Social network analysis shows direct evidence for social transmission of tool use in wild chimpanzees. PLoS Biol 12(9): e1001960

Hohmann G, Fruth B (2003) Culture in bonobos? Between-species and within-species variation in behavior. Curr Anthropol 44(4):563–571

Holzhaider JC, Hunt GR, Gray RD (2010) The development of pandanus tool manufacture in wild New Caledonian crows. Behaviour 147(5):553–586

Hopper LM, Spiteri A, Lambeth SP, Schapiro SJ, Horner V, Whiten A (2007) Experimental studies of traditions and underlying transmission processes in chimpanzees. Anim Behav 73 (6):1021–1032

Hopper LM, Lambeth SP, Schapiro SJ, Whiten A (2008) Observational learning in chimpanzees and children studied through 'ghost' conditions. Proc R Soc Lond B Biol Sci 275 (1636):835–840

Hopper LM, Schapiro SJ, Lambeth SP, Brosnan SF (2011) Chimpanzees' socially maintained food preferences indicate both conservatism and conformity. Anim Behav 81(6):1195–1202

Hopper LM, Holmes AN, Williams LE, Brosnan SF (2013) Dissecting the mechanisms of squirrel monkey (*Saimiri boliviensis*) social learning. PeerJ 1:e13

Hopper LM, Kurtycz LM, Ross SR, Bonnie KE (2015) Captive chimpanzee foraging in a social setting: a test of problem solving, flexibility, and spatial discounting. PeerJ 3:e833

Hoppitt W, Laland KN (2008) Social processes influencing learning in animals: a review of the evidence. Adv Study Behav 38:105–165

Hoppitt W, Laland KN (2013) Social learning: an introduction to mechanisms, methods, and models. Princeton University Press, Princeton

Horner V, Proctor D, Bonnie KE, Whiten A, de Waal FB (2010) Prestige affects cultural learning in chimpanzees. PLoS One 5(5):e10625

Horner V, Whiten A (2005) Causal knowledge and imitation/emulation switching in chimpanzees (Pantroglodytes) and children (Homo sapiens). Anim Cogn 8(3):164–181

Hrubesch C, Preuschoft S, van Schaik C (2009) Skill mastery inhibits adoption of observed alternative solutions among chimpanzees (*Pan troglodytes*). Anim Cogn 12(2):209–216

Hunt GR, Gray RD (2003) Diversification and cumulative evolution in New Caledonian crow tool manufacture. Proc R Soc Lond B Biol Sci 270(1517):867–874

Jaeggi AV, Dunkel LP, Van Noordwijk MA, Wich SA, Sura AA, Van Schaik CP (2010) Social learning of diet and foraging skills by wild immature Bornean orangutans: implications for culture. Am J Primatol 72(1):62–71

Janik VM, Slater PJ (1997) Vocal learning in mammals. Adv Study Behav 26:59–100

Katz M, Lachlan RF (2003) Social learning of food types in zebra finches (*Taenopygia guttata*) is directed by demonstrator sex and feeding activity. Anim Cogn 6(1):11–16

Kawai M (1965) Newly-acquired pre-cultural behavior of the natural troop of Japanese monkeys on Koshima Islet. Primates 6(1):1–30

Kendal RL, Coolen I, van Bergen Y, Laland KN (2005) Trade-offs in the adaptive use of social and asocial learning. Adv Study Behav 35:333–379

Kendal R, Hopper LM, Whiten A, Brosnan SF, Lambeth SP, Schapiro SJ, Hoppitt W (2015) Chimpanzees copy dominant and knowledgeable individuals: implications for cultural diversity. Evol Hum Behav 36(1):65–72

Kenward B, Weir AA, Rutz C, Kacelnik A (2005) Behavioural ecology: tool manufacture by naive juvenile crows. Nature 433(7022):121–121

Koops K, Visalberghi E, van Schaik CP (2014) The ecology of primate material culture. Biol Lett 10(11):20140508

Krützen M, Mann J, Heithaus MR, Connor RC, Bejder L, Sherwin WB (2005) Cultural transmission of tool use in bottlenose dolphins. Proc Natl Acad Sci USA 102(25):8939–8943

Krützen M, Willems EP, van Schaik CP (2011) Culture and geographic variation in orangutan behavior. Curr Biol 21(21):1808–1812

Kundey SM, German R, De Los Reyes A, Monnier B, Swift P, Delise J, Tomlin M (2012) Domestic dogs'(Canis familiaris) choices in reference to agreement among human informants on location of food. Anim Cogn 15(5):991–997

Laland KN (2004) Social learning strategies. Anim Learn Behav 32(1):4–14

Laland KN, Janik VM (2006) The animal cultures debate. Trends Ecol Evol 21(10):542–547

Laland KN, Williams K (1998) Social transmission of maladaptive information in the guppy. Behav Ecol 9(5):493–499

Langergraber KE, Boesch C, Inoue E, Inoue-Murayama M, Mitani JC, Nishida T, Pusey A, Reynolds V, Schubert G, Wrangham RW, Wroblewski E, Vigilant L (2011) Genetic and 'cultural' similarity in wild chimpanzees. Proc R Soc B 278:408–416

Leca J-B, Gunst N, Huffman MA (2007) Japanese macaque cultures: inter-and intra-troop behavioural variability of stone handling patterns across 10 troops. Behaviour 144(3):251–281

Lefebvre L, Giraldeau L-A (1994) Cultural transmission in pigeons is affected by the number of tutors and bystanders present. Anim Behav 47(2):331–337

Logan CJ, Breen AJ, Taylor AH, Gray RD, Hoppitt WJ (2016) How New Caledonian crows solve novel foraging problems and what it means for cumulative culture. Learn Behav 44(1):18–28

Lonsdorf EV (2006) What is the role of mothers in the acquisition of termite-fishing behaviors in wild chimpanzees (Pan troglodytes schweinfurthii)? Anim Cogn 9(1):36–46

Lonsdorf EV, Eberly LE, Pusey AE (2004) Sex differences in learning in chimpanzees. Nature 428 (6984):715–716

Luncz LV, Boesch C (2014) Tradition over trend: neighboring chimpanzee communities maintain differences in cultural behavior despite frequent immigration of adult females. Am J Primatol 76 (7):649–657

Marshall-Pescini S, Whiten A (2008) Chimpanzees (Pan troglodytes) and the question of cumulative culture: an experimental approach. Anim Cogn 11(3):449–456

Matsuzawa T, Biro D, Humle T, Inoue-Nakamura N, Tonooka R, Yamakoshi G (2001) Emergence of culture in wild chimpanzees: education by master-apprenticeship. In: Matsuzawa T (ed) Primate origins of human cognition and behavior. Springer, Tokyo, pp 557–574

Mesoudi A, Whiten A (2008) The multiple roles of cultural transmission experiments in understanding human cultural evolution. Philos Trans R Soc B Biol Sci 363(1509):3489–3501

Molleman L, van den Berg P, Weissing FJ (2014) Consistent individual differences in human social learning strategies. Nat Commun 5:3570

Musgrave S, Morgan D, Lonsdorf E, Mundry R, Sanz C (2016) Tool transfers are a form of teaching among chimpanzees. Sci Rep 6:34783

Neadle D, Allritz M, Tennie C (2017) Food cleaning in gorillas: Social learning is a possibility but not a necessity. PLoS One 12(12):e0188866

Nicol CJ, Pope SJ (1999) The effects of demonstrator social status and prior foraging success on social learning in laying hens. Anim Behav 57(1):163–171

Nishida T, Matsusaka T, McGrew WC (2009) Emergence, propagation or disappearance of novel behavioral patterns in the habituated chimpanzees of Mahale: a review. Primates 50(1):23–36

Perry S, Baker M, Fedigan L, Gros-Louis J, Jack K, MacKinnon KC, Manson JH, Panger M, Pyle K, Rose L (2003) Social conventions in wild white-faced capuchin monkeys. Curr Anthropol 44(2):241–268

Pike TW, Laland KN (2010) Conformist learning in nine-spined sticklebacks' foraging decisions. Biol Lett 6:466–468
Pongracz P, Miklosi A, Kubinyi E, Topal J, Csanyi V (2003) Interaction between individual experience and social learning in dogs. Anim Behav 65(3):595–603
Price EE, Wood LA, Whiten A (2017) Adaptive cultural transmission biases in children and nonhuman primates. Infant Behav Dev 48(Pt A):45–53. https://doi.org/10.1016/j.infbeh.2016.11.003
Raihani NJ, Ridley AR (2008) Experimental evidence for teaching in wild pied babblers. Anim Behav 75(1):3–11
Reader SM, Laland KN (2001) Primate innovation: sex, age and social rank differences. Int J Primatol 22(5):787–805
Reader SM, Laland KN (2003) Animal innovation, vol 10. Oxford University Press, Oxford
Rendell L, Fogarty L, Hoppitt JE, Morgan TJH, Webster MM, Laland KN (2011) Cognitive culture: theoretical and empirical insights into social learning strategies. Trends Cogn Sci 15(2):68–76
Robbins MM, Ando C, Fawcett KA, Grueter CC, Hedwig D, Iwata Y, Lodwick JL, Masi S, Salmi R, Stoinski TS, Todd A, Vercellio V, Yamagiwa J (2016) Behavioral variation in gorillas: evidence of potential cultural traits. PLoS One 11(9):e0160483
Santorelli CJ, Schaffner CM, Campbell CJ, Notman H, Pavelka MS, Weghorst JA, Aureli F (2011) Traditions in spider monkeys are biased towards the social domain. PLoS One 6(2):e16863
Sanz CM, Morgan DB (2013) Ecological and social correlates of chimpanzee tool use. Philos Trans R Soc B Biol Sci 368(1630):20120416
Sasaki T, Biro D (2017) Cumulative culture can emerge from collective intelligence in animal groups. Nat Commun 8:15049
Schuppli C, Forss SI, Meulman EJ, Zweifel N, Lee KC, Rukmana E, Vogel ER, van Noordwijk MA, van Schaik CP (2016) Development of foraging skills in two orangutan populations: needing to learn or needing to grow? Front Zool 13(1):43
Schwab C, Bugnyar T, Schloegl C, Kotrschal K (2008) Enhanced social learning between siblings in common ravens, Corvus corax. Anim Behav 75(2):501–508
Slater P (1986) The cultural transmission of bird song. Trends Ecol Evol 1(4):94–97
Tennie C, Call J, Tomasello M (2010) Evidence for emulation in chimpanzees in social settings using the floating peanut task. PLoS One 5(5):e10544–e10544
Tennie C, Call J, Tomasello M (2012) Untrained chimpanzees (Pan troglodytes schweinfurthii) fail to imitate novel actions. PLoS One 7:e41548
Thornton A, McAuliffe K (2006) Teaching in wild meerkats. Science 313(5784):227–229
Thornton A, Raihani NJ (2010) Identifying teaching in wild animals. Learn Behav 38(3):297–309
Tomasello M, Davis-Dasilva M, CamaK L, Bard K (1987) Observational learning of tool-use by young chimpanzees. Hum Evol 2(2):175–183
Tomasello M, Call J, Warren J, Frost GT, Carpenter M, Nagell K (1997) The ontogeny of chimpanzee gestural signals: a comparison across groups and generations. Evol Commun 1 (2):223–259
Vale GL, Davis SJ, van de Waal E, Schapiro SJ, Lambeth SP, Whiten A (2017) Lack of conformity to new local dietary preferences in migrating captive chimpanzees. Anim Behav 124:135–144
van de Waal E, Whiten A (2012) Spontaneous emergence, imitation and spread of alternative foraging techniques among groups of vervet monkeys. PLoS One 7(10):e47008
van de Waal E, Renevey N, Favre CM, Bshary R (2010) Selective attention to philopatric models causes directed social learning in wild vervet monkeys. Proc Biol Sci 277(1691):2105–2111
van de Waal E, Krützen M, Hula J, Goudet J, Bshary R (2012) Similarity in food cleaning techniques within matrilines in wild vervet monkeys. PLoS One 7(4):e35694
van de Waal E, Borgeaud C, Whiten A (2013) Potent social learning and conformity shape a wild primate's foraging decisions. Science 340(6131):483–485
van de Waal E, Bshary R, Whiten A (2014) Wild vervet monkey infants acquire the food-processing variants of their mothers. Anim Behav 90:41–45

van de Waal E, van Schaik CP, Whiten A (2017) Resilience of experimentally seeded dietary traditions in wild vervets: Evidence from group fissions. Am J Primatol 79(10). https://doi.org/10.1002/ajp.22687

van Leeuwen EJ, Haun DB (2013) Conformity in nonhuman primates: fad or fact? Evol Hum Behav 34(1):1–7

van Leeuwen EJC, Kendal RL, Tennie C, Haun DBM (2015) Conformity and its look-a-likes. Anim Behav 110:e1–e4

van Schaik CP, Ancrenaz M, Borgen G, Galdikas B, Knott CD, Singleton I, Suzuki A, Utami SS, Merrill M (2003) Orangutan cultures and the evolution of material culture. Science 299 (5603):102–105

Voelkl B, Huber L (2000) True imitation in marmosets. Anim Behav 60(2):195–202

Watson CF, Caldwell CA (2009) Understanding behavioral traditions in primates: Are current experimental approaches too focused on food? Int J Primatol 30(1):143–167

Watson SK, Townsend SW, Schel AM, Wilke C, Wallace EK, Cheng L, West V, Slocombe KE (2015a) Vocal learning in the functionally referential food grunts of chimpanzees. Curr Biol 25 (4):495–499. https://doi.org/10.1016/j.cub.2014.12.032

Watson SK, Townsend SW, Schel AM, Wilke C, Wallace EK, Cheng L, West V, Slocombe KE (2015b) Reply to Fischer et al. Curr Biol 25(21):R1030–R1031

Watson SK, Reamer LA, Mareno MC, Vale G, Harrison RA, Lambeth SP, Schapiro SJ, Whiten A (2017) Socially transmitted diffusion of a novel behavior from subordinate chimpanzees. Am J Primatol 79(6). https://doi.org/10.1002/ajp.22642

Watson SK, Lambeth SP, Schapiro SJ, Whiten A (2018) Chimpanzees prioritise social information over pre-existing behaviours in a group context but not in dyads. Anim Cogn 24:1–12

Whitehead H, Rendell L (2014) The cultural lives of whales and dolphins. University of Chicago Press, Chicago

Whiten A (2005) The second inheritance system of chimpanzees and humans. Nature 437 (7055):52–55

Whiten A (2011) The scope of culture in chimpanzees, humans and ancestral apes. Philos Trans R Soc B Biol Sci 366(1567):997–1007

Whiten A, Goodall J, McGrew WC, Nishida T, Reynolds V, Sugiyama Y, Tutin CE, Wrangham RW, Boesch C (1999) Cultures in chimpanzees. Nature 399(6737):682–685

Whiten A, Goodall J, McGrew WC, Nishida T, Reynolds V, Sugiyama Y, Tutin CE, Wrangham RW, Boesch C (2001) Charting cultural variation in chimpanzees. Behaviour 138 (11):1481–1516

Whiten A, Horner V, Litchfield CA, Marshall-Pescini S (2004) How do apes ape? Anim Learn Behav 32(1):36–52

Whiten A, Horner V, De Waal FB (2005) Conformity to cultural norms of tool use in chimpanzees. Nature 437(7059):737–740

Whiten A, Caldwell CA, Mesoudi A (2016) Cultural diffusion in humans and other animals. Curr Opin Psychol 8:15–21

Wood LA, Kendal RL, Flynn EG (2013) Whom do children copy? Model-based biases in social learning. Dev Rev 33(4):341–356. https://doi.org/10.1016/j.dr.2013.08.002

Wrangham RW, Koops K, Machanda ZP, Worthington S, Bernard AB, Brazeau NF, Donovan R, Rosen J, Wilke C, Otali E, Muller MN (2016) Distribution of a chimpanzee social custom is explained by matrilineal relationship rather than conformity. Curr Biol 26(22):3033–3037

Yamamoto S, Humle T, Tanaka M (2013) Basis for cumulative cultural evolution in chimpanzees: social learning of a more efficient tool-use technique. PLoS One 8(1):e55768

The Zone of Latent Solutions and Its Relation to the Classics: Vygotsky and Köhler

Eva Reindl, Elisa Bandini, and Claudio Tennie

It is most difficult for chimpanzees to imitate anything, unless they themselves understand it.
Wolfgang Köhler, *The Mentality of Apes*, p. 157

Primates cannot be taught [...] through imitation [...], because they have no zone of proximal development.
Lev Vygotsky, *Mind in Society*, p. 88

Abstract In 2009, Tennie et al. proposed the theory of the Zone of Latent Solutions (ZLS), defined as the range of behaviors (solutions) an individual of a species can invent independently, i.e., which it can acquire without any form of social learning. By definition, species limited to their ZLS are unable to innovate and/or transmit behavioral traits outside their ZLS, i.e., they lack traits which go beyond the level of the individual—traits resulting from a gradual cultural evolution over successive transmission events ["cumulative culture", Boyd and Richerson (Proc Br Acad 88:73–93, 1996)]. However, this does not exclude an influence of social learning on the population frequency of these behaviors: social learning can facilitate the (ultimately individual) acquisition of latent solutions and thus speed up and consolidate their spread within a population. Cultures—defined as behaviors at least influenced by social learning—are therefore still possible (and often, even likely). Here, we elaborate on the ZLS account and relate it to the theories of Vygotsky (*Mind in Society: Development of Higher Psychological Processes*. Harvard University Press, 1978) who studied the role

E. Reindl (✉)
Department of Psychology and Neuroscience, University of St Andrews, St Andrews, UK

E. Bandini
Department for Early Prehistory and Quaternary Ecology, University of Tübingen, Tübingen, Germany

C. Tennie
Department for Early Prehistory and Quaternary Ecology, University of Tübingen, Tübingen, Germany

School of Psychology, University of Birmingham, Birmingham, UK

L. D. Di Paolo et al. (eds.), *Evolution of Primate Social Cognition*, Interdisciplinary Evolution Research 5, https://doi.org/10.1007/978-3-319-93776-2_15

of social learning in human culture. We argue that the ZLS is a missing phylogenetic "baseline" of Vygotsky's Zone of Actual Development (ZAD). Vygotsky's neglect of a need for a human "baseline ZAD" (namely, a human ZLS) may have been due to his (mis-)interpretation of Köhler's (*The Mentality of Apes*. Liveright, 1925) work on great ape behaviors: Köhler used his observations on individual chimpanzees to draw conclusions about the chimpanzee species as a whole, stating that chimpanzees can only copy what they could have invented themselves, thus coming close to the ZLS concept. Vygotsky—studying the range of behaviors individuals could achieve independently—seemingly did not see that Köhler was suggesting a species "baseline" (ZLS), upon which Vygotsky's idea of an individual's ZAD could develop. As a result, Vygotsky also did not see the need for a ZLS for his own study subjects: humans. Yet, there is no reason to assume that humans lack a ZLS, and in fact we present evidence for a human ZLS for tool-use behaviors.

Keywords Lev Vygotsky · Wolfgang Köhler · Zone of Latent Solutions · Cumulative culture · Tool use · Zone of Proximal Development · Individual learning

1 The Zone of Latent Solutions (ZLS)

Culture—behaviors that are acquired and transmitted (partly) via some form of social learning—is not restricted to humans but has been reported in a variety of species, such as chimpanzees (Whiten et al. 1999, 2001), orangutans (van Schaik et al. 2003, 2009), capuchin monkeys (Perry 2011), and cetaceans (Whitehead and Rendell 2014). Despite surface similarities between human and non-human great ape (hereafter great ape) cultures, the diversity and complexity of human culture remain unprecedented (Boyd et al. 2011; Pagel and Mace 2004). Humans may well be unique in their ability to produce cumulative culture (Galef 1992; Boyd and Richerson 1996), i.e., culture whose complexity and/or efficiency has increased over transmission events to such an extent that it cannot be invented by individuals anymore [also known as the ratchet effect, Tomasello (1999) and Tennie et al. (2009)].

To explain the occurrence of cumulative culture in humans and its suspected absence in great apes, Tennie et al. (2009) put forward the Zone of Latent Solutions (ZLS) theory. This account assumes that many animal species (including humans) possess a range of behaviors (latent solutions) which can potentially be reinnovated by each able-bodied and typically developed individual on its own, i.e., whose behavioral form does not need to be learned from others. High-fidelity social transmission mechanisms such as imitation or imitation-based teaching are not needed for latent solutions to be acquired [but they likely play an enabling and facilitating role in the acquisition of cumulative culture, Dean et al. (2014), Tennie et al. (2018), Wasielewski (2014); but see Caldwell and Millen (2009)]. Although individual learning is the main force behind the form of latent solutions, this does not exclude the possibility of social learning influencing the likelihood of an individual reinnovation of a latent solution: low-fidelity social learning mechanisms such as stimulus enhancement or exposure (i.e., artifacts left behind by conspecifics) can catalyze the individual innovation of a latent solution and therefore speed up the spread of the behavior within the population (Bandini and Tennie 2017; Neadle et al. 2017; Tennie et al. 2009). Thus, social learning can (and often does) influence the frequency of the behavior in the population. Nevertheless, the spread of latent solutions in a population is fueled by individual reinnovations, not by (social) copying. In other words, behaviors within the ZLS are the product of both genetic predispositions and bouts of individual learning in relation to the environment. A useful metaphor is one in which different objects roll down a hill (representing cultural traits developing over time). Latent solutions are like *stones* rolling down the hill—they "gather no moss" (or here: gather no snow). These stones can kick loose other stones along the way, leading to whole avalanches of stones. However, the *form* of the stones does not change; in the end, there may be more or less stones, but their form remained unchanged along the way. Cumulative culture, on the other hand, is like a snowball rolling down a snowy hill, accumulating (and sometimes losing) snow on their surfaces, thus changing its form along the way.

The cultural (tool-use) behaviors of wild great apes are argued to lie within the respective species' ZLS (Tennie et al. 2009). For instance, chimpanzees and bonobos have been observed to swallow rough-surfaced leaves without chewing, perhaps for self-medication (Wrangham and Nishida 1983). In order to learn this technique, a great ape is unlikely to depend on observing and copying others' actions (in this example, note that the actions are mostly unobservable, involving muscles hidden from view, such as movements of the tongue). Rather, exposure to the ecological materials and the problem space may be enough to induce leaf swallowing. In this case, the behavior would be deemed a latent solution.

Evidence for the claim that great ape tool-use behaviors lie within the ZLS comes from latent solution experiments, in which culturally naïve individuals (those who have never seen the behavior) are provided with the required raw materials (e.g., food rewards, sticks) and tested for the spontaneous invention of the behavior in question, i.e., of the behavior for which there is a cultural claim. For instance, if the behavior involves a long walk to gather a stick prior to using it in a special, supposedly "culturally learned" way, then the latent solution test does not need to recreate the "long walk" aspect. What matters is that the stick is used by naïve subjects in the same way (the behavioural form of the tool use) as in the target population. With this approach, if at least two independent, naïve subjects spontaneously invent the behavior in question, the ZLS theory regards this behavior to fall within the species' ZLS, i.e., to be a latent solution [for relatively more complex behaviors (such as nutcracking) the "single-case ZLS standard" (Bandini and Tennie 2017) applies and which only requires a single reinnovation].

One might ask in how far one can infer from the observation of a couple of naïve individual apes tested in zoos that a certain behavior is potentially reinnovatable by all able-bodied and motivated members of their species, including their wild counterparts whose social and environmental experience differs in many ways from zoo-housed individuals. Does this mean that the ZLS is entirely genetically determined? We do not claim this to be the case: the expression of latent solutions depends on both genetic and environmental effects in complex ways. For example, if we find several zoo-housed chimpanzees spontaneously cracking nuts independently from each other, we would predict such a capacity in wild chimpanzees as well, ultimately allowed by their genetic similarity. However, this does not mean that environmental effects play no role: even though the specifics of the environments of zoo-housed and wild chimpanzees differ, they have in common that they provide the individual chimpanzees with opportunities to learn about affordances (Gibson 1979), i.e., about ways in which the individual may interact with its environment. In contrast, we would expect chimpanzees raised in deprived conditions, e.g., those who have not had the opportunity to handle any objects at all, to perform worse (including null performances).

Of course, there may also be differences in social contexts between captive and wild animals. Again, socially deprived captive apes are not a suitable population to study, it is well known that their behavioral range is restricted (Menzel et al. 1970). Apes that are not socially deprived should be tested instead. Thus, the (lack of) social environment can restrict the range of ZLS behaviors expressed in captive individuals. However, latent solution tests are robust against such limitations: the actual reinnovation of a target behavior by naïve subjects represents positive data with a straightforward interpretation (i.e., social learning is *not* necessary to explain the behavioral form). And so, while social factors (as well as potential lack of physical stimuli in the environment) must be considered as an alternative explanation in cases where a latent solution test finds no reinnovation, a positive finding automatically guards against such problems.

Using latent solution experiments, food washing and wheat placer mining in Japanese macaques (Allritz et al. 2013), nettle feeding in gorillas (Tennie et al. 2008; Masi 2011), algae-scooping in chimpanzees (Bandini and Tennie 2017), and nut cracking in capuchin monkeys (Visalberghi 1987) have been clearly identified as latent solutions, i.e., these behaviors reappeared in naïve subjects. The literature also contains studies whose initial focus was not on latent solutions but whose outcomes are nevertheless meaningful with regard to the ZLS hypothesis. Studies exploring physical cognitive skills of a species or those investigating subsequent social learning of the target behaviors have often inadvertently tested captive and target-behavior-naïve subjects for wild-type behaviors. As long as some of the subjects reinnovated the behavior spontaneously, the behavior can be regarded as a latent solution. In this way, several studies can be reclaimed as latent solution tests and these indeed found evidence for latent solutions [e.g., leaf swallowing in chimpanzees, Huffman and Hirata (2004), later directly tested by Menzel et al. (2013); tool-use behaviors in orangutans, such as "Dip Stick," Lehner et al. (2010); and nut cracking in chimpanzees, Marshall-Pescini and Whiten (2008)]. The latent solution approach is also closely related to intentional controlled rearing experiments and such studies provided evidence for spontaneous simple tool manufacture in crows (Kenward et al. 2005) or stick tool use in woodpecker finches (Tebbich et al. 2001).

The ZLS theory not only claims that great ape cultural behaviors lie within the respective species' ZLS, but it also hypothesizes that great apes are unable—at least when left among themselves, i.e., without enculturation and/or training by humans—to learn behaviors lying beyond their ZLS. Experimental evidence indeed suggests that great apes do not imitate behavioral forms that are new to them (Tennie et al. 2012).[1] Thus, Tennie et al. (2009) presented what we call a ZLS-only account for the great ape

[1]Of course it is possible that some untrained or unenculturated great apes sometimes imitate—or in other ways learn something new. Current evidence suggests that this is rather unlikely and it is even less likely that such learning (reaching beyond the ZLS) underlies great ape "wild cultures."

behavioral repertoire: all great ape behavioral forms—including their "wild cultures"—might be explainable solely through latent solutions. Therefore, as long as convincing evidence is missing that representative—i.e., wild—great apes are able to learn new behavioral forms, we defend this strong formulation of the ZLS hypothesis (i.e., the ZLS-only account) for great apes, namely, that great apes are "stuck" inside their ZLS. Yet, even if it was shown that great apes copy behaviors outside their ZLS, this would not negate the ZLS hypothesis for all species. Rather, it would place great apes in a similar ZLS category as humans: into one where the ZLS potentially explains only part of their behavioral repertoire. Tennie et al. (2009) presented this strong ZLS-only claim for great apes partly also to invite critique. A ZLS-only claim should therefore, in theory, be very easy to falsify. So far, this has not happened in well-controlled experiments.

In order to further test the ZLS-only account, Tennie et al. (2009) presented great apes (and human children) with a task in which wooden wool had to be formed into a loop which could be used to pull a platform with a reward into reach. The authors found that great apes neither spontaneously manufactured a loop nor learned how to make one after a social demonstration—supporting the ZLS-only hypothesis for the apes. In contrast, one child spontaneously made a loop, so the authors concluded that loop production is a behavior within the ZLS of humans (of that age), but not that of great apes. Tennie et al. (2009) thus showed that the ZLS is not restricted to non-human animals, i.e., that humans also possess a ZLS. The crucial difference is that humans are not restricted by their ZLS in a way that great apes seem to be.

In contrast to great apes, whose cultural repertoire seems to be limited by their ZLS (ZLS-only account), modern humans have succeeded in moving beyond their ZLS through the production of cumulative culture (Tennie et al. 2009; ZLS-plus account[2]). Our ability to accumulate cultural traits is usually explained by a set of potentially uniquely human cognitive and motivational factors, such as the frequent use of imitative learning, active imitation-based teaching, language, enhanced mind-reading abilities, and a capacity for shared intentionality and cooperative actions (Boyd et al. 2011; Csibra and Gergely 2011; Kline 2015; Tennie et al. 2009; Tomasello 1999). In addition, it may be that humans alone are capable of frequent copying of several sources of information at once (e.g., actions and environmental results), which may have allowed the human lineage to increase their copying fidelity sufficiently to produce cumulative culture (Acerbi and Tennie 2016)—which in turn supported further increases in copying fidelity. However, human cumulative culture is unlikely to have arisen "from zero." We argue that the ZLS represents the ontogenetic and phylogenetic baseline on which cumulative culture was built. However, this does not imply that the ZLS is and was always fixed (though its upper limit is fixed for any individual at a given time and a given population). Instead, it can change as a result of (ongoing) evolution (see, e.g., Tennie and Hedwig 2009; Pradhan et al. 2012).

[2]Admittedly, these terms are somewhat cumbersome. But the topic at hand is somewhat complex, and the terms should also reflect this. Other terms may be used instead, if these are preferred [e.g., ZLS constrained/unconstrained or ZLS-only versus "ZLS + ZPD" or "ZLS + OCI" (for OCI, see below)].

Having elaborated on the ZLS theory, we aim to relate it to classic theoretical and empirical work, namely, to the approaches of German psychologist and chimpanzee researcher Wolfgang Köhler (1887–1967) and Russian psychologist Lev Vygotsky (1896–1934). We focus on three main publications: Köhler's (1925) *The Mentality of Apes*, in which he describes his research on chimpanzees' spontaneous problem-solving abilities (including tool use); Vygotsky's (1978) *Mind in Society: Development of Higher Psychological Processes*, which describes his theory about the Zones of Proximal and Actual Development; and Luria and Vygotsky's (1930) *Ape, Primitive Man, and Child: Essays in the History of Behavior*, in which Vygotsky presents his interpretation of Köhler's work.

2 Köhler's Thinking: Precursor of the ZLS Theory

As director of the chimpanzee research station of the Prussian Academy of Sciences on the island of Tenerife (1914–1920), Köhler carried out a series of experiments on chimpanzee problem-solving to determine whether chimpanzees' behaviors could be described as "intelligent" (Köhler 1925, p. 2) and to identify "what the chimpanzee is able to achieve by his unaided efforts" (ibid.). Many of Köhler's studies involved placing food outside the subjects' immediate bodily reach and providing them with simple materials such as boxes or sticks to allow the rewards to be retrieved. Köhler demonstrated that chimpanzees, contrary to previous belief, were capable of solving such problems unaided. He suggested that the chimpanzees had used "insight" learning to do so (Köhler 1925, p. 190).

The chimpanzees in Köhler's experiments were most likely naïve to the tests and tools given to them (at least at the start of the testing series). Thus, one of his main findings was that these solutions were innovated by the chimpanzees on their own: independently of each other and of human influence, without social learning and teaching being necessary.[3] The chimpanzees were also neither forced nor trained to show these tool behavioral forms. Köhler thus concluded that even though the "animals may be granted opportunities which they would not enjoy in their native African forests, we shall always in these experiments be observing the natural chimpanzee, and not any artificial product" (Köhler 1925, p. 69). By aiming to investigate what behaviors the chimpanzees were able to invent on their own, Köhler's experiments came close to modern latent solution tests.

In addition, Köhler observed spontaneous tool use in chimpanzees outside his experiments. Within a 2-year period, he observed the chimpanzees spontaneously perform a variety of tool behaviors. For example, the animals used sticks as weapons, levers, or digging tools to gather ants and water, and they also used leaves for wiping their bodies (Köhler 1925, pp. 73–87). Köhler suggested that it was

[3]However, note that the chimpanzees were born in the wild (Teuber 1994), so we cannot fully exclude the possibility of cultural descent of these behaviors in these particular subjects.

unlikely that "these achievements were largely due to their captive state" (1925, p. 67) as the objects in the chimpanzees' enclosure were "hardly more numerous and diverse than the products of nature in the forests of the Cameroons" (ibid.).

One similarity between the ZLS theory and Köhler's thinking is the focus on behaviors on a species level: Köhler applied insights he gained from observing a (small) group of chimpanzees to the cognitive abilities of chimpanzees in general. Similarly, the ZLS theory also draws conclusions on a species level: behaviors invented by even a few single naïve chimpanzees in latent solution tests are considered to fall within the ZLS of the entire species (Tennie et al. 2009).

In addition, Köhler and the ZLS theory both state that chimpanzees have limited imitative capabilities and that the role of imitation in the acquisition of novel behaviors in chimpanzees is only small. The ZLS theory claims that behavioral traits lying outside a species' ZLS—i.e., cumulative culture—can only be acquired via high-fidelity social transmission, e.g., imitation (Tennie et al. 2009). The ZLS hypothesis thus agrees with the notion that great apes' lack of cumulative culture stems from their deficit of high fidelity social learning skills (such as imitative skills, including imitation-based teaching [and possibly also from their lack of imagination; Dean et al. (2012), Legare and Nielsen (2015), Nagell et al. (1993), Tennie et al. (2009, 2012), and Tomasello (1999); but see Boesch (1991) and Whiten et al. (2009)]. Köhler reached similar conclusions, stating that imitation in chimpanzees is rare and "when it does occur, the situation, as well as its solution, must lie just about within the bounds set for spontaneous solutions" (1925, p. 222). Köhler argued that chimpanzees would only imitate behavior which they themselves understand (see introductory quote; as is mostly the case for humans—except for some mimicry perhaps); however, in contrast to humans, chimpanzees were limited in their understanding of others' behaviors, especially "when one of these apes sees another executing intelligent new actions quite unusual among chimpanzees" (1925, p. 308). In sum, Köhler anticipated the ZLS-only account by concluding that chimpanzees were unable to acquire and produce behavior outside of their own species' spontaneous capacities.

3 Vygotsky's Zone of Proximal Development

At the same time as Köhler was carrying out his research on individually learned behaviors in chimpanzees, another influential psychologist, Lev Vygotsky (and his students) focused on the role of social learning for the development of behavior, investigating another species: humans. To explain how an individual's behavioral repertoire could be extended by learning novel skills from adults and peers, Vygotsky developed his Zone of Proximal Development (ZPD) hypothesis. His emphasis on social learning as the driver behind the acquisition of behaviors which individuals could not invent on their own is again reflected in Tennie et al.'s (2009) ZLS-plus account.

Vygotsky undertook extensive empirical work on children, aiming to illuminate the development of higher cognitive functions, such as memory, attention, or the use of signs and symbols. Vygotsky's approach revolutionized the field of psychology: contrary to contemporary belief, he proposed that human higher cognition was not the result of isolated maturation processes. Instead, these capacities would undergo a developmental process dependent on input from social interactions (including teaching) and culture (e.g., language and tools). This idea is the precursor of the modern "ontogenetic cultural intelligence hypothesis" [in short: OCI; see Herrmann et al. (2007), Tennie and Over (2012), and van Schaik and Burkart (2011)].

Vygotsky investigated these developmental processes experimentally by presenting children of different ages with problem-solving tasks that varied in difficulty, where each task allowed for a range of solutions (including no solution at all). In this way Vygotsky gained insight about the development of mental functions and their interaction with social learning. For example, he found that young children who had not yet gained access to cultural solutions to certain problems—e.g., using symbols as external memory aid—tended to fail the related tasks. The solutions which were initially beyond what children could innovate on their own would gradually become incorporated into the growing children's cognitive repertoire—as he hypothesized: via social learning (and teaching).

Based on observations like these, Vygotsky (1978) formulated his ZPD concept. This concept has been highly influential in subsequent psychological and educational research (with 585 search results for "Zone of Proximal Development" in Web of Science, July 2018). However, its counterpart—the Zone of Actual Development (ZAD)—is less well known (with only three search results for "Zone of Actual Development" in Web of Science, July 2018), even though Vygotsky (1978) himself regarded it as equally important: "[A] child's mental development can be determined only by clarifying its two levels: the actual developmental level and the zone of proximal development" (p. 87). We concur with Vygotsky that the ZAD is a very important concept. In addition, we regard the ZAD as being closely related—but not equal—to the concept of the ZLS.

The ZAD is the range of behaviors, skills, and mental functions which an individual—at a set developmental stage—can achieve independently, without any guidance from others. It is the "result of certain already completed developmental cycles" (Vygotsky 1978, p. 85) and the result of past cultural learning and thus also includes skills which have previously been beyond what an individual could achieve independently. Thus, it can include skills learned through imitation or being taught. And so, a current ZAD will often include skills which were previously in a different zone, namely, the ZPD. The ZPD can be thought of as the range of skills an individual can achieve only with social help, i.e., "under adult guidance or in collaboration with more capable peers" (Vygotsky 1978, p. 86). These skills are therefore still in the stages of maturation. However, the ZPD is not unbounded: "[A] person can imitate only that which is within her developmental level" (Vygotsky 1978, p. 88), i.e., individuals will be unable to imitate or be taught anything that is beyond their current ZPD. For instance, even the best teacher will not be able to teach higher mathematics to a child who just learned how to add numbers (at least not in the short term). Thus, the ZPD can be viewed as a stage in the process of a skill turning from something that is beyond an individual's capacity (even in the presence of teaching) into a part of the ZAD.

In order for traits within the ZPD to become incorporated into the ZAD—i.e., for cognitive development to take place—learning must take place. To Vygotsky, learning in humans has a "specific social nature" (1978, p. 88): learners usually receive help such as scaffolding by and collaboration with adults and peers (more generally: teaching). In addition, learners possess the capacity and motivation to learn via imitation. Together, these factors allow humans to acquire novel skills that lie beyond their current scope of spontaneous achievements.[4]

Social learning plays an almost analogous role in both Tennie et al.'s (2009) and Vygotsky's theories. Tennie et al. (ibid.) suggest that—on a species level—high-fidelity social learning and high fidelity teaching allow the human behavioral repertoire to be extended beyond the ZLS through the acquisition of cumulative culture (ZLS-plus account). On the strict level of the individual, Vygotsky argues that social learning can extend the ZAD through the acquisition of behaviors within the ZPD. In this way, the ZLS account is very similar to Vygotsky's account—but simply more focused on population-level outcomes.

Vygotsky differed from Köhler not only in his focus on cultural (as opposed to independent) learning of cultural traits but also because despite his interest in individual differences he was much less interested in species-typical behavior. According to Vygotsky, every human individual possesses a ZAD and a ZPD, and these zones can and do vary in content and size between individuals. The ZAD and ZPD capture individuals' current and future cognitive capacities and can be used to describe individual learning trajectories and differences between individuals.

Maybe because of Vygotsky' focus on individual social learning, he never asked for the human case what Köhler asked for great apes, namely, what kinds of skills humans as a species would be able to innovate without social learning. Answers to this question would have likely helped him identify the extent to which the cognitive skills he studied depend on social learning. Yet, Vygotsky did not look for such a baseline ZAD, i.e., a phylogenetic basis of an individual's ZAD—or how we would like to call it: a ZLS. Yet, there is the need to assume a species-level ZAD (ZLS): one can only build upon already existing traits when one grows up.

In fact, Luria and Vygotsky (1930) emphasized the importance of understanding the phylogeny of a given trait (in addition to studying its ontogeny) in order to fully understand the development of the trait. Vygotsky rightly regarded the comparison of humans to other primates as a means to explore these phylogenetic roots of human behavior. Yet, even though he emphasized the need for species comparisons, Vygotsky did not adapt his individual-based ZAD/ZPD concept to a broader species level, thus missing the anticipation of the ZLS-plus account as well as the chance to incorporate Köhler's ideas into his own theory.

[4]For a modern and extended version of this argument, see Kim Sterelny's (2011) *Apprentice Learning Model*. Generally, this can be subsumed again under OCI.

We propose that the ZLS concept—by treating behavioral traits as capacities on a species level—is the missing piece in Vygotsky's theory, providing the phylogenetically derived "baseline" of the ZAD. These baseline skills develop through a complex interplay of cognitive skills, genetic predispositions, and (low-fidelity) social learning—all influenced by past biological evolution and interaction of these factors [see van Schaik and Pradhan (2003); also known as "phylogenetic cultural intelligence hypothesis," or in short PCI; see Tennie and Over (2012)].

4 Köhler in the Eyes of Vygotsky

We argue that Vygotsky (1) agreed with Köhler's interpretation of the chimpanzees' behaviors and (2) successfully applied his ZAD/ZPD hypothesis to the chimpanzee case but (3) failed to see that Köhler was close to the idea of a ZLS, and so failed to incorporate such a concept into his own theory.

First, Vygotsky agreed with Köhler's (and our) interpretation that the chimpanzees' solutions in Köhler's tool tasks could be explained by individual effort (Luria and Vygotsky 1930). Vygotsky (1978) further agreed that "primates can use imitation to solve only those problems that are of the same degree of difficulty as those they can solve alone" (p. 88).

What is novel is Vygotsky's attempt to explain the chimpanzees' behavior within his own theoretical framework. For this, he transferred his human-based ZAD/ZPD theory across species and interpreted Köhler's finding of the restricted imitative abilities of chimpanzees as follows: "[P]rimates cannot be taught [...] through imitation, nor can their intellect be developed, because they have no zone of proximal development" (ibid.). From this one might infer that Vygotsky believed that chimpanzees, too, possess a ZAD but that they are restricted to it because they—in contrast to humans—lack a ZPD.

However, Vygotsky did not see—or explicitly acknowledge—that Köhler's aim was to generalize across individuals to draw conclusions about the entire chimpanzee species. Vygotsky seems to have overlooked that Köhler was hinting at an additional concept which would be complementary to the ZAD: a baseline ZAD (a ZLS). Maybe if Vygotsky had identified this point, he might have incorporated such a concept into his own theory and thus would have applied it to his own study subjects: human children. However, at the time, it was not known that social learning could take place without high-fidelity transmission. It was also unknown that a species could possess culture without showing cumulative culture. This might explain why Vygotsky did not think about a phylogenetic baseline underlying human cultural learning. However, he could have inferred the existence of a ZLS from the following: only a minority of Köhler's tasks were solved by all chimpanzees. In most cases, only a couple of individuals innovate a solution. Crucially, however, Köhler (1925) nevertheless suggested that the behavior in question could be claimed to be within the mental capacities of *all* chimpanzees (this interpretation is consistent with the interpretation the ZLS theory would make).

Combining his theory with Köhler's ideas, Vygotsky could have argued (as we do here) that chimpanzees as a species are capable of spontaneously using a variety of tools for a variety of purposes (ZLS). He could also have stated that this does not imply that every chimpanzee always immediately shows every single behavior: the individuals' capacities are latently there, but may require the individual being at an appropriate developmental stage (e.g., in terms of fine motor abilities), or may require trial-and-error learning, and/or the right motivation and context in order to be expressed etc. (this motivation can come from seeing others engage with the target objects and in general through several channels of low-fidelity social learning[5]). The ZLS theory and Köhler do not require every individual to always show a given behavior for it to count as a latent solution. Both theories thus acknowledge variance—in a complex interplay between genetic and environmental inheritance (together with catalyzing effects of low-fidelity social learning).

5 Interim Summary: Contrasting the ZAD with the ZLS

In sum, the ZLS is separate from, but underlying, the ZAD (Fig. 1). In species lacking a ZPD (as seems to be the case for great apes though the jury is still out), the maximum range of an individual's ZAD will be limited by the ZLS (ZLS-only

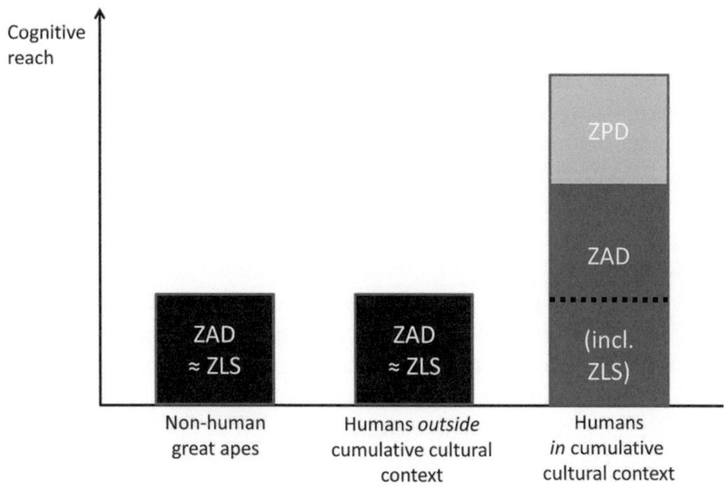

Fig. 1 The relative cognitive reaches of humans and non-human great apes with and without cumulative culture (cumulative culture in non-human great apes not depicted as its existence is questionable)

[5]At times, there might also be ontogenetic stages after which the animals are no longer interested in engaging with the task and/or in developing different approaches to solve it.

account). In the human case, the ZLS is only the starting point for the ZAD, whose behavioral range can be extended by the acquisition of traits lying outside the ZAD (i.e., within the ZPD) via high-fidelity social transmission (ZLS-plus account). For humans, the ZPD (enabled by our species' ability and motivation for high-fidelity transmission) is a necessary additional zone to the ZAD/ZLS—in fact without it, human behavior might not look much different from that of other apes [a view which contrasts starkly with that of Morin (2016)].

6 Why We Need a ZLS Theory

As Vygotsky seems to have never considered a ZLS concept for humans, he consequently overlooked the *need* for a ZLS; in fact, in the tool-use domain, he even suggested that spontaneous tool use in children was "practically zero" (Luria and Vygotsky 1930, p. 114). However, this would suggest that at some point after the split of the hominin and non-human great ape lineages at 14 million years ago (Tavaré et al. 2002) our ancestors lost all basic tool-use abilities from their cognitive repertoire, i.e., they would have lost (most of) their ZLS.[6] We think this is unlikely, especially given the enormous adaptive value of flexible tool use (Shumaker et al. 2011). Furthermore, evidence has now accumulated about tool-use behaviors which young children can invent on their own (Piaget 1952; Willatts 1984), and so the need to postulate a ZLS for humans became ever more apparent.

The ZLS-plus account predicts that humans, too, have a ZLS. Thus, it postulates that humans are not born as blank slates, but instead already possess a baseline of latent cognitive skills, anatomy, genetic predispositions, and general motivations that will channel them toward baseline behavioral response to various situations. Furthermore, it predicts that these latent solutions will come about sooner or later in each typically developing individual, unless he or she is socially and/or physically deprived and/or lacking the motivation (including social motivation) to engage in the behavior. However, it is at times possible that situational constraints or past experiences may negatively affect the expression of such behaviors (e.g., if the ZLS allows for more than one way to solve a task and the individual/population has already hit upon one solution, then additional solutions may be rare or suppressed).

The ZLS is the first and non-cultural version of the human ZAD, upon which a potential for development through social learning exists [ZPD, or, perhaps somewhat modernized, OCI, Tennie and Over (2012)]. In this respect, the ZLS theory is in contrast to Vygotsky's (Luria and Vygotsky 1930) claim that children were incapable of independently showing any form of tool-use behaviors.

[6]This hypothesis is not that far-fetched: in other domains, human reliance on cumulative culture has had deep impacts on our appearance and skills. For instance, our ability to cook food has led to shrinkage of our digestive organs, eventually making humans dependent on this cultural form of "pre-digesting" food (Henrich 2015; Wrangham 2009; Zink and Lieberman 2016).

The ZLS theory also allows for a second specific prediction: it expects behaviors argued to lie within a species' ZLS to occur spontaneously in latent solution tests in at least some target-naïve individuals of the species.

By creating such a baseline of skills across species, the ZLS theory facilitates the understanding of the evolution of cognition in general. The ZLS account also provides a framework for exploring the phylogenetic cultural intelligence roots (ZLS) of ontogenetic cultural intelligence (Tennie and Over 2012). Lastly, it raises the question of what is included in the human ZLS (Reindl et al. 2016) and what instead is a product of cumulative culture.

7 Empirical Evidence for a Human ZLS

In order to identify behaviors that fall within the human ZLS, the same logic as the one used in research with non-human animals applies. The goal is to explore what kinds of behaviors individuals can innovate without copying from others (Tennie and Hedwig 2009). However, latent solution experiments in humans are more challenging as humans are embedded in a rich cultural environment from birth. There have been occasional attempts to deprive humans from these cultural influences (the so-called forbidden experiment, Shattuck 1980), but not only do the findings remain anecdotal but these highly unethical attempts should also never be repeated. An ethically valid approach is testing very young children with novel tasks which they are *unlikely* to have encountered before. This method is similar to the one applied in great ape studies: the goal is not to deprive an individual of social contact and experience, but to test ecologically valid subjects in behaviors they naturally have not encountered previously.

Given the complexity and diversity of human tools, it would seem sensible to assume that all forms of human tool use must be acquired by social learning. Indeed, it has been suggested that learning of even "simple" forms of tool use, such as using a spoon, is a prolonged process of several stages, requiring significant social scaffolding by adults (Connolly and Dalgleish 1989; McCarty et al. 1999). Similarly, Luria and Vygotsky (1930) argued that tool behaviors in young children, such as spoon use, wiping with a towel, or using a plate, must be the product of imitation. However, this conclusion seems odd, especially given that Vygotsky believed in the continuity of cognitive skills between humans and great apes. He even described the tool behaviors observable in other animals as the "embryonic form[s]" (ibid., p. 32) of human tool use. However, arguing that great apes show "rudimentary versions" (ibid., p. 4) of human tool use—and likely do so spontaneously and without imitation—it is surprising that Vygotsky assumed that humans themselves did not possess these rudimentary behaviors.

Studies on children's spontaneous tool-use abilities are rare. Piaget (1952) showed that infants start using tools at the end of their first year of life: 9- to 11-month-olds spontaneously pull objects such as cloths, strings, and hooks to obtain out-of-reach objects, but spatial contact between tool and target seems to be necessary for their success (Bates et al. 1980). With 2 years, children are able to use tools on spatially separate objects (Brown 1990; Rat-Fischer et al. 2012). Finally, 3-year-olds have been found to spontaneously use and make hooks from pipe cleaners to retrieve a bucket from a bottle (Beck et al. 2011; Nielsen et al. 2014). This evidence tentatively suggests that the human ZLS is not zero.

To further explore human spontaneous tool behaviors, we conducted the first *explicit* latent solution experiment with children between 2 and 3.5 years on a test battery of 12 problem-solving tasks (Reindl et al. 2016). To identify candidate tool behaviors which possibly lie within the human ZLS, we based our tasks on cultural tool behaviors observed in wild chimpanzees and orangutans (van Schaik et al. 2009; Whiten et al. 2001). It was this "ape tool" approach that increased the ecological validity of the tasks, i.e., making them less influenced from a human perspective on problem-solving—and thus also less likely to have been encountered before by the children. Furthermore, it also allows for species comparisons to be made on the contents of the ZLS (and thus, for a phylogenetic analysis).

Children showed spontaneous tool use in the vast majority (11 of 12) of the ape tasks, indicating that nearly all of the studied tool behaviors lie within the scope of what children can invent on their own. Thus, we provided direct evidence for the presence of a human ZLS for tool behaviors. The ZLS theory also allows for predictions to be made based on cognitive cladistics (Byrne 1995). Reindl et al. (2016) found a large overlap in the spontaneous tool behaviors between modern humans, chimpanzees, and orangutans. From this, two hypotheses can be derived: first, since orangutans are the great ape species phylogenetically least related to us and chimpanzees the closest, the ZLS theory predicts via phylogenetic relations that other great apes more closely related to us (e.g., bonobos—to be tested in the future) will also share these tool-use skills. Second, applying similar reasoning, we conclude that the studied tool behaviors are likely shared by the last common ancestor of humans and great apes, living around ~14 million years ago (Tavaré et al. 2002).

8 Conclusion

We argued that Köhler's ideas about chimpanzee tool behaviors came very close to the notion of a chimpanzee ZLS. However, while focusing on his group of captive chimpanzees, Köhler did not attempt to explain cultural behaviors of chimpanzees in the wild—and neither could he, as he was unaware that these cultures existed. Vygotsky can be viewed as one of the first proponents within psychology of the ontogenetic cultural intelligence hypothesis (stating that cognition might be influenced by cultural forces); however, he missed to see both the need and existence of a ZLS in humans. The ZLS theory (Tennie et al. 2009) combines similar insights as those expressed by these two classic psychological approaches, while also extending them in crucial ways. First, the ZLS theory can explain culture in wild great ape populations [including not only currently living great apes but also extinct forms including also early hominins, Tennie et al. (2016)]: cultures that derive from and in "ZLS-only species" are often restricted to a species' ZLS. In the latter, low-fidelity social learning facilitates the individual reinnovation of a behavior, resulting in homogeneous behaviors (latent solutions) within a population. In addition, past realizations of latent solutions in a population as well as genetic and environmental factors also influence the likelihood of latent solutions being expressed and transmitted in a population. The ZLS-plus account applies to humans; it predicts a ZLS for our own species—and indeed we provided evidence for a human ZLS for tool-use behaviors (Reindl et al. 2016, see also Herrmann et al. 2007).

The ZLS theory reinforces the view that human culture is unique among great apes. As we continue to find more features that were previously thought to define culture in other species, the ability to produce cumulative culture may be key in explaining why human culture has advanced so much more dramatically (Boyd et al. 2011; Henrich 2015). Continued research into the ZLS of humans and non-humans will hopefully provide knowledge about the interplay of cognition and culture and will contribute to the interpretation of the work of the two quintessential psychologists Vygotsky and Köhler.

References

Acerbi A, Tennie C (2016) The role of redundant information in cultural transmission and cultural stabilization. J Comp Psychol 130:62–70. https://doi.org/10.1037/a0040094

Allritz M, Tennie C, Call J (2013) Food washing and placer mining in captive great apes. Primates 54:361–370. https://doi.org/10.1007/s10329-013-0355-5

Bandini E, Tennie C (2017) Spontaneous reoccurrence of "scooping", a wild tool-use behaviour, in naïve chimpanzees. PeerJ 5:e3814. https://doi.org/10.7717/peerj.3814

Bates E, Carlson-Luden V, Bretherton I (1980) Perceptual aspects of tool using in infancy. Infant Behav Dev 3:127–140. https://doi.org/10.1016/S0163-6383(80)80017-8

Beck SR, Apperly IA, Chappell J, Guthrie C, Cutting N (2011) Making tools isn't child's play. Cognition 119(2):301–306. https://doi.org/10.1016/j.cognition.2011.01.003

Boesch C (1991) Teaching among wild chimpanzees. Anim Behav 41:530–532

Boyd R, Richerson PJ (1996) Why culture is common but cultural evolution is rare. Proc Br Acad 88:73–93

Boyd R, Richerson PJ, Henrich J (2011) The cultural niche: why social learning is essential for human adaptation. Proc Natl Acad Sci USA 108:10918–10925

Brown AL (1990) Domain-specific principles affect learning and transfer in children. Cogn Sci 14:107–133. https://doi.org/10.1207/s15516709cog1401_6

Byrne RW (1995) The thinking ape: evolutionary origins of intelligence. Oxford University Press, New York

Caldwell CA, Millen AE (2009) Social learning mechanisms and cumulative cultural evolution: is imitation necessary? Psychol Sci 20(12):1478–1483. https://doi.org/10.1111/j.1467-9280.2009.02469.x

Connolly K, Dalgleish M (1989) The emergence of a tool-using skill in infancy. Dev Psychol 25 (6):894–912. https://doi.org/10.1037/0012-1649.25.6.894

Csibra G, Gergely G (2011) Natural pedagogy as evolutionary adaptation. Philos Trans R Soc Lond B Biol Sci 366:1149–1157

Dean LG, Kendal RL, Schapiro SJ, Thierry B, Laland KN (2012) Identification of the social and cognitive processes underlying human cumulative culture. Science 335:1114–1118. https://doi.org/10.1126/science.1213969

Dean LG, Vale GL, Laland KN, Flynn E, Kendal RL (2014) Human cumulative culture: a comparative perspective. Biol Rev 89:284–301. https://doi.org/10.1111/brv.12053
Galef GB (1992) The question of animal culture. Hum Nat 3:157–178
Gibson JJ (1979) The ecological approach to visual perception. Houghton Mifflin, Boston
Henrich J (2015) The secret of our success: how culture is driving human evolution, domesticating our species, and making us smart. Princeton University Press, Princeton
Herrmann E, Call J, Hernández-Lloreda MV, Hare B, Tomasello M (2007) Humans have evolved specialized skills of social cognition: the cultural intelligence hypothesis. Science 317:1360–1366. https://doi.org/10.1126/science.1146282
Huffman MA, Hirata S (2004) An experimental study of leaf swallowing in captive chimpanzees: insights into the origin of a self-medicative behavior and the role of social learning. Primates 45:113–118
Kenward B, Weir AAS, Rutz C, Kacelnik A (2005) Tool manufacture by naive juvenile crows: the use of twigs by these birds to coax out hidden food seems to be an instinctive skill. Nature 433:121
Kline MA (2015) How to learn about teaching: an evolutionary framework for the study of teaching behavior in humans and other animals. Behav Brain Sci 38:1–71
Köhler W (1925) The mentality of apes. Liveright, New York. German edition: Köhler W (1921) Intelligenzprüfungen am Menschenaffen. Springer, Berlin
Legare CH, Nielsen M (2015) Imitation and innovation: the dual engines of cultural learning. Trends Cogn Sci 19:688–699
Lehner SR, Burkart JM, van Schaik CP (2010) An evaluation of the geographic method for recognizing innovations in nature, using zoo orangutans. Primates 51:101–118
Luria AR, Vygotsky L (1930) Ape, primitive man, and child: essays in the history of behaviour. Harvester Wheatsheaf, London
Marshall-Pescini S, Whiten A (2008) Social learning of nut-cracking behavior in East African sanctuary-living chimpanzees (*Pan troglodytes schweinfurthii*). J Comp Psychol 122:186–194
Masi S (2011) Differences in gorilla nettle-feeding between captivity and the wild: local traditions, species typical behaviours or merely the result of nutritional deficiencies? Anim Cogn 14:921–925. https://doi.org/10.1007/s10071-011-0457-7
McCarty ME, Clifton RK, Collard RR (1999) Problem solving in infancy: the emergence of an action plan. Dev Psychol 35(4):1091–1101. https://doi.org/10.1037//0012-1649.35.4.1091
Menzel EW Jr, Davenport RK, Rogers CM (1970) The development of tool using in wild-born an restriction-reared chimpanzees. Folia primatol 12:273–283
Menzel C, Fowler A, Tennie C, Call J (2013) Leaf surface roughness elicits leaf swallowing behavior in captive chimpanzees (*Pan troglodytes*) and bonobos (*P. paniscus*), but not in gorillas (*Gorilla gorilla*) or orangutans (*Pongo abelii*). Int J Primatol 34:533–553. https://doi.org/10.1007/s10764-013-9679-7
Morin O (2016) How traditions live and die. Oxford University Press, New York
Nagell K, Olguin RS, Tomasello M (1993) Processes of social learning in the tool use of chimpanzees (*Pan troglodytes*) and human children (*Homo sapiens*). J Comp Psychol 107:174–186
Neadle D, Allritz M, Tennie C (2017) Food cleaning in gorillas: social learning is a possibility but not a necessity. PLoS One 12(12):e0188866
Nielsen M, Tomaselli K, Mushin I, Whiten A (2014) Exploring tool innovation: a comparison of Western and Bushman children. J Exp Child Psychol 126:384–394
Pagel M, Mace R (2004) The cultural wealth of nations. Nature 428:275–278
Perry S (2011) Social traditions and social learning in capuchin monkeys (*Cebus*). Philos Trans R Soc Lond B Biol Sci 366:988–996
Piaget J (1952) The origins of intelligence in children. International University Press, New York
Pradhan GR, Tennie C, van Schaik CP (2012) Social organization and the evolution of cumulative technology in apes and hominins. J Hum Evol 63:180–190. https://doi.org/10.1016/j.jhevol.2012.04.008

Rat-Fischer L, O'Regan JK, Fagard J (2012) The emergence of tool use during the second year of life. J Exp Child Psychol 113:440–446. https://doi.org/10.1016/j.jecp.2012.06.001
Reindl E, Beck SR, Apperly IA, Tennie C (2016) Young children spontaneously invent wild great apes' tool use behaviors. Proc R Soc B 283(1825):20152402. https://doi.org/10.1098/rspb.2015.2402
Shattuck R (1980) The forbidden experiment: the story of the wild boy of Aveyron. Kodansha International, Pennsylvania
Shumaker RW, Walkup KR, Beck BB (2011) Animal tool behavior: the use and manufacture of tools by animals. Johns Hopkins University Press, Baltimore
Sterelny K (2011) From hominins to humans: how sapiens became behaviourally modern. Philos Trans R Soc Lond B Biol Sci 366:809–822. https://doi.org/10.1098/rstb.2010.0301
Tavaré S, Marshall CR, Will O, Soligo C, Martin RD (2002) Using the fossil record to estimate the age of the last common ancestor of extant primates. Nature 416:726–729
Tebbich S, Taborsky M, Fessl B, Blomqvist D (2001) Do woodpecker finches acquire tool-use by social learning? Proc R Soc B 268(1482):2189–2193
Tennie C, Hedwig D (2009) How latent solution experiments can help to study differences between human culture and primate traditions. In: Potocki E, Krasinski J (eds) Primatology: theories, methods and research. Nova, New York
Tennie C, Over H (2012) Cultural intelligence is key to explaining human tool use. Behav Brain Sci 35:242–243. https://doi.org/10.1017/S0140525X11001968
Tennie C, Hedwig D, Call J, Tomasello M (2008) An experimental study of nettle feeding in captive gorillas. Am J Primatol 70:584–593
Tennie C, Call J, Tomasello M (2009) Ratcheting up the ratchet: on the evolution of cumulative culture. Philos Trans R Soc Lond B Biol Sci 364:2405–2415. https://doi.org/10.1098/rstb.2009.0052
Tennie C, Call J, Tomasello M (2012) Untrained chimpanzees (*Pan troglodytes schweinfurthii)* fail to imitate novel actions. PLoS One 7:1–19
Tennie C, Braun DR, Premo LS, SP MP (2016) The island test for cumulative culture in paleolithic cultures. In: Haidle MN, Conard NJ, Bolus M (eds) The nature of culture, Vertebrate paleobiology and paleoanthropology. Springer, Netherlands
Tennie C, Caldwell C, Dean L (2018) Cumulative culture. In: Callan H (ed) International encyclopedia of anthropology. Wiley, Oxford
Teuber ML (1994) The founding of the Primate Station, Tenerife, Canary Islands. Am J Psychol 107(4):551–582
Tomasello M (1999) The cultural origins of human cognition. Harvard University Press, Cambridge
van Schaik CP, Burkart JM (2011) Social learning and evolution: the cultural intelligence hypothesis. Philos Trans R Soc Lond B Biol Sci 366:1008–1016
van Schaik CP, Pradhan GR (2003) A model for tool-use traditions in primates: implications for the coevolution of culture and cognition. J Hum Evol 44:645–664. https://doi.org/10.1016/S0047-2484(03)00041-1
van Schaik CP, Ancrenaz M, Borgen G et al (2003) Orangutan cultures and the evolution of material culture. Science 299:102–105. https://doi.org/10.1126/science.1078004
van Schaik CP, Ancrenaz M, Djojoasmoro R et al (2009) Orangutan cultures revisited. In: Wich SA, Utami Atmoko SS, Mitra Setia T, van Schaik CP (eds) Orangutans: geographic variation in behavioral ecology and conservation. Oxford University Press, New York
Visalberghi E (1987) Acquisition of nut-cracking behavior by 2 capuchin monkeys (*Cebus apella*). Folia Primatol 49:168–181
Vygotsky L (1978) Mind in society: development of higher psychological processes. Harvard University Press, Cambridge
Wasielewski H (2014) Imitation is necessary for cumulative cultural evolution in an unfamiliar, opaque task. Hum Nat 25:161–179
Whitehead H, Rendell LE (2014) The cultural lives of whales and dolphins. University of Chicago Press, Chicago

Whiten A, Goodall J, McGrew WC et al (1999) Cultures in chimpanzees. Nature 399:682–685. https://doi.org/10.1038/21415

Whiten A, Goodall J, McGrew WC et al (2001) Charting cultural variation in chimpanzees. Behaviour 138:1481–1516. https://doi.org/10.1163/156853901317367717

Whiten A, McGuigan N, Marshall-Pescini S, Hopper LM (2009) Emulation, imitation, over-imitation and the scope of culture for child and chimpanzee. Philos Trans R Soc Lond B Biol Sci 364:2417–2428. https://doi.org/10.1098/rstb.2009.0069

Willatts P (1984) The stage-IV infant's solution of problems requiring the use of supports. Infant Behav Dev 7:125–134. https://doi.org/10.1016/S0163-6383(84)80053-3

Wrangham RW (2009) Catching fire: how cooking made us human. Basic Books, New York

Wrangham RW, Nishida T (1983) *Aspilia* spp. leaves: a puzzle in the feeding behavior of wild chimpanzees. Primates 24:276–282

Zink KD, Lieberman DE (2016) Impact of meat and lower palaeolithic food processing techniques on chewing in humans. Nature 531:500–503. https://doi.org/10.1038/nature16990

Minimal Cognitive Preconditions on the Ratchet

Elizabeth Renner and Tadeusz Zawidzki

Abstract *H. sapiens* stands out from other primates along many social dimensions; however, none seems as prominent and important as the capacity of our species for cumulative cultural evolution or, as Tomasello calls it, "the ratchet." Although other primate species show evidence of cultural variation, there is little evidence of cumulative cultural evolution, i.e., the gradual accumulation, modification, and refinement of traditions and skills over historical time, in any primate species other than our own. This is clearly an extremely significant component of the human phenotype, responsible for our unparalleled cultural, social, political, and technological achievements. However, it remains extremely controversial what sorts of cognitive capacities are necessary to trigger cumulative cultural evolution and whether any currently proposed candidates are really distinctive of humans. Furthermore, the ratchet raises a bootstrapping problem: before complex skills and technologies are present and necessary for biological success, there appear to be few advantages to high-fidelity social learning; however, without such high-fidelity social learning, it is unclear how traditions capable of generating complex skills and technologies could arise in the first place. In this chapter, we survey relevant empirical research in comparative and developmental psychology, integrating it with a novel theoretical analysis of the bootstrapping problem to defend a hypothesis about the minimal cognitive preconditions on the ratchet.

Keywords Ratchet · Cumulative cultural evolution · Primate social learning · Primate innovation · Sterelny · Social niche construction

E. Renner
Center for the Advanced Study of Human Paleobiology, Department of Anthropology, The George Washington University, Washington, DC, USA

Psychology, University of Stirling, Stirling, UK
e-mail: elizabeth.renner@stir.ac.uk

T. Zawidzki (✉)
Department of Philosophy, The George Washington University, Washington, DC, USA
e-mail: zawidzki@gwu.edu

L. D. Di Paolo et al. (eds.), *Evolution of Primate Social Cognition*, Interdisciplinary Evolution Research 5, https://doi.org/10.1007/978-3-319-93776-2_16

1 Introduction

There is a broad consensus in comparative cognitive science that "the ratchet" sets humans apart from other animals. "The ratchet," or "cumulative cultural evolution," refers to a manifest fact about human cultural products, including technologies and skills: they typically change radically over relatively short periods of historical time, as later generations replicate and then innovate on the products of earlier generations. There is evidence of this in the human archaeological record going back at least 100,000 years (McBrearty and Brooks 2000). But there is very little evidence of this in contemporary populations of non-human animals.

Explanations of this apparent discontinuity in the animal kingdom garner much less consensus. One area of controversy concerns the cognitive capacities required to give rise to and maintain cumulative culture. Another concerns to what extent these capacities are exclusive to humans. A third concerns the phylogenetic roots of the ratchet: what distinctive features of hominin ecology explain this unusual adaptive response?

In this chapter, we address all three controversies. Section 2 is an overview of the problem, highlighting various alternatives that have been proposed in response to the three controversies. One lesson we draw from this overview is that no current proposal adequately responds to a "bootstrapping" problem concerning cumulative cultural evolution: the sorts of cognitive capacities required for cumulative cultural evolution are adaptive only in contexts where cumulative cultural evolution is already present (Boyd and Richerson 1996). Section 3 reviews the empirical evidence concerning similarities and differences in social learning and innovation capacities between humans and non-humans. One lesson we draw from this evidence is that it is very difficult to identify specifically cognitive differences between humans and non-humans that can explain why we have the ratchet, while others do not. This sets the stage for our arguments in Sect. 4. Following some suggestions of Kim Sterelny (2012), we argue that the ratchet was likely triggered by social and motivational differences between our ancestors and other species. The advantage of this view is that it suggests ways of solving the two puzzles highlighted above: bootstrapping and the apparent lack of specifically cognitive differences between humans and non-humans relevant to cumulative cultural evolution.

2 Cumulative Cultural Evolution: The State of Play

Most theories of cumulative cultural evolution focus on the cognitive capacities, presumably implemented in the brains of individuals, which appear to be necessary for it. It is plausible that cumulative cultural evolution is impossible without two components: efficient, high-fidelity replication and innovation. Consider two generations: G_a (ancestral) and G_o (offspring). It is plausible that a technology or skill, S_a, possessed by members of G_a can evolve cumulatively into a technology or skill, S_o, possessed by members of G_o, only under the following conditions: (1) members of G_o have some method for efficiently and faithfully replicating S_a from observing it executed by members of G_a, and (2) members of G_o have the capacity to innovatively tweak S_a, in ways that yield S_o. These constraints embody the commonsense maxim that there is no use in re-inventing the wheel. It is important for faithful replication to be highly efficient: if members of G_o spend too much time and effort replicating S_a, there will not be enough time and energy to tinker with it in ways that improve it, yielding S_o. This highly intuitive decomposition of the ratchet has framed most discussions of it.

This framing of cumulative cultural evolution gives rise to the following research agenda. First, we must identify human social learning and innovation capacities, including general socio-cognitive capacities that might facilitate them, like theory of mind, and show them to be engaged in cumulative cultural evolution. Second, we must, through comparative research, establish that these social learning, innovation, and associated capacities are unique to humans. Otherwise, it remains a mystery why the ratchet occurs only in human populations. Third, we must investigate the fossil and archaeological records with a view to identifying features unique to the human lineage that, plausibly, could have helped select for these capacities.

Most discussions of cumulative cultural evolution fit this template. For example, in their discussions of the ratchet over the last 25 years, Tomasello and colleagues have stressed the human capacity for imitation, arguing that it is necessary to achieve the transmission fidelity required for the ratchet and that it is unique to humans (Tomasello et al. 1993a, 2005). Although Tomasello and colleagues have written less about the unique features of the human niche that likely selected specifically for imitation, they do argue that it is a by-product of uniquely human capacities to "read minds," probably selected for other reasons (Ibid.). One problem with this approach is that it takes for granted something that remains controversial: that only humans truly imitate and that only humans read minds in the relevant sense.

More recently, other theorists have refined this strategy by focusing on a specific kind of imitation that appears truly unique to humans: overimitation. Following the pioneering discussions of Tomasello and colleagues, imitation is usually conceived in terms of two components: acquiring new goals from observing others and acquiring new means to those goals from observing others. Initially, Tomasello and colleagues argued that, although many species can acquire new goals from observing others and new means to old goals from observing others, only humans can acquire new means to new goals from observing others. That is, non-humans can acquire a new goal from observing a conspecific and then learn their own means to that goal or acquire new means to goals they already have from observing a conspecific, but only humans can acquire both a new goal and the means to achieve it from observing a conspecific. However, Horner and Whiten (2005) showed that the latter capacity is not unique to humans; chimpanzees are capable of acquiring both new goals and the means to achieve them from observing others. The same experiment showed something else as well: when chimpanzees are aware of a method for achieving a goal they observe another achieve that is more efficient than the method employed by this model, then they immediately shift to this more efficient method, unlike human children. That is, human children persist in employing the method employed by the model to achieve the new goal, even when they know of a more efficient method of achieving that goal. This is overimitation, and it appears to be, as far as we know, unique to humans.

The recent discovery of overimitation appears highly relevant to explaining cumulative cultural evolution. It is clearly a mechanism of efficient and extremely high-fidelity replication. It is unsurprising, therefore, that many recent discussions focus on it. For example, Nielsen and Tomaselli (2010) argue that, although overimitation appears irrational, since the imitator is needlessly using the less efficient of two known methods to accomplish a goal, it likely played an extremely important role in human evolution, making possible the ratchet effect. Lyons (2009) argues that, specifically, the need to acquire skill at constructing and using complex technologies explains why overimitation evolved only in the human lineage. Here is the idea. Many complex, multi-part technologies and techniques require engaging in sequences of multiple, precisely executed steps, the roles of which are opaque relative to the ultimate goal. Consider flint knapping. Expert knappers must engage in multiple steps with no obvious connection to the ultimate goal of fashioning a blade with which to cut. If these steps are not executed precisely, the ultimate goal of fashioning a cutting blade is compromised. A novice learning from an expert cannot identify a more efficient knapping technique without first mastering the complex, multi-step technique of the expert. Overimitation seems ideally suited to solve this problem. Overimitating novices would automatically copy the precise method of the expert, with no thought of how it contributes to the ultimate goal or whether there is a more efficient method possible. This efficient, faithful replication is the first component of the ratchet. Once techniques are mastered, innovation can be used to tweak them, thereby making possible cumulative cultural evolution.

Not everyone accepts Lyons' proposal. Recently, a number of theorists have linked overimitation to ritual, group affiliation, and other conventional behaviors (Over and Carpenter 2012, 2013; Legare and Souza 2012; Legare et al. 2015; Legare and Nielsen 2015). Learning ritual poses an opacity problem similar to that posed by learning complex techniques and technologies. One must acquire complex behaviors, with no obvious connection to goals, from observing experts. Based on recent evidence from human children, these theorists argue that overimitation, due to the efficiency and extreme fidelity it makes possible, is used mainly for acquiring conventional behaviors, while the acquisition of technological skill triggers more innovative responses.

Thus, there are two recent, competing approaches to explaining the ratchet. Both agree that overimitation is a uniquely human capacity used in social learning. But they disagree regarding the functions of this capacity. Lyons argues that overimitation is necessary for the acquisition of opaque techniques for mastering technological competence. Over and Carpenter and Legare and colleagues argue that it is necessary for the acquisition of rituals and other conventional behaviors. It seems, therefore, that the key disagreement concerns phylogeny, specifically, which features unique to hominin prehistorical ecology selected for overimitation, giving rise to the ratchet. On one view, this had to do with the importance of acquiring competence in the construction and use of complex technologies. On the other view, it had to do with the importance of mastering rituals and other conventional behaviors required for group identification and cohesion.

There are a number of problems with both of these approaches. The hypothesis invoking complex technology faces a bootstrapping problem (Boyd and Richerson 1996). Complex technology requires the ratchet to even exist. That is, in order for a particular technology to go from simple to complex, at least one individual must be capable of both copying and improving upon the simple design (instantiating "the ratchet"). In order for the new, complex artifact to continue to exist, at least one other local individual in each generation must be capable of copying this new, complex design. However, at this early point in the lifetime of the ratchet, there is no selection pressure for high-fidelity copying of artifacts. Hence, complex technology cannot constitute the initial feature of hominin ecology that selected for the cognitive capacities that give rise to the ratchet. Boyd and Richerson (1996) suggest that other features of complex hominin social arrangements selected for advanced social cognition, e.g., theory of mind, which then yielded the ratchet as a kind of by-product. That is, once hominin offspring were adept at reading the minds of adults, they could imitate their behavior more faithfully, before innovating upon it. But this suggestion overstates the differences between human and non-human primate social cognition. As we argue in Sect. 3, there is evidence that non-human primate socio-cognitive capacities are sufficient to give rise to a ratchet. Something else must explain the uniqueness of human, cumulative cultural evolution. The suggestion that ritual and other conventional, group-affiliative behaviors played an extremely important role in human prehistory is promising; however, it runs into a timing problem. On most views, ritual-mediated group bonding has selective value only in demographically dense socio-ecologies, where there is a lot of group competition and individuals have to be able to signal affiliation with their groups (Powell et al. 2009). But the archaeological record shows evidence of complex technology, seemingly requiring some kind of cumulative cultural evolution, long before any evidence of demographic densities likely to give rise to intense group competition (Ibid.). Thus, overimitation, or other cognitive capacities likely to yield cumulative cultural evolution, had to be in place long before learning rituals would have been adaptive.

Besides these problems of phylogenetic timing, there is a problem raised by research in comparative social cognition. Although it is true that there is little evidence of non-human overimitation, there seems to be no cognitive reason for this. A variety of non-human species show evidence of all the capacities that appear to be required for overimitation: they can learn from each other, identify each other's goals, and mimic the methods others use to achieve goals. We now turn to a review of the empirical evidence in comparative social cognition relevant to explaining the ratchet. In Sect. 4, we argue that, together with the foregoing arguments, this evidence shows that the ratchet likely cannot be explained in terms of cognitive adaptations expressed in individual brains. Following some suggestions of Sterelny (2012), we argue that, instead, the ratchet is a product of unique social motivations and social circumstances that made possible uniquely human capacities for coordination on cooperative projects.

3 Human vs. Non-human Social Cognition, Social Learning, and Innovation: The Evidence

Accumulating evidence from a variety of non-human species attests to the presence of abilities that are relevant to cumulative cultural evolution. These abilities include both social learning (needed to transmit cultural objects or behaviors to others) and innovation (needed to create new behaviors or objects or to alter existing ones).

Social learning is widely taxonomically distributed. Various forms of social learning have been documented in insects, fish, reptiles, birds, and mammals. Although psychologists disagree about whether social learning is mechanistically distinct from, e.g., individual learning (Heyes 2012), the important question here is simply the extent to which non-human species are capable of transmitting information between individuals and the fidelity with which they are able to do so. As the group of species most closely related to humans, we focus here on non-human primates.

Social learning is often divided into various types, based on the fidelity with which a model's actions or goals are copied. Stimulus enhancement and local enhancement are considered low-fidelity forms: these occur when a model's presence or action draws an individual's attention to an object or a location. In this case, an individual does not learn how to solve a particular problem but simply that there is something of interest about an object or location. Emulation, a higher-fidelity form of social learning, occurs when an individual idiosyncratically copies a goal of (and therefore, if the goal is achieved, the result obtained by) a model. Imitation, often considered the highest-fidelity form of copying, occurs when an individual acquires both the goal and the method of a model.

Because imitation has been a theorized requirement for the ratchet, it is important to determine whether non-human primates can learn via this process. Distinct lines of evidence point to non-human primates' ability to engage in the separate behaviors that make up acts of imitation.

First, chimpanzees and macaque monkeys are able to detect the goals of other individuals. In one paradigm used to demonstrate this, one chimpanzee had access to an array of objects, and another (related) chimpanzee in the next room was presented one of two tasks that require different tools to solve. Among the objects in the first chimpanzee's array was a tool necessary to solve the second chimpanzee's problem. When chimpanzees could see which problem was facing their partner, they more frequently provided the useful tool than a tool that was non-functional in this context (Yamamoto et al. 2012). In a separate study in which a human carer "dropped" and reached toward an object, young chimpanzees retrieved the object (Warneken and Tomasello 2006), although these individuals did not help in other types of situations when the carer requested it. In a paradigm similar to ones more commonly used with human infants, macaque monkeys were shown videos of a human demonstrator engaging in a goal-directed action: reaching over a barrier to grasp a toy. When shown a video of the experimenter reaching toward the toy with no barrier present, the monkeys looked significantly longer at the video in which the demonstrator used an inefficient route (as if she still had to reach over the barrier) than the video in which the demonstrator used an efficient, direct route (Rochat et al. 2008). In such paradigms, longer looking time is interpreted to indicate surprise, and this performance is similar to that of human infants in a related task. Taken together, these findings show that non-human primates are sensitive to the goals of other individuals, conspecific or otherwise.

Second, chimpanzees can copy the goals or results achieved by models, even when they must use an idiosyncratic method to do so. In the floating peanut task, an individual is presented with a tube containing a peanut (or other desirable, lightweight object) that is not reachable with the fingers. The tube is fixed to a substrate so that the object cannot be obtained by turning the tube upside down. One solution to this task is to add water to the tube until the object floats into reach. When shown a demonstration of a human solving this task by adding water to the tube via a water bottle, some chimpanzees subsequently gathered water with their mouths and spat it into the tube (Tennie et al. 2010). This type of response was described as emulation—achieving a demonstrated goal via a novel method. A different study used a task consisting of a box with a panel that can be slid to the left or the right in order to access a reward. Chimpanzees were shown various types of demonstration, including one in which a model chimpanzee did not touch the box, but the experimenter, from out of sight, discreetly slid the panel open using fishing line. When an apparatus appears to solve itself, this is known as a "ghost" demonstration (Hopper 2010). In this case, the chimpanzee model did not visibly perform any actions but did reap the benefits of the ghost demonstration by retrieving the reward. After seeing a ghost demonstration, most observer chimpanzees copied the direction of motion that they had seen when they opened the panel themselves, although they had not seen anyone physically manipulate the task (Hopper et al. 2008). The idiosyncratic solutions in these tasks, which chimpanzees had not seen demonstrated, indicate that some individuals have both the ability and the motivation to copy the goals of another, even if the means are not demonstrated directly.

Finally, some captive chimpanzees are able to copy (fairly simple) combinations of actions and goals that are demonstrated to them by human or chimpanzee models. Many experiments associated with imitative learning use so-called artificial fruits. These generally consist of a box-like apparatus protected by some kind of defense, such as doors or bolts, which must be defeated in order to obtain a food item stored inside. This method mimics an ecological problem (the need to extract fruits, nuts, or seeds from protective casings when foraging) but presents never-before-seen objects to the individuals tested, to prevent prior experience from affecting performance. In one such study, chimpanzee models from different groups were trained to open such an artificial fruit using one of two alternative methods (lift the door or slide the door to the side). A second member from each group was able to observe one of the models solve the problem and then given an opportunity to solve it. A third, naïve chimpanzee from each group was able to watch the interaction of the second chimpanzee with the "fruit" and then given a chance to solve it and so on for five or six "generations." Thus, the solution was transmitted to multiple members of each group; most group members copied the method they had seen used by the model a majority of the time (Horner et al. 2006). A similar pattern emerged in an open diffusion study in which all group members in different groups were able to watch differently trained demonstrators solve a task and obtain a food reward by using a stick to either lift a handle of an apparatus or poke inside the apparatus to dislodge a blockage (Whiten et al. 2005).

The ability to copy both actions and goals is not confined to chimpanzees. Orangutans (Stoinski and Whiten 2003) and gorillas (Stoinski et al. 2001), in similar tasks, learned how to remove the defenses of "artificial fruits," though generally not as successfully as the chimpanzee groups studied. Capuchin monkeys (Dindo et al. 2008) and vervets (van de Waal et al. 2013) have also learned to open artificial fruits from their conspecifics.

Further evidence of non-human primates' imitative ability comes from "do-as-I-do" paradigms, in which individuals are instructed to copy a model's body movements or actions relating to objects. Tomasello et al. (1993b) used a do-as-I-do paradigm with a variety of everyday objects. They tested young human children, enculturated apes (chimpanzees and bonobos brought up in close contact with humans and taught symbolic systems), and mother-reared apes. Young children and enculturated apes performed similarly on this task, and these groups outperformed mother-reared apes. This result points to a key role of experience (versus simply ability) in whether individuals engage in copying behaviors.

Imitation in non-human primates is not limited to three-dimensional tasks. In a paradigm known as the "simultaneous chain," participants must touch pictures in a predetermined order on a touchscreen computer. The spatial locations of pictures are shuffled from trial to trial so that participants cannot use spatial cues to remember the order of the pictures; instead, they must use the contents of the pictures to seriate them. Macaque monkeys were able to learn four-item sequences by watching another monkey in an adjacent enclosure enter the correct sequence (Subiaul et al. 2004). This result indicates that the ability to learn from others extends even to fairly abstract tasks.[1]

A further piece of evidence that some non-human species have all the cognitive capacities relevant to the ratchet is the acquisition by wild chimpanzees of certain complex foraging skills, including termite fishing and nut cracking (Galef 2013). These skills take years to master and appear to have some social learning components, even if we cannot strictly say that they are learned by imitation because no individual does them correctly on the first try. This pattern of skill acquisition is not

[1]Not every experiment examining whether non-human primates can imitate has evidenced this ability. The results of the studies summarized here simply show that such copying is not outside the cognitive abilities of some members of these species.

dissimilar to that displayed by humans when learning numerous technical skills or rituals. When acquired by social learning, these skills involve, first, learning somewhat abstract representations and goals (versus specific movements) from others and, second, engaging in individual learning to realize those representations or goals. Galef (2013) uses the examples of alpine skiing and tai chi; novices cannot learn how to be a good skier or how to do tai chi by simply watching an expert. Rather, they learn what a good skier looks like on the way down the slope, or that it is possible to be highly skilled in tai chi, and must acquire the relevant behaviors through individual practice. Any skill that is possible or even easy to do well on the first attempt may not be amenable to such improvement through practice, and the comparative payoff of learning such a skill socially versus individually would be low. But for a skill that requires practice for mastery, social learning is likely to be the optimal (if not the only) method of acquisition. Therefore, skills that require practice or expertise are more likely to be affected by the ratchet than skills that do not. Chimpanzees' complex foraging skills, acquired via some combination of social learning and individual practice, follow the pattern of skills that are subject to cumulative culture in humans.

There appears to be no question, then, that non-human primates are capable of behaviors required to fuel the first component of the ratchet: social learning. What of the second component, innovation?

Although innovation is difficult to study systematically, especially in wild populations, aggregation of individual reports of novel behaviors can give an estimate of the low end of the possible innovation rate. Reader and Laland (2002) located 533 innovation events in the literature on wild and captive primates. This figure represents only the groups studied by humans and only fairly recent data. More innovations, therefore, can be expected to have occurred in groups unobserved by humans and in the time before modern primatology focused academic interest on non-human primates.

In addition to the innovations directly observed by humans, some past innovations can be inferred from present circumstances. Cultures, or socially transmitted variations in foraging methods and social behaviors between different social groups of the same species, have been documented in wild chimpanzees (Whiten et al. 1999), orangutans (van Schaik et al. 2003), and capuchin monkeys (Panger et al. 2002; Perry et al. 2003). In many cases, this variation cannot be explained by local ecologies (such as the presence or absence of a certain type of prey or nut species). In each local group of these wild populations, at some point in the past, an individual must have created a novel behavior (either from scratch or based on an existing behavior), which the other members of that group then adopted; in other groups, different novel behaviors similarly developed and spread. Wild primate cultures often display such geographic variation across multiple behaviors. The present variation in cultural practices between groups of wild primates attests to multiple past innovation events.

As a third strand of evidence regarding innovation, laboratory researchers have "encouraged" some captive individuals to engage in or invent new behaviors. This is done by presenting individuals with a task that becomes more difficult to solve on subsequent presentations. Indeed, some non-human great ape individuals have been able to successfully solve increasingly difficult versions of these tasks. In one paradigm, members of all four non-human great ape species were presented with a grape at the bottom of a vertical tube (Manrique et al. 2013). In the first stage, because there was a hole near the bottom of the tube, the fingers of one hand could be used to extract the grape. The hole in the tube in the next stage was higher, and apes had to lift a handle attached to the base of the tube to raise the grape several centimeters in order to access it with their fingers; most subjects were successful in retrieving grapes at this stage. The hardest task involved a tube with no hole, and the only effective method of retrieving the grape was to "shoot" it upwards using the handle attached to the base, a solution that animals had to discover on their own. Around half of the individuals in this study were able to retrieve grapes in this difficult version of the task (Manrique et al. 2013). In another study, orangutans were first given a similar apparatus: a wide vertical tube with syrup in the bottom (Lehner et al. 2011). Initially, syrup could be retrieved simply by putting fingers into the tube. The orangutans were next given a narrower tube into which their fingers could not fit, and they dipped tools, such as paper, vegetables, or leafy twigs, into the tube to access the syrup. In the final and most difficult condition, favored tools (leafy twigs) were removed, so other tools had to be used. Several orangutans combined techniques from previous stages of the task to come up with efficient ways to retrieve the syrup, such as squashing paper down into the tube with a stick and, after syrup had stuck to it, retrieving it using the stick (Lehner et al. 2011).[2]

Some would argue that not all innovations have equivalent effects. That is, in the ratchet, innovations must occur, not just from scratch, but rather as improvements to existing skills, technologies, or practices (Charbonneau 2015). There is currently little direct evidence that non-human primates are capable of innovations that build on existing traditions. However, Dean and colleagues (2014) consider several extant traditions in wild primate groups that may be variations that have arisen via innovation from previous versions of these traditions. For example, one group of capuchins that initially engaged in bond-testing rituals such as hand sniffing developed a new ritual of eye poking that may be a derivative form (Perry 2011). Boesch (2003) considers three candidate behaviors to have arisen via successive innovations in wild chimpanzees. These are nut cracking, with the trait potentially evolving from nuts held in the hand (initial state) to nuts cracked on anvils with stone hammers (current state); ectoparasite manipulation, with the initial innovation of removing parasites and placing them on leaves for inspection and subsequent ones of folding and cutting the leaves with a fingernail (at one site) or using a pile of leaves for inspection (at another); and well digging, with the trait potentially evolving from digging in dried water beds during dry periods to access water (initial state) to digging in wet periods near water sources, possibly to avoid parasites (current state). Although Dean and colleagues (2014) doubt that any of these traits arose via

[2]n.b., success is not universal in all such "forced innovation" tasks. Marshall-Pescini and Whiten (2008) and Dean et al. (2012) found very low levels of performance in their respective tasks.

innovation on existing traditions, the matter is not settled. Long-term studies of wild primate groups have the potential to clarify whether specific traditions have arisen as variations on initial behavior patterns.[3]

Field and experimental work, then, provides evidence that non-human primates are capable of innovating a variety of new behaviors to solve various types of problems and socially learning both social and technological behaviors. Hence, there is evidence that some non-human primates have the cognitive capacities we identified earlier as necessary for cumulative cultural evolution: high-fidelity imitation and innovation. However, this phenomenon has not been observed in non-human primate populations. This is the crux of our puzzle. It suggests that the key to explaining the rise of cumulative cultural evolution in the hominin lineage lies elsewhere than in the cognitive capacities of individuals. Next, we propose and explore an alternative approach: identifying distinctively human social and motivational factors relevant to cumulative cultural evolution.

4 Social and Motivational Roots of the Ratchet

It seems that explaining the human uniqueness of the ratchet encounters a number of puzzles. First, it is not clear how influential hypotheses focused on overimitation accommodate the facts of human phylogeny: Lyons' technological skill acquisition hypothesis faces a phylogenetic bootstrapping problem, while the various ritual-focused hypotheses face problems of phylogenetic timing. Second, there is evidence that contemporary non-human primates have the cognitive capacities required for the ratchet. Thus, the human uniqueness of the ratchet, and its phylogenetic roots, remain mysterious. We think that part of the problem is an excessive focus on the cognitive capacities required for the ratchet. In this section, we suggest a reconceptualization of the ratchet, along lines suggested by Kim Sterelny (2012), which avoids this excessive focus on cognitive capacities, thereby pointing the way to resolving the problems posed by bootstrapping, phylogenetic timing, and the social learning and innovation capacities of contemporary non-human primates.

In his conceptually innovative book on human evolution, *The Evolved Apprentice* (2012), Kim Sterelny offers an alternative to mainstream conceptualizations of human cognitive evolution. According to most evolutionary psychologists, the

[3]A reviewer mentions evidence that even human children do not show capacities for innovation until relatively late in development: 7 years of age (Beck et al. 2011; Cutting et al. 2011, 2014). But the kind of innovation needed to give rise to the ratchet is relatively minimal: just copying with less than perfect fidelity. All that is needed is a source of variation. And there is plenty of evidence that young children and chimpanzees often copy methods of achieving new goals with less than perfect fidelity. In fact, Legare et al. (2015) provide evidence that children as young as four already innovate in imitative contexts marked as instrumental rather than conventional. And the whole point of the Horner and Whiten (2005) study is that chimpanzees are more likely to adopt more efficient methods for accomplishing a goal, i.e., more likely to innovate, than human children.

evolution of distinctively human behavioral traits should be modeled roughly as the gradual accumulation of neurally implemented cognitive modules, dedicated to specific domains. The human mind is likened to a "Swiss army knife," with different components serving different functions that had significant adaptive value at some point in human prehistory (Barkow et al. 1992). For example, there might be a module dedicated to social cognition, employing so-called theory of mind; another module dedicated to knowledge of the natural environment, employing so-called folk biology; another module dedicated to manipulating physical objects, employing so-called folk physics; and yet another module dedicated to detecting and responding to norm violations, e.g., Cosmides and Tooby's "cheater detection module" (1992). The idea is that there were constant, adaptively relevant ecological features in human prehistory, roughly the Pleistocene, and these constitute our so-called environment of evolutionary adaptedness (Tooby and Cosmides 1990). These ecological features selected for genes coding for the construction of bits of brain tissue that implement the cognitive modules identified by evolutionary psychologists. According to Sterelny, this is a fundamentally misguided picture of human cognitive evolution.

Sterelny notes that many of the cognitive domains most central to human evolutionary success do not show the kind of long-term stability necessary to select for genes coding for specific, neurally implemented cognitive modules. The social domain is a good example. It is likely that styles of behavior and normative regimes varied greatly across human prehistory, as they have during recorded history. Even the natural environment is likely to have been relatively unstable, not just due to climatic variation but also, and importantly for Sterelny, due to human effects on the natural environment. For example, foraging strategies that were effective at one epoch and location in human prehistory likely were not effective at other epochs or locations, partly due to the effects of human foraging, e.g., driving certain prey species to extinction. Furthermore, many of the most important behaviors likely involved drawing simultaneously on capacities for dealing with multiple domains. For example, foraging likely required skillful manipulation of the social, natural, and physical domains, since hunting parties are social entities, employing physical technologies, to hunt or extract parts of the natural environment, i.e., other species. So, the sorts of behaviors that likely played important roles in human survival in prehistory required flexible skills at integrating knowledge about multiple domains, not compartmentalized, domain-specific modules.

Instead of piecemeal accumulation of cognitive modules hypothesized by the "Swiss army knife" model, Sterelny proposes the following. Human prehistory consisted of complex feedback loops, in which the effects of certain behaviors on the social, natural, and physical environments of our precursors led to the construction of new niches, which selected for new, multi-domain skills, leading to further effects on the environment, in a constant, dynamic spiral. Hominins managed these unstable dynamics by gradually developing skills at learning from each other, i.e., essentially, the ratchet. But these skills were not primarily the products of cognitive modules constructed from naturally selected genetic blueprints. Rather, they were by-products of changing social circumstances, especially more stable populations with denser demographics, and relatively minor tweaks of socially relevant motivations.

Here is the basic hominin evolutionary arc that Sterelny defends. The first step involved relatively minor modifications to basic great ape behavior patterns in more cooperative directions. Sterelny argues that the first hominins to diverge from the lineage we share with chimpanzees cooperated on predator defense, scavenging, and care for offspring. But even such minor forms of cooperation so altered the hominin niche that new selection pressures emerged, favoring more cooperation and, especially, cross-generational information sharing. Successful cooperative scavenging and predator defense required the preservation and accumulation of knowledge about local flora and fauna, coordination techniques, and, eventually, technological expertise. This required the cross-generational transmission of information. Fortunately, successful cooperative foraging and predator defense also led to better nutrition and lower adult mortality. This allowed juveniles to mature at a slower rate, protected and nourished by whole groups of adults, rather than just their immediate progenitors. This made possible improved cross-generational information sharing, which led to even better cooperative foraging and predator defense, again lowering adult mortality, improving nutrition, and hence enhancing conditions necessary for even better cross-generational information sharing. In time, this dynamic triggered major demographic changes, yielding much denser and more diverse populations. These changes continued to support improved cross-generational information transfer, but they also gave rise to new challenges, especially interaction with strangers and, more broadly, dealing with increasingly heterogeneous social environments, as groups fissioned, and a division of labor led to increasing specialization.

Notice that the key changes in this phylogenetic story do not involve cognitive innovations. The changes are entirely social and motivational. Early hominins were distinct in motivations to cooperate on predator defense, scavenging, and care for offspring. This relatively minor motivational tweak led to dramatic alterations in the social niche: more efficient coordination on cooperative projects was now rewarded, and social environments became more stable and more densely populated. This selected for better social learning: inclusive fitness benefitted when offspring acquisition of skills at coordination was somehow scaffolded. Initially this need not have required specifically cognitive adaptations. Sterelny (2012) speculates that relatively minor tweaks to motivations, e.g., adult tolerance of offspring observation of and participation in coordinative activities, may have been enough to improve social learning using already extant individual cognitive capacities of the kind discussed above: offspring would have more opportunities to internalize abstract behavioral representations and goals involved in coordination and then practice them on their own. As coordination became more effective through more efficient social learning, hominin populations became more successful foragers and hunters, reaping nutritional boons that made groups even more stable and dense in redundant, adult social models. These changes in the social niche also made possible ever more protracted childhoods, as increasing numbers of adults were available for alloparenting and protection against predation. With more adult social models, and longer juvenile periods, social learning was enhanced once again, leading to consequent improvements in foraging, hunting, and nutrition, in a positive feedback loop.

On this view, distinctively human motivations and social facts are the key drivers of the ratchet. Only populations motivated to coordinate on cooperative endeavors, possessing a sufficient density of adult social models, with motivations to at least tolerate offspring observation of and participation in coordinative activities and to support protracted childhoods devoted to social learning through alloparenting, are likely to yield a ratchet. Once these social and motivational facts are on the scene, cognitive capacities that we share with non-human primates can yield the ratchet.

Note that, on this view, the bootstrapping problem raised above is entirely avoided. We need not posit a special domain, like skill at constructing and using complex technologies, to explain phenomena like overimitation, required for the ratchet. Instead, the ratchet is a by-product of motivational and social factors that are rewarded for other reasons, e.g., better predator defense, foraging, and care of offspring. Once such factors are in place, the ratchet can become self-sustaining, in virtue of yielding better coordination on cooperative endeavors, which yield better foraging, predator defense, and offspring care, which, in turn, further feed these motivational and social factors in a positive feedback loop. Once such a way of life is stable and self-sustaining, there may then be selection for further minor tweaks to motivational and cognitive factors that make the ratchet more efficient. For example, social learning can be improved if offspring come with innate motivations to copy precise sequences of adult behavior and better cognitive capacities to parse such behavior. Complex technologies would be a natural by-product of such capacities. Furthermore, once population densities reach levels in which ritual and other mechanisms of group identification and cohesion are beneficial, there would be another useful function for enhanced social learning mechanisms, like overimitation.

Thus, Sterelny's story succeeds in avoiding the puzzles raised by other accounts of the ratchet that focus exclusively on allegedly unique, human cognitive capacities. Bootstrapping is not a problem for it, since it shows how improved social learning can gradually emerge as a by-product of social and motivational enablers and consequences of coordination on cooperative activities. Phylogenetic timing is likewise not an issue, since efficient social learning would be rewarded very early on in populations reliant on coordination on cooperative activities. Finally, the presence of cognitive capacities seemingly sufficient for the ratchet in contemporary non-human primates is no longer mysterious. The human uniqueness of the ratchet derives from our unique cooperative motivations and social factors, like population densities and protracted childhoods, rather than from unique cognitive capacities. Indeed, the historical archaeological record provides a very strong argument for this perspective: there are cases of human populations experiencing significant declines in technological sophistication, i.e., apparent reversals of the ratchet, due to radical changes in demographics like massive declines in density, not to any cognitive deficits. One classic example is the ancestral Aboriginal population in Tasmania (Henrich 2004).

5 Conclusion

We began with two broad puzzles about the human uniqueness of the ratchet: phylogenetic timing and contemporary non-human capacities for social learning and innovation. The most influential hypotheses about the phylogeny of the ratchet encounter problems of timing. The idea that the ratchet was required to acquire tool making and using expertise encounters a bootstrapping problem: the ratchet needs to be in place before there are complex technologies to master. The idea that the ratchet was required to acquire ritual necessary for group identification and cohesion seems at odds with the timing available from the archaeological record: tools requiring the ratchet appear to predate the kinds of demographics in which ritual would be rewarded by hundreds of thousands of years. Furthermore, there is growing evidence that non-human primates (at least) have all the cognitive capacities presupposed by the ratchet.

Kim Sterelny's (2012) hypothesis that human evolution is the result of a positive feedback loop, in which coordination on cooperative endeavors was rewarded and then altered social and natural niches in ways that selected for ever more sophisticated coordination on cooperative endeavors, has the potential to explain the human uniqueness of the ratchet, without raising puzzles about phylogenetic timing, and the contemporary distribution of social learning and innovation capacities among primates. On Sterelny's story, motivations to cooperate and learn from one another, together with social factors like population densities sufficient to support alloparenting and protracted childhoods, as well as redundant social models, arose early in the hominin lineage, and gave rise to a self-sustaining ratchet.

References

Barkow JH, Cosmides L, Tooby J (eds) (1992) The adapted mind: evolutionary psychology and the generation of culture. Oxford University Press, New York

Beck SR, Apperly IA, Chappell J, Guthrie C, Cutting N (2011) Making tools isn't child play. Cognition 119:301–306

Boesch C (2003) Is culture a golden barrier between human and chimpanzee? Evol Anthropol 12: 82–91. https://doi.org/10.1002/evan.10106

Boyd R, Richerson PJ (1996) Why culture is common but cultural evolution is rare. Proc Br Acad 88:73–93

Charbonneau M (2015) All innovations are equal, but some more than others: (re)integrating modification processes to the origins of cumulative culture. Biol Theory 10:322–335. https://doi.org/10.1007/s13752-015-0227-x

Cosmides L, Tooby J (1992) Cognitive adaptations for social exchange. In: Barkow JH et al (eds) The adapted mind: evolutionary psychology and the generation of culture. Oxford University Press, New York, pp 163–228

Cutting N, Apperly IA, Beck SR (2011) Why do children lack the flexibility to innovate tools? J Exp Child Psychol 109:497–511

Cutting N, Apperly IA, Chappell J, Beck SR (2014) The puzzling difficulty of tool innovation: why can't children piece their knowledge together? J Exp Child Psychol 125:110–117

Dean LG, Kendal RL, Schapiro SJ, Thierry B, Laland KN (2012) Identification of the social and cognitive processes underlying human cumulative culture. Science 335:1114–1118. https://doi.org/10.1126/science.1213969

Dean LG, Vale GL, Laland KN, Flynn E, Kendal RL (2014) Human cumulative culture: a comparative perspective. Biol Rev 89(2):284–301

Dindo M, Thierry B, Whiten A (2008) Social diffusion of novel foraging methods in brown capuchin monkeys (*Cebus apella*). Proc R Soc B 275:187–193. https://doi.org/10.1098/rspb.2007.1318

Galef BG (2013) Imitation and local enhancement: detrimental effects of consensus definitions on analyses of social learning in animals. Behav Process 100:123–130. https://doi.org/10.1016/j.beproc.2013.07.026

Henrich J (2004) Demography and cultural evolution: how adaptive cultural processes can produce maladaptive losses – the Tasmanian case. Am Antiq 69(2):197–214

Heyes C (2012) Simple minds: a qualified defence of associative learning. Philos Trans R Soc B 367:2695–2703. https://doi.org/10.1098/rstb.2012.0217

Hopper LM (2010) 'Ghost' experiments and the dissection of social learning in humans and animals. Biol Rev 85:685–701. https://doi.org/10.1111/j.1469-185X.2010.00120.x

Hopper LM, Lambeth SP, Schapiro SJ, Whiten A (2008) Observational learning in chimpanzees and children studied through 'ghost' conditions. Proc R Soc B 275:835–840. https://doi.org/10.1098/rspb.2007.1542

Horner V, Whiten A (2005) Causal knowledge and imitation/emulation switching in chimpanzees (*Pan troglodytes*) and children (*Homo sapiens*). Anim Cogn 8:164–181

Horner V, Whiten A, Flynn E, de Waal FBM (2006) Faithful replication of foraging techniques along cultural transmission chains by chimpanzees and children. Proc Natl Acad Sci USA 103(37):13878–13883. https://doi.org/10.1073/pnas.0606015103

Legare CH, Nielsen M (2015) Imitation and innovation: the dual engines of cultural learning. Trends Cogn Sci 19(11):688–699

Legare CH, Souza AL (2012) Evaluating ritual efficacy: evidence from the supernatural. Cognition 124:1–15

Legare CH, Wen NJ, Herrmann PA, Whitehouse H (2015) Imitative flexibility and the development of cultural learning. Cognition 142:351–361

Lehner SR, Burkart JM, van Schaik CP (2011) Can captive orangutans (*Pongo pygmaeus abelii*) be coaxed into cumulative build-up of techniques? J Comp Psychol 125:446–455. https://doi.org/10.1037/a0024413

Lyons DE (2009) The rational continuum of human imitation. In: Pineda JA (ed) Mirror neuron systems: the role of mirroring processes in social cognition. Humana Press, New York, pp 77–106

Manrique HM, Völter CJ, Call J (2013) Repeated innovation in great apes. Anim Behav 85: 195–202. https://doi.org/10.1016/j.anbehav.2012.10.026

Marshall-Pescini S, Whiten A (2008) Chimpanzees (*Pan troglodytes*) and the question of cumulative culture: an experimental approach. Anim Cogn 11:449–456. https://doi.org/10.1007/s10071-007-0135-y

McBrearty S, Brooks AS (2000) The revolution that wasn't: a new interpretation of the origin of modern human behavior. J Hum Evol 39:453–563. https://doi.org/10.1006/jhev.2000.0435

Nielsen M, Tomaselli K (2010) Overimitation in Kalahari Bushman children and the origins of human cultural cognition. Psychol Sci 21(5):729–736

Over H, Carpenter M (2012) Putting the social into social learning: explaining both selectivity and fidelity in children's copying behavior. J Comp Psychol 126(2):182–192

Over H, Carpenter M (2013) The social side of imitation. Child Dev Perspect 7(1):6–11

Panger MA, Perry S, Rose L, Gros-Louis J, Vogel E, Mackinnon KC, Baker M (2002) Cross-site differences in foraging behavior of white-faced capuchins (*Cebus capucinus*). Am J Phys Anthropol 119:52–66

Perry S (2011) Social traditions and social learning in capuchin monkeys (*Cebus*). Philos Trans R Soc B 366:988–996. https://doi.org/10.1098/rstb.2010.0317

Perry S, Baker M, Fedigan L, Gros-Louis J, Jack K, MacKinnon KC, Manson JH, Panger M, Pyle K, Rose L (2003) Social conventions in wild white-faced capuchin monkeys: evidence for traditions in a neotropical primate. Curr Anthropol 44:241–268

Powell A, Shennan S, Thomas MG (2009) Late Pleistocene demography and the appearance of modern human behavior. Science 324:1298–1301

Reader SM, Laland KN (2002) Social intelligence, innovation, and enhanced brain size in primates. Proc Natl Acad Sci USA 99:4436–4441. https://doi.org/10.1073/pnas.062041299

Rochat MJ, Serra E, Fadiga L, Gallese V (2008) The evolution of social cognition: goal familiarity shapes monkeys' action understanding. Curr Biol 18:227–232. https://doi.org/10.1016/j.cub.2007.12.021

Sterelny K (2012) The evolved apprentice. MIT, Cambridge

Stoinski TS, Whiten A (2003) Social learning by orangutans (*Pongo abelii* and *Pongo pygmaeus*) in a simulated food-processing task. J Comp Psychol 117:272–282. https://doi.org/10.1037/0735-7036.117.3.272

Stoinski TS, Wrate JL, Ure N, Whiten A (2001) Imitative learning by captive Western lowland gorillas (*Gorilla gorilla gorilla*) in a simulated food-processing task. J Comp Psychol 115: 272–281. https://doi.org/10.1037/0735-7036.115.3.272

Subiaul F, Cantlon JF, Holloway RL, Terrace HS (2004) Cognitive imitation in rhesus macaques. Science 305:407–410. https://doi.org/10.1126/science.1099136

Tennie C, Call J, Tomasello M (2010) Evidence for emulation in chimpanzees in social settings using the floating peanut task. PLoS One 5(5):e10544. https://doi.org/10.1371/journal.pone.0010544

Tomasello M, Kruger AC, Ratner HH (1993a) Cultural learning. Behav Brain Sci 16:495–552

Tomasello M, Savage-Rumbaugh S, Kruger AC (1993b) Imitative learning of actions on objects by children, chimpanzees, and enculturated chimpanzees. Child Dev 64:1688–1705

Tomasello M, Carpenter M, Call J, Behne T, Moll H (2005) Understanding and sharing intentions: the origins of cultural cognition. Behav Brain Sci 28:675–735

Tooby J, Cosmides L (1990) The past explains the present: emotional adaptations and the structure of ancestral environments. Ethol Sociobiol 11(4–5):375–424

van de Waal E, Claidière N, Whiten A (2013) Wild vervet monkeys copy alternative methods for opening an artificial fruit. Anim Cogn 18:617–627. https://doi.org/10.1007/s10071-014-0830-4

van Schaik CP, Ancrenaz M, Borgen G, Galdikas B, Knott CD, Singleton I, Suzuki A, Utami SS, Merrill M (2003) Orangutan cultures and the evolution of material culture. Science 200: 102–105. https://doi.org/10.1126/science.1078004

Warneken F, Tomasello M (2006) Altruistic helping in human infants and young chimpanzees. Science 311:1301–1303. https://doi.org/10.1126/science.1121448

Whiten A, Goodall J, McGrew WC, Nishida T, Reynolds V, Sugiyama Y, Tutin CEG, Wrangham RW, Boesch C (1999) Cultures in chimpanzees. Nature 399:683–685. https://doi.org/10.1038/21415

Whiten A, Horner V, de Waal FBM (2005) Conformity to cultural norms of tool use in chimpanzees. Nature 437:737–740. https://doi.org/10.1038/nature04047

Yamamoto S, Humle T, Tanaka M (2012) Chimpanzees' flexible targeted helping based on an understanding of conspecifics' goals. Proc Natl Acad Sci USA 109:3588–3592. https://doi.org/10.1073/pnas.1108517109

Emulation, (Over)imitation and Social Creation of Cultural Information

Laura Desirèe Di Paolo and Fabio Di Vincenzo

Abstract The creation of cultural information by humans is an ability that requires to compound together different factors. Although information needs to be transmitted faithfully enough so to prevent errors, space must be left to create innovations at the same time.

Individual trial and error is the principal source of innovations among all primate species, especially in emulative contexts, but it does not explain the quantity, quality or rapidity of human cultural production. On the other hand, imitation and (over) imitation explain quite well faithful transmission and error control but do not explain the creation of cultural novelties nor the ratchet effect of human culture. To explain these latter components, we need a combination of trial and error in emulative contexts and (over)imitation. Here we suggest that this combination of the ability in creating innovations and transmitting them faithfully occurred for the first time during the Palaeolithic. In that time frame, we can detect the establishment of imitation as the main social learning strategy in the genus *Homo*. Adopting a niche construction (henceforth NC) paradigm, we propose that this combination became a social characteristic of *Homo sapiens* which ontogenetically happens when children reach the school age in modern humans.

Keywords Innovation · Monkey · Ape · Human · Hunter-gatherer · Palaeolithic · Human evolution

L. D. Di Paolo (✉)
Lichtenberg-Kolleg—Institute for Advanced Study at the Georg-August-Universität Göttingen, Göttingen, Germany

F. Di Vincenzo
Department of Environmental Biology, Sapienza Università di Roma, Roma, Italy

L. D. Di Paolo et al. (eds.), *Evolution of Primate Social Cognition*, Interdisciplinary Evolution Research 5, https://doi.org/10.1007/978-3-319-93776-2_17

1 Introduction

Individual trial and error is the main source of innovation among non-human primates. In everyday life, individuals discover new solutions for finding food, mating, sleeping comfortably, etc., by themselves. However, a large part of these new (and useful) solutions are randomly lost, or do not spread within the social group. One of the reasons for such low transmission rates is that innovations are not (always) coupled with faithful social learning mechanisms. Only by appearing jointly they can acquire cultural relevance, permitting reliable social preservation and accumulating cross-generational information (Legare and Nielsen 2015; Shipton and Nielsen 2015). Since this joint action depends on the interweaving of genetic, ecological, social and cognitive factors, the differential contributions of these factors to the cultural behaviours exhibited by primates can determine differences in the cultural strategies adopted by each species.

All primates display a wide array of cognitive abilities that allow them to manage their *specific forms* of cultural information. Furthermore they can shape their individual behaviour focusing on the others' behaviour through different strategies of observational and social learning (henceforth SL) and depending on the differences in the learner's attentional focus (e.g. Tomasello 1996, 1999). This focus can be directs on the object (or its affordances) used by a conspecific (*stimulus enhancement*), on the outcome of a procedure (*emulative learning*) or on the procedure itself and the steps through which the model reaches a particular goal (*imitative learning*) (Byrne and Russon 1998). It is also possible to distinguish different modalities of cultural knowledge and skill acquisition among primates, grounding this distinction in the preferred acquisition channel adopted by the animals.

For instance, capuchin monkeys (*Sapajus apella*) adopt standardised procedures in nut-cracking behaviour. They learned or developed that specific behaviour by observing and interacting with conspecifics, thanks to the material outcomes of their actions (tools and places for collecting them, places for food processing, resulted waste, etc.), within favourable ecological conditions (e.g. Koops et al. 2014; Fragaszy 2012; Visalberghi and Fragaszy 2012). These monkeys offer a clear case of a socially stimulated tool-use (cultural) tradition as they learnt significant food processing techniques from one another. However, in this case, ecological elements work as specific constrains in the acquisition of nut-cracking techniques, for example, which kind of stone to use, where to collect it and how to choose the best location for displaying the behaviour, the age and the weight of the individuals. This behaviour can be considered an *ecologically based (potentially cumulative) culture*, in which any form of accumulation of innovations is possible and it can largely depend on (and is highly constrained by) the subsistence environment of the species. This assumption has recently been proven by the presence of flakes made by the capuchin monkeys, as an intentional result of the primary intention of damaging stones (Proffitt et al. 2016).

The case of chimpanzees (*Pan troglodytes*) is different: in this species, true socially driven, imitative learning seems to be stimulated by opaque causal conditions. Even though they prefer individual trial and error within the *emulative* learning context in problem-solving tasks (individual strategies), these animals adopt a procedurally oriented process (*imitative* learning) as they do not realise the causal connection guiding the demonstrator's behaviour (Whiten et al. 2004; Horner and Whiten 2005; Buttelmann et al. 2007, but see 2013).[1] They selectively choose emulation or imitation, and they exhibit their cognitive capacity shifting from an *ecologically based culture* to a true *socially based (potentially cumulative) culture*. Being restricted to the objects involved in the situation, they learn either how to solve in autonomy the task (emulation) or to reproduce all the procedures adopted by the demonstrator (imitation). However, in chimpanzees, these two strategies are never combined, and individuals use single strategies alternatively and in different situations. It is this inability in combining strategies which, in our opinion, strongly limits the possibility for this species to establish the *ratchet effect* (*sensu* Tomasello 1999) which characterises human cumulative culture.

We can assume that the same strong separation is shown by humans, particularly during the early stages of infancy. In contrast to chimpanzees, humans switch between individual strategies and pedantic copying, particularly at a younger age and in relation to their social context (Shimpi et al. 2013). Given the correlation between imitation of unnecessary actions and acquisition of cultural practices (Nielsen et al. 2012), young humans seem to imitate and over-imitate much more in the presence of adult (competent) demonstrators, paying particular attention to adults' actions (McGuigan et al. 2011; Harris 2012; Wood et al. 2012, 2013a, b). In one study, 12-, 18-, and 24-month-olds were left watching an adult retrieving a toy from a closed box by disengaging a switch located on the front of the box (Nielsen 2006). The box could be easily opened by hand; however, the demonstrator complicated the action by using a miscellaneous object to operate the switch. Twenty-four-month-olds (unlike 12-month-olds) persevered in copying the inefficient procedure proposed by the model (i.e. using the object), often failing the opening of the box. Subsequent studies have shown that this tendency to copy all of the actions used by an adult when solving a novel task increases with age, even if adults copy less efficiently than children (McGuigan et al. 2011).

On the other hand, they act more individualistically in a socially stimulated play context (Nielsen 2011), safely experimenting with their individual potential in play and pretend behaviours via trial and error, in what can be named *playful emulation*. Later in development, together with the acquisition of linguistic proficiency, they can combine these two strategies: by means of (over)imitative behaviour, children acquire both the fundamental competences in the tool-use domain and the deep cognitive organisation that will raise the scaffolding for the efficient spread of new inventions. In human children, innovations are still the product of individual trial and

[1]Additionally, chimpanzees seem to be particularly prone to display imitation if they have had extensive contact with humans (e.g. Russon and Galdikas 1993; Hobaiter and Byrne 2010; Horner and Whiten 2005; Whiten et al. 2005; Tennie et al. 2012). The probable reason is that, within the human context, they could have been often surrounded by opaque tasks, triggering imitation more often than among conspecifics or in nature.

error (mainly within an emulative context), but they can be generated and accumulated more easily compared to what happens in apes, because they grow upon those socio-cognitive and procedural pillars that are provided by (over)imitation. Additionally, and this is the main difference with chimpanzees and other great apes which we claim to be specific to human SL/creativity, this combination turns the imitative chain (i.e. the *procedure*) into an outcome per se, something that can be modified, and *innovated upon itself*, by turning the imitative procedure (in which passages for reaching the goal are known) into an emulative procedure (in which only the goal or the intention is known). In this framework, the preference for (over)imitation accorded by children is not only connected with the transmission of procedures; it also supports their rapid invention. Thus, we claim, since human exploitation of learning strategies is very unique, compared to their closest evolutionary relatives, humans can be considered purely *social innovators*.

2 Between Innovation and Transmission

For modern humans, transmitting innovations is not such a big deal. Making inventions can be difficult but, once made, they spread quickly since humans are equipped with very faithful mechanisms of social transmission. Looking at extant hunter-gatherer societies, infants and toddlers up of 5 years of age apprehend pivotal information, mostly vertically, from adult individuals, whereas horizontal and oblique learning occur later, in middle and late childhood. It is only from 5 to 12 years of age that children of those societies start to apprehend information horizontally (Högberg et al. 2015). The well-known scarcity of *direct teaching* (involving *helping* and *correcting*, *showing* and *explaining*: Högberg et al. 2015) in extant hunter-gatherer societies suggests that transmission happens by means of SL strategies, such as imitation and emulation, and children up to 5 years of age are particularly prone to imitate and pedantically (over)imitate in situations involving tool use (Berl and Hewlett 2015; Nielsen and Tomaselli 2010). Under these conditions, it is more likely that horizontal learning would be favoured in a cognitive framework already "shaped" by childhood (over)imitation.

The cognitive development in our species seems to be driven by two tendencies: to (1) (over)imitate and to (2) individually produce innovations. These two do not run side by side from the beginning; in fact, some authors even claim they work *in tandem* (Legare and Nielsen 2015), but they combine at some point during development, probably only around 5 years of age, when children become prone to acquiring information horizontally. It is in fact at this age that children become socially and technically competent on their own and able to learn from their peers (having already acquired, and embodied, basic skills during the previous phases of development). Actually, in children younger than 5 years, innovative abilities cannot go beyond the inventive level of the individual (Tennie et al. 2014).

One explanation of the pedantic (over)imitative tendency of children is their propensity to judge all elements of the procedure shown by an adult as causally relevant (Lyons et al. 2011) even if other scholars (e.g. Kenward 2012; Kenward et al. 2011) reported that children over-imitate even when they know the causal connections between actions and outcomes. Growing up, proven to be competent on their own and skilled in establishing causal relationships, children understand that not all sequences are relevant for, or causally connected to, the aimed goal. Able to keep the goal in their memory background and to include a different midterm goal, they can now turn their attention to procedures, innovating them or adapting old ones to new situations. The true novelty here is that—in order to obtain this procedural innovation—humans exploit a sort of emulative strategy *on top of* the procedures that they have previously learnt or embodied via pedantic (over)imitation.

In this way, in humans, there is no need to create an expensive and evolutionary new SL strategy from scratch, because an ontogenetic combination of those already possessed works pretty well by turning them into two elements of a very productive "learning-inventing" strategy, which also combines physical, procedural and social cues. In more detail, this combination makes possible the interruption of, and consequent, innovation upon imitative chains. It also makes possible the production of a wide array of new and different outcomes, starting from a stemmed (probably well-trained) procedure, and rearrangement of operational sequences and technological knowledge hierarchically rather than linearly (Andersson 2011). This largely contributes to reducing cognitive effort and time spent inventing or performing new procedures, introduced to reach a new goal or outcome. More importantly, this strategy reduces the possibility of running into *fatal errors* associated with new, untrained and unexplored procedures.

In this type of *emulation 2.0*, what modern humans do is not to perform the complete reconstruction of the learned procedure but making innovations *within* this structure and, more significantly, *by means* of this structure. Actually, the particular combination realised by *emulation 2.0* works as a coarse filter *before* ratcheting. On one hand, it prevents (or strongly reduces) the severe mistakes that are usually associated with non-trained procedures (and in general with non-experts' activities), while from the other, it permits minor errors or variants to pass this sieve, thereby becoming a main source of variations and innovations. Thus, these new variants can be safely and easily incorporated (both cognitively and productively) into the extant well-trained procedures supplied by imitation.

We claim that this combination, or synthesis, of the cognitive structure of pedantic (over)imitation and emulation allows the kind of culture humans exploit, characterised by the "ratchet effect". In fact, imitation does not only offer a faithful channel for the transmission of innovations, but also, as childish over-imitation, it actively constrains and shapes the cognitive mechanisms of embodiment and prediction which allow error avoidance (sensu Clark 2013) and increase efficiency and parsimony in both time expenditure and memory effort. Meaning that the *embodiment* (acquisition) of basic skills (including know-how and rules), through pedantic training, so repetitive that at first sight it seems to have nothing to do with creativity, is necessary not only for becoming proficient in a particular field of application but even for becoming a competent innovator.

This model is similar to that proposed by Johnson-Laird (1988), in his multistage causal explanation of human creativity, in which both internal (e.g. cognitive, developmental) and external (e.g. environmental, social) constraints (*criteria*, according to Johnson-Laird's terminology) interact in the production and spread of novelties. Our proposal here extends Johnson-Laird's model in a NC perspective. In this sense, the sociocultural environment not only "chooses" individuals' novelties and best procedures, but by doing this, it determines which selective pressures will modify the distribution of the frequencies of the "internal" (genotypic) constraints—propensity to (over)imitate, for instance—within the population. This allows for the fast and correct acquisition and execution of procedures and the production of new outcomes. These pressures increased dramatically during the Palaeolithic, due to the shifting of the early *Homo* into a new trophic (and social) niche, boosting in this way the cognitive development of our own genus. It was in this period that humans, contrarily to their evolutionary relatives, started to actively produce innovations, becoming able to store more data inside and outside the brain. From an ontogenetic point of view, the strong dependence of human infants and toddlers on the social group (on the mother in particular) for a prolonged time in respect, for example, to chimpanzees (namely, altriciality) releases the early stages of ontogeny (infancy) from any immediate application of acquired knowledge while allowing a creative rearrangement of learnt information. In such a view, the creativity characterising modern humans' industries is not only a matter of the cognitive capacity required to produce many innovations, but it is grounded in the opportunity to have a control on correctness, generativeness and functional reliability of the innovations at each stage of their production. This control (or expertise) is ultimately provided by the tight link it has with pedantic (over)imitative tendencies. This synthesis is offered by *emulation 2.0*, and in our view, it theoretically explains and allows for the evolution of the ratchet effect and the high quality of standard human innovations and as such is one of the reasons why culture has assumed such a momentous relevance in human evolution, particularly in *Homo sapiens*.

3 Pedantic Imitation Makes You an Expert Toolmaker

Some scholars (e.g. Caldwell and Millen 2009; Caldwell et al. 2012; Pradhan et al. 2012; Caldwell 2014; Zwirner and Thornton 2015; Morin 2016) claimed that imitation is not essential for cultural transmission and that reliability on cultural transmission—avoiding the loss of significant information—does not require pedantic copying of all passages but only that the final result will be achieved (Boesch 1997; for a critique of a unitary method of cultural transmission, see Sterelny 2012). With regard to this objection, we agree that pedantic (over)imitation is probably not the only way to reach high-fidelity cultural transmission. In fact, in order to obtain a very low information dropout rate, different channels—not necessarily social—can be exploited (Boesch 1997; Sterelny 2012; Zwirner and Thornton 2015), and some animal species can show a high degree of reliable transmission of information even without imitation, as we showed above being the case for capuchin monkeys. However, it is a quite trivial observation that human culture is not comparable in terms of number of innovations and transmission integrity in respect to that of the other primate species. And since the only really relevant difference consists in human children's preference for imitation, stubbornly copying a model's actions in both causally opaque and clear circumstances (Horner and Whiten 2005; Lyons et al. 2007; Whiten et al. 2009; Hopper et al. 2012; Legare and Nielsen 2015), this has to be related to the kind of culture humans developed during their evolution. In other words, in order to explain the specific feature of human-like culture, we must assume the existence of an evolutionary link between cultural *production* and faithful *transmission* of inventions and new procedures, through the preferential use of (over)imitative channels. Exploitation of these channels can in fact give an account of both faithful transmission *and* social innovation. Actually, imitative behaviour is already a clear social fact, depending not only on the ability of the imitator to parse, memorise and carefully reproduce a whole behavioural sequence (Byrne 2002, 2003) but also on his cognitive capacity, as *learner*, to pay attention to the action of another member of his social group. In this sense, a basic *knowledge about conspecifics*, therefore a basic ability to memorise and keep track of the social relationships of other group members (Seyfarth and Cheney 2015) and predict their behaviour on the base of such knowledge (Bates and Byrne 2015), goes strictly together with imitation. The cognitive existence of a social group, built upon a basic ability to take into consideration triadic social interactions of conspecifics, cognitively liberates individuals from choosing (or elaborating) the best methodology, offering them the possibility to copy the *most common* behaviour (Boyd and Richerson 2005). This assumes that the most common behaviour was useful enough to pass untouched through the tangles of selective pressures.

This latter assumption seems to be corroborated by the large amount of archaeological evidence suggesting that (over)imitation was triggered in our evolutionary lineage at the beginning of the Palaeolithic by tool-making activities, for which naive individuals needed to not only learn technical skills but also the correct sequences for producing tools (*operative chains*) (e.g. Boëda et al. 1990; Boëda 1994; Texier and Roche 1995). If at the very beginning the usefulness of tools initiated imitative learning—giving the chance to naive young individuals to deconstruct the procedural sequence in order to reproduce it more carefully without putting too much pressure on working memory—later on, the typical bifacial shape of the handaxes of the early Palaeolithic of Mode 2 (*Acheulean*) suggests that selective pressures (within both the physical and social environments) became much stronger (e.g. Shipton 2010). Bifacial tools have in fact remained highly conservative in shape since their earliest appearance in the archaeological record 1.75 Ma ago in Africa (Asfaw et al. 1991; Dominguez-Rodrigo et al. 2005; Beyene et al. 2013). That their shape was unaltered during their spread in Eurasia until 0.2 Ma (Dennell 2009) suggests that imitation became very much *pedantic* during the early Palaeolithic.

Even so, scholars have pointed out that repetitiveness of the shape of these tools does not mean stasis or stagnation in producing innovations (e.g. Goren-Inbar et al.

2008; Shipton et al. 2009; Shipton 2010). On the contrary, Acheulean (or Mode 2) history offers evidence for a certain degree of accumulation of innovations, mainly as improvements and refinements in the manufacture of bifaces (progressively thinner and sharper) (Shipton et al. 2009; Shipton 2010) or in the exploitation of *platform preparation* (Stout et al. 2014). It is also arguable that Mode 2 increased the hierarchical organisation of the knapping actions, with individuals adopting different and more sophisticated techniques during the time, such as the soft hammer or the earliest development of Levallois techniques (Goren-Inbar et al. 2008). In synthesis, Mode 2 demonstrates for the first time in evolutionary history a peculiar ability to consider the general shape and function of tools, irrespective of minor variants or modifications. This attention to the general shape can be considered as a consequence of the irrevocability of errors in tool-making activities and knapping sequences (Winton 2005; Stout and Chaminade 2012) from one side and the intrinsic limits of working memory in early *Homo* (Shipton 2010) to the other. The developed of (over)imitation can be seen, thus, as a way to embodied early in age and with a reduced cognitive effort, procedures and demonstrators in order to guarantee the correct execution of the complex hierarchically nested operative chains for producing tools (Di Vincenzo 2011; Di Vincenzo and Manzi 2012).

The early Palaeolithic therefore offers a complex scenario in which imitation and innovation interplay in an even more positive way than previously suggested (Legare and Nielsen 2015), and in our opinion, the role which imitation assumes with respect to innovation is clear. The former may in fact have been the best option for obtaining reliable and functionally precise tools, and only its early development and exploitation would have supplied children with the pivotal information and skills necessary for becoming, once grown up, expert toolmakers. That is, *the more they imitated as children, the more they would have known as adults*. Following this rule and stressing the role of imitation in learning, early *Homo* shaped the cognitive structure of their descendants. For example, scaffolding the cognitive capacity which allowed the later *Homo sapiens* to become able to produce in the tool making a set of outcomes different in shape and function starting from the same stem (undifferentiated) operative chain (Andersson 2011), they produced innovative tools as "variations on the same theme".

It is this interaction between imitation, over-imitation and innovation during the Palaeolithic that created the typical human-like cultural environment in the first place (Shipton and Nielsen 2015, 2018). And it is by NC that the environment then became selective for this ontogenetic combination of imitation and innovation, which we assume to be a typical trait of *Homo sapiens*.

NC stresses the *reciprocal role* (Laland and O'Brien 2011) that individual and social behaviours and the ecological context have in shaping evolutionary pressures. The tool-making activities of early *Homo* changed the learning environment of new generations, thereby selectively modifying those pressures that later acted upon them as a species. Those modifications went in the direction of making surroundings more stable and *predictable*. The developments in technology and culture aided early humans in successfully buffering the crude effects of changes and fluctuations in the physical world, very different from the type of modifications other species make to their surroundings (e.g. Odling-Smee et al. 2003). Tool-making activities and the selective pressures they put on SL tightly connected the two main aspects related to the development of human cognition, the social aspect and the "extended" one (the proficient relationship between cognition and artefacts *sensu* Clark and Chalmers 1998, 2010). In this connected and more predictable environment (at least in terms of regular access to vital resources), the fitness value of SL increased in terms of adaptiveness. Within such an environment, individual learning, being error-prone, is much too costly (Boyd and Richerson 2005), while imitation increases the average fitness of imitators (Castro and Toro 2004). In this way, a combined socio-physical environment, stable and predictable, together with the possibility to carefully learn motor skills and improve them during an extended period of ontogenetic development resulted in humans developing a ratchet-effect culture.

However, ecological changes and climatic instability were in some way "necessary" to combine social and physical environments in the first place in order to successfully deal with them and making the surroundings more predictable by means of cumulative culture. We are going to look at this supposed paradox in the next paragraph.

4 The Importance of the Ecological Context

Palaeoclimatic and palaeoanthropological evidence strongly suggests that at the end of the Pliocene, pulsed climate instability and loss of suitable habitat produced increasing rates of speciation and adaptation within the lineage of early African hominins (*Australopithecus*) (Vrba 1995; Plummer et al. 2009; Potts 2012; Antón et al. 2014; Maslin et al. 2015). Some species became proficient consumers of C4 plants and low-quality foods, which required the acquisition of physical and dental specialisations (Cerling et al. 2011; Sponheimer et al. 2013; but see Wood and Strait 2004), whereas others, e.g. members of the newly emerging genus *Homo*, undertook opportunistic behaviours. This split between different species/genera is associated with the first appearance of an archaeological record alongside the earliest members of the human genus around 2.6 Ma (Villmoare et al. 2015). These early *Homo* were able to produce tools by means of other tools [*tool making*], a skill lacking in their ancestors (Semaw et al. 1997). This change of perspective, from the use of physical objects as tools to the production and modification of tools, could be seen as the behavioural correlate of the phenomenon of *encephalisation* (the non-allometric increase of cerebral volumes) which strongly characterises the evolution of the genus *Homo* (e.g. Bruner et al. 2003; Dunbar 1998, 2003).

The existence of a clear relationship between variation in brain metrics in fossil hominins and increasing behavioural complexity is hard to establish. However, a positive correlation between variation in brain size and cultural innovations is documented in extant primates (Reader and Laland 2001, 2002). Furthermore, evidence based on the study of fossil endocasts shows that with respect to the australopiths, a cortical reorganisation occurred early in the genus *Homo* (Zollikofer and Ponce De León 2013), progressively involving the differential growth of parietal lobes (e.g. Bruner et al. 2003; Kyriacou and Bruner 2011) and cortical areas functionally related to visuospatial integration, category recognition, memory and praxis, bridging the somatosensory and visual cortex (e.g. Bruner and Iriki 2016).

This is very important considering that a growing brain is a metabolically expensive organ (Aiello and Wheeler 1995; Pontzer et al. 2016), which started to require very high-quality nutrients for its development, particularly fatty acids (*ω3* and *ω6*) essential for supporting the activity of neural membranes and normal brain development (e.g. Brenna and Carlson 2014). The problem is that primates do not produce essential fatty acids autonomously and therefore need to acquire them either through their diet or from mother to offspring during pregnancy and breastfeeding. Even if some scholars argued that the savannah was not a "good" environment for gathering such nutrients (e.g. Cunnane et al. 1993), recent studies on the nutritional contents of extant African ruminants suggest that both energy and fatty acids could actually have been available to early humans (Cordain et al. 2001). By means of scavenging or hunting activities, early and later *Homo* could have had access to marrow and cerebral tissue of animals, supplying first the necessary energy and second, the required fatty acids (Cordain et al. 2001). Scavenging would also have opened up a new, more meat-oriented, ecological niche for early humans (Hernandez Fernández and Vrba 2006; Zink and Lieberman 2016), in which the increased availability of proteins and lipids, including high-quality nutrients, then allowed for and triggered the encephalisation process.

If this is true, early *Homo* would have needed to exploit and produce tools more extensively. Consequently, this would have progressively shaped both their society and social cognition by means of the increased selective pressures associated with efficient but cheap computational and planning abilities. This would allow for generativity, reduced time expenditure and fewer mistakes during tool making. In turn, procedural and cognitive generative abilities trigger apprenticeship learning (Rogoff 1990; Sterelny 2012) (and some authors argue also teaching: Högberg et al. 2015), by means of which the learner can capture the inner social organisation of the task by means of an environment structured by demonstrators. It is likely that within the implicitly communicative context created by the imitative channel of trans-generational transmission of motor and technical skills required to regularly access, the new (meat-oriented) trophic resource, iterative, computational, hierarchical nesting and also recursive abilities became increasingly complex (e.g. Di Vincenzo and Manzi 2012).

Furthermore, as we argued so far, a structured learning environment also triggers imitative learning, increases sharing intentionality among the members of the social group (Shipton and Nielsen 2015) and induces the same kind of high-level goal inference required in very complex social bonding and communication (Stout and Chaminade 2012). It has been pointed out that social bonding increased during this period because the new trophic niche of early *Homo* increased selection pressures on tight forms of cooperation (Sterelny 2011), making individuals completely dependent on their social group (Jeffares 2009, 2010). Human children still show this complete dependence on their social group (triadic interactions, zone of proximal development or guided participation and prolonged altriciality are common examples of this dependence), particularly in the critical period called *childhood*, which is a unique human developmental growth phase (Bogin 1988, 1990, 1998). During this prolonged infancy, which is also characterised by fast rates of brain growth and very delayed rates of synaptic maturation and axon myelinisation in the human cortex (Liu et al. 2010; Somel et al. 2014), children can acquire via imitation a first set of pivotal information (both know-how and know-what competences) momentous for their future survival. As we argued here (but for a different opinion, see Högberg et al. 2015), the acquisition of those abilities required neither explicit nor implicit teaching, instead of making them possible. By means of over-imitation, children learn to *be taught*, which means to be proper children (not infants nor simply young individuals).

5 Conclusions

In this chapter, we outlined a possible explanation not only for the presence of such a behavioural oddity as (over)imitation in human children but also its evolutionary route and meaning. If pedantic (over)imitation still exists, the reason has to be sought in its cognitive role in making humans primarily *social innovators*, even though it is not always useful or immediately beneficial in terms of fitness to the imitator.

We argued that over-imitation creates the cognitive base for making *social innovations* possible. Innovations require an individual ability to solve the problem in order to reach an intended goal, which in turn means the ability to use trial and error within an emulative context. However, being social innovations, they do not only rely on the simple ability to keep track of social interactions within a social group but also on the ability to focus attention on the actions of a particular member of the community and the objects involved. Furthermore, all of which without being distracted by the actions of other individuals or other social situations going on during the time period required for learning the procedure. We have provisionally named this ability *emulation 2.0*, stressing the reciprocity that emulation and imitation have in the production of innovations. In *emulation 2.0*, we argued, imitative learning is the key for making social innovations. Especially in the shape of pedantic and unselective (over)imitation, this ability supplies cognition with a double structure: the social mechanism of information transmission on one side and the embodiment of the technical procedures on the other. Social, technical and physical aspects of learning become so dependent on one another that they can even be perceived as a single world. Only with a long training of pedantic (over)imitation, the "creative environment" in which individual solutions can emerge has the potential to change, improve and transform with proficiency within the cognitive framework of know-how, know-what and the system of rules previously acquired.

From an evolutionary perspective, during the Palaeolithic, (over)imitation was strongly selected for being particularly useful in facilitating the transmission of procedural information in tool-making activities. It plays important roles in avoiding the high incidences of mistakes usually associated with individualistic strategies, increasing shared intentionality and strongly reducing the time required to produce tools. From this, also very strong NC perspective, procedure, demonstrator, physical objects and spaces in the world become more and more interconnected, making it actually impossible for them to be separated again.

Otherwise, separation remains in the early phases of ontogenetic development of human cognition when children are completely dependent on the social group for their survival and when they can develop their innovative abilities safely and without any direct practical purposes. In the same period, their cognition is moulded by imitation and pedantic (over)imitation, particularly triggered by procedures shown by adult demonstrators. This separation is not definitive, even though, as we claim here, the two behavioural tendencies combine at some point during development, probably when language proficiency increases. In this chapter, we do not inquire into the role language has in this story, but we suppose it is actually very much connected, both in terms of evolutionary first occurrence and developmental cognitive reconstruction in young humans.

References

Aiello LC, Wheeler P (1995) The expensive-tissue hypothesis: the brain and the digestive system in human and primate evolution. Curr Anthropol 36(2):199–221

Andersson C (2011) Paleolithic punctuations and equilibria: did retention rather than invention limit technological evolution? PaleoAnthropology 2011:243–259

Antón SC, Potts R, Aiello LC (2014) Evolution of early *Homo*: an integrated biological perspective. Science 345(6192):1236828

Asfaw B, Beyene Y, Suwa G, Walter RC, White TD, WoldeGabriel G, Yemane T (1991) The earliest Acheulean from Konso-Gardula. Nature 360(6406):732–735

Bates LA, Byrne RW (2015) Primate social cognition: what we have learned from nonhuman primates and other animals. In: Mikulincer M, Shaver PR (eds) APA handbook of personality and social psychology: vol. 1 attitudes and social cognition. American Psychological Association, Washington, DC, pp 47–78

Berl R, Hewlett B (2015) Cultural variation in the use of overimitation by the Aka and Ngandu of the Congo Basin. PLoS One 10:e0120180

Beyene Y, Katoh S, WoldeGabriel G, Hart WK, Uto K, Sudo M, Kondo M, Hyodo M, Renne PR, Suwa G (2013) The characteristics and chronology of the earliest Acheulean at Konso, Ethiopia. Proc Natl Acad Sci USA 110(5):1584–1591

Boëda E (1994) The Levallois concept: variability of methods (Monographie du CRA, 9). CNRS Editions, Paris

Boëda E, Geneste J-M, Meignen L (1990) Identification de chaînes opératoires lithiques du Paléolithique ancien et moyen. Paléo 2(1):43–80

Boesch C (1997) Three approaches for assessing chimpanzee culture. In: Russon AE, Bard KA, Parker ST (eds) Reaching into thought: the minds of the great apes. Cambridge University Press, Cambridge, pp 404–429

Bogin B (1988) Patterns of human growth. Cambridge University Press, New York

Bogin B (1990) The evolution of human childhood. Bioscience 40:16–25

Bogin B (1998) Evolutionary and biological aspects of childhood. In: Panter-Brick C (ed) Biosocial perspectives on children. Cambridge University Press, Cambridge, pp 10–44

Boyd R, Richerson PJ (2005) The origins and evolution of cultures. Oxford University Press, Oxford

Brenna JT, Carlson SE (2014) Docosahexaenoic acid and human brain development: evidence that a dietary supply is needed for optimal development. J Hum Evol 77:99–106

Bruner E, Iriki A (2016) Extending mind, visuospatial integration, and the evolution of the parietal lobes in the human genus. Quat Int 405(Part A):98–110. https://doi.org/10.1016/j.quaint.2015.05.019

Bruner E, Manzi G, Arsuaga JL (2003) Encephalization and allometric trajectories in the genus *Homo*: evidence from the Neandertal and modern lineages. Proc Natl Acad Sci USA 100 (26):15335–15340

Buttelmann D, Carpenter M, Call J, Tomasello M (2007) Enculturated chimpanzees imitate rationally. Dev Sci 10(4):F31–F38

Buttelmann D, Carpenter M, Call J, Tomasello M (2013) Chimpanzees, *Pan troglodytes*, recognize successful actions, but fail to imitate them. Anim Behav 86(4):755–761

Byrne RW (2002) Imitation of novel complex actions: what does the evidence from animals mean? Adv Study Behav 31:77–105

Byrne RW (2003) Imitation as behaviour parsing. Philos Trans R Soc Lond B Biol Sci 358 (1431):529–536

Byrne RW, Russon AE (1998) Learning by imitation: a hierarchical approach. Behav Brain Sci 21 (5):667–684

Caldwell CA (2014) Experimental studies of cumulative culture in modern humans: what are the requirements of the ratchet? In: Mesoudi A, Aoki K (eds) Learning strategies and cultural evolution during the palaeolithic. Springer, Tokyo, pp 145–154

Caldwell CA, Millen AE (2009) Social learning mechanisms and cumulative cultural evolution is imitation necessary? Psychol Sci 20(12):1478–1483

Caldwell CA, Schillinger K, Evans CL, Hopper LM (2012) End state copying by humans (*Homo sapiens*): implications for a comparative perspective on cumulative culture. J Comp Psychol 126 (2):161

Castro L, Toro MA (2004) The evolution of culture: from primate social learning to human culture. Proc Natl Acad Sci USA 101(27):10235–10240

Cerling TE, Wynn JG, Andanje SA, Bird MI, Korir DK, Levin NE, Mace W, Macharia AN, Quade J, Remien CH (2011) Woody cover and hominin environments in the past 6 [thinsp] million years. Nature 476(7358):51–56

Clark A (2013) Whatever next? Predictive brains, situated agents, and the future of cognitive science. Behav Brain Sci 36(3):181–204

Clark A, Chalmers D (1998) The extended mind. Analysis 58(1):7–19

Clark A, Chalmers D (2010) Memento's revenge: the extended mind, extended. In: Menary R (ed) The extended mind. MIT Press, Cambridge, MA

Cordain L, Watkins BA, Mann NJ (2001) Fatty acid composition and energy density of foods available to African hominids. Evolutionary implications for human brain development. World Rev Nutr Diet 90:144–161

Cunnane SC, Harbige LS, Crawford MA (1993) The importance of energy and nutrient supply in human brain evolution. Nutr Health 9(3):219–235

Dennell R (2009) DISCUSSION 2: Transitions: behavioral change in the early pleistocene. In: Sourcebook of Paleolithic transitions. Springer, New York, pp 229–234

Di Vincenzo F (2011) Toward a neuro-archaeology of the faculty of language. In: Evoluzione, mente e linguaggio Atti del IV Convegno del CODISCO (2010), Roma, pp 255–266

Di Vincenzo F, Manzi G (2012) Social learning and origin of the language faculty by means of natural selection. J Anthropol Sci 91:23–52

Dominguez-Rodrigo M, Pickering TR, Semaw S, Rogers MJ (2005) Cutmarked bones from Pliocene archaeological sites at Gona, Afar, Ethiopia: implications for the function of the world's oldest stone tools. J Hum Evol 48(2):109–121

Dunbar RIM (1998) The social brain hypothesis. Evol Anthropol 6:178–190

Dunbar RIM (2003) The social brain: mind, language, and society in evolutionary perspective. Annu Rev Anthropol 32:163–181

Fernández MH, Vrba ES (2006) Plio-Pleistocene climatic change in the Turkana Basin (East Africa): evidence from large mammal faunas. J Hum Evol 50(6):595–626

Fragaszy D (2012) Community resources for learning: how capuchin monkeys construct technical traditions. Biol Theory 6(3):231–240

Goren-Inbar N, Sharon G, Alperson-Afil N, Laschiver I (2008) The Acheulean massive scrapers of Gesher Benot Ya'aqov—a product of the biface chaîne opératoire. J Hum Evol 55(4):702–712

Harris PL (2012) Trusting what you're told: how children learn from others. Harvard University Press, Cambridge, MA

Hobaiter C, Byrne RW (2010) Able-bodied wild chimpanzees imitate a motor procedure used by a disabled individual to overcome handicap. PLoS One 5(8):e11959

Högberg A, Gärdenfors P, Larsson L (2015) Knowing, learning and teaching – how *Homo* became docens. Camb Archaeol J 25(4):847–858

Hopper LM, Marshall-Pescini S, Whiten A (2012) Social learning and culture in child and chimpanzee. In: de Waal FB, Ferrari PF (eds) The primate mind: built to connect with other minds. Harvard University Press, Cambridge, MA, pp 99–118

Horner V, Whiten A (2005) Causal knowledge and imitation/emulation switching in chimpanzees (*Pan troglodytes*) and children (*Homo sapiens*). Anim Cogn 8(3):164–181

Jeffares B (2009) The evolution of technical competence: strategic and economic thinking. In: ASCS09: Proceedings of the 9th Conference of the Australasian Society for Cognitive Science, pp 162–169

Jeffares B (2010) The co-evolution of tools and minds: cognition and material culture in the hominin lineage. Phenomenol Cogn Sci 9(4):503–520

Johnson-Laird PN (1988) The computer and the mind: an introduction to cognitive science. Harvard University Press, Cambridge, MA

Kenward B (2012) Over-imitating preschoolers believe unnecessary actions are normative and enforce their performance by a third party. J Exp Child Psychol 112:195–207

Kenward B, Karlsson M, Persson J (2011) Over-imitation is better explained by norm learning than by distorted causal learning. Proc R Soc B Biol Sci 278:1239–1246

Koops K, Visalberghi E, van Schaik CP (2014) The ecology of primate material culture. Biol Lett 10(11):20140508

Kyriacou A, Bruner E (2011) Special issue: innovation and the evolution of human behavior brain evolution, innovation, and endocranial variations in fossil hominids. PaleoAnthropology 130:143

Laland KN, O'Brien MJ (2011) Cultural niche construction: an introduction. Biol Theory 6 (3):191–202

Legare CH, Nielsen M (2015) Imitation and innovation: the dual engines of cultural learning. Trends Cogn Sci 19(11):688–699

Liu C, Teng Z-Q, Santistevan NJ, Szulwach KE, Guo W, Jin P, Zhao X (2010) Epigenetic regulation of miR-184 by MBD1 governs neural stem cell proliferation and differentiation. Cell Stem Cell 6(5):433–444

Lyons DE, Young AG, Keil FC (2007) The hidden structure of overimitation. Proc Natl Acad Sci USA 104(50):19751–19756

Lyons DE, Damrosch DH, Lin JK, Macris DM, Keil FC (2011) The scope and limits of overimitation in the transmission of artefact culture. Philos Trans R Soc Lond B Biol Sci 366 (1567):1158–1167

Maslin MA, Shultz S, Trauth MH (2015) A synthesis of the theories and concepts of early human evolution. Philos Trans R Soc Lond B Biol Sci 370(1663):20140064

McGuigan N, Makinson J, Whiten A (2011) From over-imitation to super-copying: adults imitate causally irrelevant aspects of tool use with higher fidelity than young children. Br J Psychol 102 (1):1–18

Morin O (2016) How traditions live and die. Oxford University Press, Oxford
Nielsen M (2006) Copying actions and copying outcomes: social learning through the second year. Dev psychol 42(3):555–565
Nielsen M (2011) Imitation, pretend play, and childhood: essential elements in the evolution of human culture. J Comp Psychol 126(2):170
Nielsen M, Tomaselli K (2010) Over-imitation in Kalahari Bushman children and the origins of human cultural cognition. Psychol Sci 21:729–736
Nielsen M, Moore C, Mohamedally J (2012) Young children overimitate in third-party contexts. J Exp Child Psychol 112:73–83
Odling-Smee FJ, Laland KN, Feldman MW (2003) Niche construction: the neglected process in evolution. Princeton University Press, Princeton
Plummer TW, Ditchfield PW, Bishop LC, Kingston JD, Ferraro JV, Braun DR, Hertel F, Potts R (2009) Oldest evidence of toolmaking hominins in a grassland-dominated ecosystem. PLoS One 4(9):e7199
Pontzer H, Brown MH, Raichlen DA, Dunsworth H, Hare B, Walker K, Luke A, Dugas LR, Durazo-Arvizu R, Schoeller D, Plange-Rhule J, Bovet P, Forrester TE, Lambert EV, Thompson ME, Shumaker RW, Ross SR (2016) Metabolic acceleration and the evolution of human brain size and life history. Nature 533(7603):390–392
Potts R (2012) Environmental and behavioral evidence pertaining to the evolution of early *Homo*. Curr Anthropol 53(S6):S299–S317
Pradhan GR, Tennie C, van Schaik CP (2012) Social organization and the evolution of cumulative technology in apes and hominins. J Hum Evol 63(1):180–190
Proffitt T, Luncz LV, Falotico T, Ottoni EB, de la Torre I, Haslam M (2016) Wild monkeys flake stone tools. *Nature* 539(7627):85–88
Reader SM, Laland KN (2001) Primate innovation: sex, age and social rank differences. Int J Primatol 22(5):787–805
Reader SM, Laland KN (2002) Social intelligence, innovation, and enhanced brain size in primates. Proc Natl Acad Sci 99(7):4436–4441
Rogoff B (1990) Apprenticeship in thinking: cognitive development in social context. Oxford University Press, Oxford
Russon AE, Galdikas BM (1993) Imitation in free-ranging rehabilitant orangutans (*Pongo pygmaeus*). J Comp Psychol 107(2):147
Semaw S, Renne P, Harris JWK, Feibel CS, Bernor RL, Fesseha N, Mowbray K (1997) 2.5-million-year-old stone tools from Gona, Ethiopia. Nature 385(6614):333–336
Seyfarth RM, Cheney DL (2015) Social cognition. Anim Behav 103:191–202
Shimpi PM, Akhtar N, Moore C (2013) Toddlers' imitative learning in interactive and observational contexts: the role of age and familiarity of the model. J Exp Child Psychol 116(2):309–323
Shipton C (2010) Imitation and shared intentionality in the Acheulean. Camb Archaeol J 20 (2):197–210
Shipton C, Nielsen M (2015) Before cumulative culture. Hum Nat 26(3):331–345
Shipton C, Nielsen M (2018) The acquisition of biface knapping skill in the Acheulean. In: Di Paolo LD, Di Vincenzo F, De Petrillo F (eds) Evolution of primate social cognition. Springer, Cham
Shipton C, Petraglia MD, Paddayya K (2009) Inferring aspects of Acheulean sociality and cognition from lithic technology. In: Adams B, Blades BS (eds) Lithic materials and palaeolithic societies. Wiley, Chichester, pp 219–231
Somel M, Liu X, Khaitovich P (2014) Human brain evolution: transcripts, metabolites and their regulators. Nat Rev Neurosci 14(2):112–127
Sponheimer M, Alemseged Z, Cerling TE, Grine FE, Kimbel WH, Leakey MG, Lee-Thorp JA, Kyalo Manthi F, Reed KE, Wood BA, Wynn JG (2013) Isotopic evidence of early hominin diets. Proc Natl Acad Sci 110(26):10513–10518
Sterelny K (2011) From hominins to humans: how sapiens became behaviourally modern. Philos Trans R Soc Lond B Biol Sci 366(1566):809–822
Sterelny K (2012) The evolved apprentice. MIT Press, Cambridge, MA
Stout D, Chaminade T (2012) Stone tools, language and the brain in human evolution. Philos Trans R Soc Lond B Biol Sci 367(1585):75–87

Stout D, Apel J, Commander J, Roberts M (2014) Late Acheulean technology and cognition at Boxgrove, UK. J Archaeol Sci 41:576–590

Tennie C, Call J, Tomasello M (2012) Untrained chimpanzees (*Pan troglodytes schweinfurthii*) fail to imitate novel actions. PLoS One 7(8):e41548

Tennie C, Walter V, Gampe A, Carpenter M, Tomasello M (2014) Limitations to the cultural ratchet effect in young children. J Exp Child Psychol 126:152–160

Texier PJ, Roche H (1995) The impact of predetermination on the development of some Acheulean chaînes opératoires. Evolución Humana en Europa y los Yacimientos de la. Sierra de Atapuerca 2:403–420

Tomasello M (1996) Do apes ape? In: Heyes CM, Galef BG (eds) Social learning animals: the roots of culture. Academic, London, pp 319–346

Tomasello M (1999) The cultural origins of human cognition. Harvard University Press, Cambridge, MA

Villmoare B, Kimbel WH, Seyoum C, Campisano CJ, DiMaggio EN, Rowan J, Braun DR, Arrowsmith JR, Reed KE (2015) Early *Homo* at 2.8 Ma from Ledi-Geraru, Afar, Ethiopia. Science 347(6228):1352–1355

Visalberghi E, Fragaszy D (2012) Learning how to forage: socially biased individual learning and "niche construction" in wild capuchin monkeys. In: de Waal F, Ferrari PF (eds) The priemate mind. Harvard University Press, Cambridge, MA, pp 81–98

Vrba ES (1995) Paleoclimate and evolution, with emphasis on human origins. Yale University Press, New Haven

Whiten A, Horner V, Litchfield CA, Marshall-Pescini S (2004) How do apes ape? Learn Behav 32 (1):36–52

Whiten A, Horner V, De Waal FBM (2005) Conformity to cultural norms of tool use in chimpanzees. Nature 437(7059):737–740

Whiten A, McGuigan N, Marshall-Pescini S, Hopper LM (2009) Emulation, imitation, overimitation and the scope of culture for child and chimpanzee. Philos Trans R Soc Lond B Biol Sci 364(1528):2417–2428

Winton V (2005) An investigation of knapping-skill development in the manufacture of palaeolithic handaxes. In: Roux V, Brill B (eds) Stone knapping: the necessary conditions for a uniquely hominin behaviour. McDonald Institute for Archaeological Research, Cambridge, pp 109–116

Wood B, Strait D (2004) Patterns of resource use in early *Homo* and *Paranthropus*. J Hum Evol 46 (2):119–162

Wood LA, Kendal RL, Flynn EG (2012) Context-dependent model-based biases in cultural transmission: children's imitation is affected by model age over model knowledge state. Evol Hum Behav 33(4):387–394

Wood LA, Kendal RL, Flynn EG (2013a) Copy me or copy you? The effect of prior experience on social learning. Cognition 127(2):203–213

Wood LA, Kendal RL, Flynn EG (2013b) Whom do children copy? Model-based biases in social learning. Dev Rev 33(4):341–356

Zink KD, Lieberman DE (2016) Impact of meat and Lower Palaeolithic food processing techniques on chewing in humans. Nature 531(7595):500–503

Zollikofer CPE, de León MSP (2013) Pandora's growing box: inferring the evolution and development of hominin brains from endocasts. Evol Anthropol 22(1):20–33

Zwirner E, Thornton A (2015) Cognitive requirements of cumulative culture: teaching is useful but not essential. Sci Rep 5:16781

The Acquisition of Biface Knapping Skill in the Acheulean

Ceri Shipton and Mark Nielsen

Abstract The Acheulean stone tools of *Homo erectus* and *Homo heidelbergensis* are the longest enduring of all archaeological cultures, lasting for 1.5 million years. Three competing hypotheses have been proposed to explain this longevity: that Acheulean technology lies in a zone of latent, easy to invent solutions to problems that *H. erectus* and *H. heidelbergensis* would have encountered; that there was a genetic predisposition among *H. erectus* and *H. heidelbergensis* populations to make handaxes, and perhaps other stone tools characteristic of the Acheulean; or that high-fidelity social transmission was an integral part of the behavioural repertoire of *H. erectus* and *H. heidelbergensis*. In this chapter, the first two of these hypotheses are critiqued. Experimental and anecdotal evidence from modern stone knappers is reviewed to determine how knapping expertise is acquired. The chapter concludes with reference to archaeological evidence for social transmission in the Acheulean and a model of how knapping expertise is and was acquired.

Keywords Imitation · Genetic determinism · Invention · Oldowan · Helical curriculum · *Homo erectus*

1 Introduction

The Acheulean is the most persistent of all archaeological cultures with its characteristic artefacts, handaxes and cleavers, produced for over 1.5 million years (Beyene et al. 2013; Haslam et al. 2011). Handaxes and cleavers are elongate stone discs, knapped on both faces of the disc, hence their collective name of bifaces. Handaxes

---

C. Shipton (✉)
Centre of Excellence for Australian Biodiversity and Heritage, The Australian National University, Canberra, ACT, Australia
e-mail: ceri.shipton@anu.edu.au

M. Nielsen
School of Psychology, University of Queensland, St Lucia, QLD, Australia

Faculty of Humanities, University of Johannesburg, Johannesburg, South Africa

L. D. Di Paolo et al. (eds.), *Evolution of Primate Social Cognition*, Interdisciplinary Evolution Research 5, https://doi.org/10.1007/978-3-319-93776-2_18

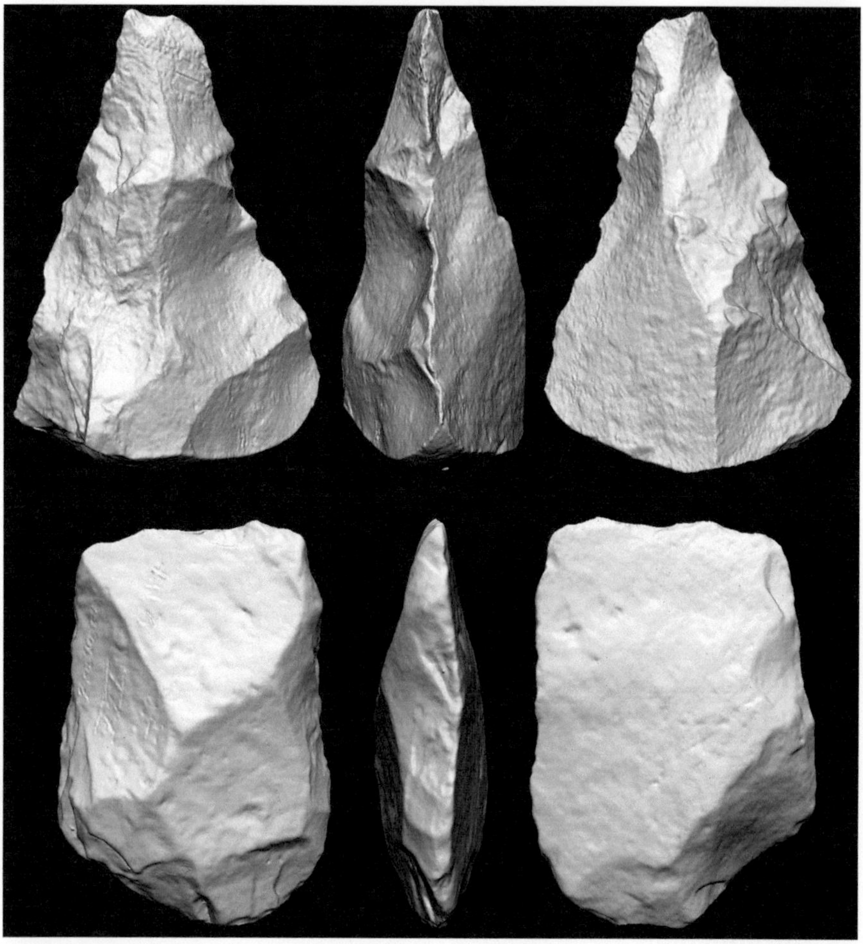

Fig. 1 A handaxe (above) and a cleaver (below) from the site of Chirki in India. The handaxe is 118 mm long and the cleaver is 106 mm long. The handaxe is made on a chert cobble with a small patch of remnant rounded cortex visible on the butt. The cleaver is made on a flake of basalt, finished by marginal trimming, particularly of the platform and bulb of percussion which would have been the thickest parts of the flake

have a typically rounded and globular butt at one end, where they are thought to have been held (Gowlett 2006). At the other, elongate end, handaxes taper in plan view to varying degrees, with their sharp, bifacially worked cutting edge usually extending around much of their circumference (Fig. 1). Cleavers have a straight unworked edge at one end which is thought to have been the principal cutting edge, like the bit of an axe or a chisel (Movius 1948; de la Torre et al. 2014; Shipton 2016) (Fig. 1).

The Acheulean has long been assumed to be a social tradition like other archaeological cultures, but recently, this idea has been challenged from two directions. On one hand is the proposal that handaxes and cleavers lie in an easy to access 'zone of latent solutions' (ZLS) and were repeatedly reinvented (Tennie et al. 2016). On the other hand, it has been suggested that handaxe and cleaver production were in part genetically determined (Corbey et al. 2016). In this chapter, we critique these hypotheses on the grounds that independent invention and genetic control cannot explain the persistence and diversity of Acheulean biface technology, respectively. We then examine evidence for the acquisition of skill among modern knappers using both published studies and our own anecdotal observations. The Acheulean is a vast technocomplex that varies widely across time and space with several subfacies, particularly in its later stages. There is far more to Acheulean stone tools than just handaxes and cleavers, but we focus on these archetypes because both Tennie et al. (2016) and Corbey et al. (2016) make explicit reference to them, and their ubiquity across the Acheulean world makes them the most plausible candidates for all three explanations of genetic control, social transmission and latent solutions.

2 Independent Invention

Tennie et al. (2016) propose 'the Island Test' for culture in which they conduct a thought experiment to determine whether or not social transmission is necessary for the repeated appearance of artefact forms. The thought experiment involves imagining a group of hominins raised without socially transmitted knowledge of stone knapping, on an island with stone suitable for flaking and the need to cut things. Under these circumstances, Tennie et al. suggest that the Oldowan stone tool industry (the simpler predecessor of the Acheulean), and even the Acheulean, would be invented spontaneously. However, running this thought experiment in our minds does not produce these results.

Well before the recent discovery of the 3.3 million-year-old Lomekwian industry (Harmand et al. 2015), researchers noted that Oldowan tools already displayed considerable knowledge of knapping, so they were unlikely to be the earliest knapped technology (Panger et al. 2002). Studies of Oldowan artefacts over 2 million years old indicate their knappers understood appropriate platforms and angles to strike at to detach flakes; and how to maintain angles on a core so as to maximise the number of flakes that could be struck from it (de la Torre 2004; Delagnes and Roche 2005; Stout et al. 2010). This latter aspect is particularly important as it sometimes involved forgoing the easiest flake removal in favour of a strike that would allow more future removals, demonstrating an understanding of how individual actions relate to the wider manufacturing sequence. Also noteworthy is the rarity of battering on these early Oldowan artefacts, suggesting hominins knew the appropriate points and angles to strike without extensive trial and error. An experiment comparing early Oldowan flakes with those of modern novice knappers on clasts of the same material showed the Oldowan knappers used a more uniform strategy of unifacial reduction and produced thicker and more standardised flakes (Stout and Semaw 2006). Oldowan knappers were not habitually reinventing the wheel but appear instead to have been part of a knapping tradition that maintained skills.

The Acheulean must have been invented at least once, albeit very likely out of an Oldowan tradition (Gowlett 1986). Many aspects of Acheulean knapping are however counter-intuitive, so that the propagation of the Acheulean would be more easily achieved by social learning rather than independent invention. For example, even at the most basic level, to remove mass from one surface of a biface, you need to strike it on the opposite surface. In order to produce the thin bifaces characteristic of the later Acheulean, it is necessary to employ techniques such as platform faceting and raising the plane of intersection, which involve small strikes at a steep angle on the surface from which you intend to remove mass, followed by a large strike at a shallow angle on the opposite surface. To master such techniques to the point where they become efficient requires considerable practice even once the method is explained (Stout et al. 2014; Putt et al. 2014). The greater complexity of the Acheulean in comparison to the Oldowan and the later Acheulean in comparison to the earlier Acheulean does not fit with the ZLS model but suggests incipient cumulative cultural evolution at work.

An easy to invent Acheulean would result in similar forms wherever there is basic knapping. A knapping experiment in which the goal was simply to produce the largest flakes from randomised platforms resulted in several bifacial pieces, but none of these resembled a handaxe at any stage of 59 different reduction sequences (Moore and Perston 2016; see Fig. 16 for the closest example). Handaxes and cleavers are not easy to stumble upon without a specific goal in mind.

Perhaps the best evidence in support of the ZLS hypothesis is the appearance of handaxe-like forms at other times in prehistory. Bifacial knapping is the easiest way to shape a piece of stone, and the majority of ancient stone knapping industries contain types of shaped bifaces. Unsurprisingly then, bifaces that could be mistaken for handaxes do occur in later prehistory, for example, in northern Australia, or the Neolithic of north-western Europe (Brumm and Moore 2012; Moore 2003). However, the cleaver is a far more particular tool form (Tixier and Inizian 1983), and it is only known from one other time: the end of the Middle Palaeolithic of north-western Europe, in particular the Cantabrian and western Pyrenees mountains (Utrilla et al. 2015; Thiébaut et al. 2012). This is the best known candidate for the independent invention of the Acheulean, but the Vasco-Cantabrian final Middle Palaeolithic did not appear de novo; it arose out of the existing Middle Palaeolithic industry with deep roots and a piecemeal emergence out of the Acheulean over more than a hundred thousand years (Moncel et al. 2011; de la Torre et al. 2013). Even if the Vasco-Cantabrian Middle Palaeolithic cleavers were independently invented, the industry would not be mistaken for the Acheulean as it includes many other artefact types that distinguish it, such as Levallois points and formal denticulates.

To summarise, there is no evidence for the trial-and-error learning at Oldowan and Acheulean sites that would be necessary to support a scenario of repeated independent invention. Rather Oldowan and Acheulean knappers display sophisticated knowledge of knapping, including counter-intuitive methods that would seemingly be difficult to invent. Only one non-Acheulean stone knapping industry contains both handaxes and cleavers, but this contains other elements that are distinctive from the Acheulean and arises out of a long-standing knapping tradition rather than being invented de novo. We thus dispute the findings of Tennie et al.'s 'Island Test' that handaxes and cleavers are easy to invent and that this repeatedly happened during the Acheulean.

3 Genetic Control

The persistence of handaxes, and perhaps also cleavers, over the vast time span of the Acheulean is suggested by Corbey et al. (2016) to be due to these forms being under partial genetic control. We think this hypothesis is inconsistent with the diversity in the Lower Palaeolithic.

According to Corbey et al. (2016), variation in handaxe form is too little to accommodate the copying error inherent in cultural transmission. They draw support for this argument from the lower than expected size variation in archaeological handaxes (Kempe et al. 2012) when compared to outline shapes drawn on a tablet computer in an experimental transmission chain. However, this experiment does not account for the fact that handaxe size is a function not only of transmission but intrinsic constraints of hand size and strength among other factors (Gowlett and Crompton 1994). Therefore, the size of real stone handaxes should not be expected to vary as much as two-dimensional shapes drawn on a tablet. A follow-up experiment found that in reproducing three-dimensional handaxe shapes from foam blocks, there was less copying error when the parameters of social transmission were enhanced to include observation of the act of reproduction, rather than just the end product—i.e. in an imitation condition (Schillinger et al. 2015).

The lack of variation in handaxes across different environmental conditions is touted as evidence of genetic transmission (Corbey et al. 2016). However, the interregional homogeneity is overstated: handaxes are not monolithic, and there are significant regional and temporal differences in handaxe morphology (Wynn and Tierson 1990; Shipton and Clarkson 2015; Shipton and Petraglia 2010). Moreover, the relevance of a lack of variation across environments is dependent on knowing the function of handaxes and demonstrating different functional requirements for different environments. The most widely demonstrated function for handaxes is as butchery tools for large animals (e.g. Mitchell 1994), and large animal butchery seems to have been a feature of hominin behaviour across the Acheulean world (Piperno and Tagliacozzo 2001; Goren-Inbar et al. 1994; Roberts and Parfitt 1999).

It is suggested that birds' nests and bowers, widely agreed to have a strong genetic component (Frith and Frith 2004), are similar in technological complexity to handaxes (Corbey et al. 2016). But building a nest from sticks and fluffy things or a bower from sticks and shiny things is conceptually very different from handaxe knapping. The sticks are rigid to begin with, the fluffy things are fluffy, and the shiny things are shiny, and though they may be combined in intricate and impressive ways, their fundamental properties are not changed. But a lump of stone prior to knapping is not sharp, nor are the properties which make it suitable for knapping easily determinable. Corbey et al. (2016: 14) state that it was 'probably not just a simple target form, but rather a predisposition toward the basic behavioural routines involved' that was genetically determined in handaxe manufacture. But there is no one basic behavioural routine applicable to all handaxe manufacture. Depending on whether a biface is made on a large flake or a slab/cobble of stone, the process of manufacture is conceptually entirely different (Shipton et al. 2009). Even within the paradigm of flake handaxe production, a review of methods for creating large flake blanks for handaxes identifies ten distinct methods found in different regions of the Acheulean world (Sharon 2007).

The dearth of bifaces in East Asia is explained by Corbey et al. through the lack of high-quality rock for their manufacture over large swathes of the region. When such rock does become available, the genes are then expressed. However, it has been demonstrated experimentally that handaxes of the same shape can be made on a broad variety of rock types (Eren et al. 2014), and in any case, most East Asian biface assemblages do not contain genuine handaxes and cleavers (Shipton and Petraglia 2010). The rock quality explanation offered by Corbey et al. for assemblages lacking handaxes also does not explain assemblages with high-quality material lacking handaxes. At the northern fringes of Acheulean occupation in Europe, there are large assemblages without handaxes that are contemporary with handaxe-bearing assemblages further south. These 'Clactonian' assemblages date to the beginning of interglacial periods and are thought to be a pioneer phase of Acheulean (re)colonisation by small populations unable to sustain more complex handaxe technology (White 2000; Ashton 2016). Under the genetic model, there is no reason why regions and even sites with high-quality flint, which later have handaxes, should lack them for some periods of the Lower Palaeolithic.

In a study of handaxe shape, Lycett (2008) found that, consistent with the founder effect, handaxes get less variable with distance from their East African homeland, and Corbey et al. state that this pattern fits their model of genetic inheritance. The founder effect in genetics works because inheritance is particulate, and once you have lost a particular allele from a population, it is unlikely to reappear: but there is no evidence that the quantitative variation in biface shape measured by Lycett is any way particulate or discontinuous. The correlation Lycett identifies between variation and distance from East Africa is weak (<0.3), and it is noteworthy that of the assemblages with reliable age estimates, those from within Africa are older than those from outside Africa, so that this could just as easily be a chronological pattern as a geographical one.

The Movius–Schick hypothesis, which states that biface knapping was a necessary precursor to the development of Middle Palaeolithic Levallois technology, has been put forward to explain the dearth of Levallois in East Asia (Lycett 2007). Under the genetic model, there should not be a relationship between handaxe knapping and Levallois as they are propagated by different transmission mechanisms, but cumulative cultural evolution under the social transmission model explains the geographical correlation between the Acheulean and Levallois.

Cognitive change at the end of the Acheulean is key to explaining the more rapid rate of cultural change after the Acheulean in the social transmission model of handaxe conservatism. Corbey et al. attack the idea of cognitive change on the grounds that it would require convergent evolution between Neanderthals and *Homo sapiens*, but endocranial evidence indicates increases in brain size and changes in brain shape do occur in the emergence of both species (Bruner 2010; Bruner and Holloway 2010). Corbey et al. also use cultural features from Africa in the last 90,000 years (well after the end of the Acheulean) as their benchmark for evidence of cognitive change, rather than evidence from the Acheulean to Middle Palaeolithic transition, which is the period of relevance. Elsewhere we have argued that the Acheulean to Middle Palaeolithic transition is characterised by increases in hierarchical organisation, recursion and, most importantly for cultural change, generativity (Shipton et al. 2013). Throughout Corbey et al.'s article, the argument is that Acheulean handaxe variation is not consistent with modern human models of cultural transmission. However, cultural transmission in Acheulean species with less capacity for innovation should not be expected to generate the same predictions.

The reappearance of handaxe-like forms in later prehistory presents a problem for the genetic hypothesis, as later examples, including those made by our own species, should also be interpreted as genetically determined. It is argued that handaxes are fundamentally different from Middle Palaeolithic bifaces in that the former were platforms on which different working edges were made, while the latter were complete tools with integrated working edges that were maintained during resharpening (Corbey et al. 2016). However, there is no evidence to suggest that handaxes were platforms for multiple working edges rather than integrated tools, and indeed it has been shown that distinctive handaxe shapes are conserved through episodes of reduction, suggesting they were conceptualised as holistic tools (Shipton and Clarkson 2015). The relatively short-lived nature of the Middle Palaeolithic biface industries is used by Corbey et al. (2016) to suggest that it is unlikely they were under genetic control. However, some of the distinctive handaxe forms of the British Acheulean, such as twisted ovates, are similarly short-lived (Ashton 2016; Wenban-Smith 2004; Bridgland and White 2014).

A final criticism of the social transmission model offered by Corbey et al. is that in non-hominin species, genes not culture are invoked as the first resort of explanation for complex behaviour, so to invoke culture in hominins is an anthropocentric double standard. We agree with Corbey et al. that it is indeed anthropocentric, but we contend that in our immediate ancestors within the last million years, anthropocentrism is justified and more parsimonious than proposing a different transmission mechanism for knapped stone to the one which exists in the recent descendants of these hominins.

In sum, there is far too much variation in the morphology of handaxe assemblages in different times and places; between the different stones on which handaxes are made; between the manufacturing sequences through which handaxes were made; and between handaxe and non-handaxe assemblages made by closely related populations, for handaxes to have been genetically controlled. The reappearance of handaxe-like forms in several later industries requires a double standard of explanation in the genetic model for Acheulean and later hominins. It seems wholly unparsimonious to suggest that in the most recent ancestor of *Homo sapiens* and

Neanderthals, the two most social species this planet has seen, a fundamental part of their artefact repertoire was genetic and not cultural.

4 How Modern Knappers Learn

Percussive technology is known among non-human primates such as chimpanzees, macaques and capuchins. Some West African chimpanzee populations crack oil palm nuts in what has been described as a tradition (McGrew 1992), but Corbey et al. (2016) suggest that such behaviour may also be genetically controlled or alternatively it could be independently invented under conducive ecological conditions. An experiment with East African chimpanzees which do not crack nuts in the wild showed that they will learn to do so by observing a model, demonstrating genetic differences between populations do not explain the acquisition of nut-cracking behaviour (Marshall-Pescini and Whiten 2008). In addition, this study also showed that a control group provided with nuts and the materials to crack them but no model to observe did not spontaneously invent nut cracking, indicating percussive technology will not easily arise even given the right ecological circumstances.

Elsewhere we have argued that high-fidelity social learning was key to the longevity of handaxe and cleaver production (Shipton 2010; Shipton and Nielsen 2015; Nielsen 2012). Support for this position can be found among modern people learning how to knap. For example, in one of the rare ethnographic examples documented of people learning to knap large bifaces, Stout (2002) writes that stone adze makers in New Guinea work in a line commenting on the work of less experienced neighbours, reminding them where on the adze to strike and even taking over to overcome particular difficulties.

In the second half of the twentieth century, knapping was revived in many places by pioneers such as Don Crabtree, Errett Callahan and Bruce Bradley in North America, and Mark Newcomer and Francois Bordes in Europe. Having no access to other individuals who could knap, these pioneers were necessarily self-taught, trying to replicate forms that they saw in the archaeological record, in what may be described as emulation learning. It should be noted that these pioneers had the luxury of time that hominins with a spear to sharpen or a carcass to butcher would presumably not have had. Callahan describes the process (http://www.webring.org/l/rd?ring=stonetool;id=3;url=http%3A%2F%2Fwww%2Eerrettcallahan%2Ecom%2F): 'Those first 10 years were a real struggle. I had to work it all out by trial and error. I didn't even know what the questions were, much less the answers. Sometimes I'd find myself banging away for years, making one mistake after the other, trying to isolate what causes what. As slowly as evolution itself, I eventually sorted most of it out'.

Similarly Bruce Bradley (pers. comm.) states: 'I began in the 1960s on my own having never seen or heard of the process. It was entirely trial and error (a lot of error)'.

The Australian knapper Chris Clarkson (pers. comm.) has a similar story of the initial steps: 'For me, knapping began as a first year student in 1989 trying to work out how a beautiful handaxe of Bergerac flint my lecturer had brought to class had been made. I messed around with hard stones, poor quality cherts with incipient fracture planes from the creek, and inappropriately sized hammerstones of different types, including steel claw hammers. Needless to say, I almost gave up'.

As more knappers began to learn the art, they began to teach others. Callahan writes that, after the first 10 years, reading and 'hands-on instruction from the Master-level knappers' were important factors in the development of his knapping skill. Bradley states: 'In 1968 I witnessed Don Crabtree giving a knapping demonstration for about 5 min and this really started me in serious knapping... Over the years I have been fortunate to learn from and also teach many of the best knappers... I have learned from all of them but also contributed to their learning'.

In line with the above, John Lord (pers. comm.) recalls that his knapping developed by watching Newcomer, going away and practising himself for a while and then asking Newcomer for advice on overcoming specific problems. Chris Clarkson says that he 'began reading up on flintknapping, getting into Crabtree ... and Whittaker. An old grainy movie of Crabtree and Bordes making pressure blades, fluted points, Levallois and punch blades was shown in second year, and after that I was really hooked. ...In 2004 ... a student of Bruce Bradley ... showed me some valuable stuff about isolating platforms and turning edges that I knew in theory, but didn't really know too well in practice. ...The age of YouTube began around 2007, and for a while I was glued to channels ... until I had reached a point of satisfaction in my own knapping. ...I found by this point that I could obtain a great deal of information by watching even once a video of someone ... knapping, and trying it for myself'. Similarly, Whittaker (2004) writes that an average 'how I learned to knap story', including his own, would go: 'I found an arrowhead, wondered how it was made, and started experimenting. I figured out a little bit, but then got stuck. Then I met X, and he taught me enough to go on from there'.

Two factors seem to be of paramount importance in learning to knap: practice and contact with an expert. Once the pioneers had discovered an array of knapping techniques, the next generation of knappers who could learn from these early masters had a much easier time of it. In Callahan's words, 'My students are now learning in one week what it took me the first 10 years to learn on my own'.

In 1994, prior to the proliferation of instructional videos on knapping, Whittaker conducted a survey of knappers to establish how they learned (Whittaker 2004). Thirty-two respondents regarded artefacts or self-teaching as the most important factors, while 68 regarded social learning (friends, lessons, watching experts, books on knapping or attending knapping groups) as the most important. When respondents were broken down into two groups of more or less than 10 years of knapping experience, self-teaching was less important to the less experienced knappers than it was for the more experienced group (12% vs. 35% regarded it as the most important factor) (Whittaker 2004). This indicates that from its beginnings as individual learning, the acquisition of modern knapping is becoming increasingly social.

Fig. 2 Slate core, 10 mm thick, flaked bifacially using a stick as a hammer by CS's 5-year-old nephew

Young children engaging with adults in stone working were observed among modern quarry workers in the Hunsgi–Baichbal Valley in India, working with the same stone as Acheulean knappers, and it was hypothesised that knapping could begin at a very young age (Petraglia et al. 2005). An anecdotal observation from CS's experience also suggests knapping could be transmitted early in life:

> My 5 year-old nephew asked me to make him a stone arrowhead (his mother had told him that I could do this). In front of him, I took the small lump of flint that I had and reduced it in a discoidal fashion, telling him to look for any pointed flakes (needless to say he cut himself while doing this). I offered no explanation as to what I was doing because I believed it would not be something he could do at his age. About 6 weeks later he was visiting my farm again, and, without any prompting, in the exact same spot where I had been knapping I saw my nephew attempting to knap! Interestingly, he was not trying to use the remnant lump of flint or the stone hammer that I had used, as he would not have been strong enough. Instead, he was using freehand percussion to hit a slab of slate 1 cm thick, with a stick about 2 cm thick and 20 cm long. In this way he was able to remove at least nine flakes over 2 cm in maximum dimension, the largest being 6.4 cm, and he was even flaking the core bifacially. I gathered up the flakes and core he left behind after he had finished (Fig. 2).

This example indicates that very young individuals can begin learning to knap through social observation.

Several experimental studies have looked at the acquisition of knapping skill. A transmission chain experiment with naïve subjects learning Oldowan technology showed that increased social learning significantly improved the uptake of knapping ability, with imitation providing an improvement over emulation, gestural teaching providing an improvement over imitation and verbal teaching providing an improvement over gestural (Morgan et al. 2015). A handaxe knapping experiment compared novices and expert knappers in a learning condition where they were given the necessary materials and the finished artefact to copy, but were not able to view the knapping process (Geribàs et al. 2010). Participants were given as many attempts as they wished in a single session to achieve a result they were satisfied with. Novices failed to achieve the combination of acute edge angles, tilted cores and soft support of the cores that are critical to freehand percussion. This accords with the suggestion above that even the basics of freehand knapping evident in the Oldowan are difficult to acquire spontaneously. Experts rotated the core more than novices including more bifacial rotations between the faces. Novices meanwhile tried to strike the sides of blank rather than the two main faces. These deficiencies meant that the novices did not produce anything that would be archaeologically recognisable as a handaxe. This experiment supports the observation above that bifacial knapping is counter-intuitive and difficult to master.

The experience of modern knappers indicates that given enough time, it is possible to teach oneself the basics of knapping using purely emulation learning. However, even the basics of percussive technology are not easy to acquire, and the process becomes vastly easier with direct observation and active teaching. Social transmission of knapping appears to happen spontaneously when the opportunities are there. Whiten (2015) presents a 'helical curriculum' model of the social learning of complex skills: an observational episode in which the knowledge of a skill is acquired, followed by a period of practice in which the ability to enact that skill is assimilated; a new observational episode, even of an identical task, then affords new insights; and a subsequent period of practice allows the new skills to be assimilated. This process continues in a cyclical manner with the learner acquiring an increasingly complex skill repertoire. The importance of both social learning and practice in the examples presented here suggests the helical curriculum model may be apposite for the acquisition of knapping skill.

5 Conclusion

Three mechanisms have been proposed to explain the vast scale of the Acheulean: handaxes and cleavers are particularly easy to invent from scratch; they are to some extent genetically determined; and they were replicated through high-fidelity social transmission. We can find no evidence to support the genetic hypothesis, which cannot account for the variability of the Acheulean. The repeated invention hypothesis is a more serious contender as both handaxes and cleavers may have been independently invented in the final Middle Palaeolithic of north-western Europe, and the first generation of modern knappers had to reinvent the Acheulean without other knappers to copy. It should be remembered though that the north-western European final Middle Palaeolithic is currently the only known place and time that cleavers reappear and they arise in the context of a long-standing western European knapping tradition stretching back to the Acheulean. The reproduction of the Acheulean by the first modern knappers was dependent on copying artefacts in a form of emulation learning, rather than completely independent reinvention.

It is possible to recreate the Acheulean without imitation; indeed the first invention of the Acheulean cannot by definition have been learned through imitation. However, the experience of modern knappers indicates it is very difficult to learn to knap without some observation of the process. Learning through observation will also occur spontaneously if the opportunity arises. That chimpanzees acquire simple percussive technology through social learning suggests that this would also have been the default position for early *Homo*. Archaeological evidence for separate canalised knapping methods in the production of handaxes and cleavers at the Acheulean site of Isampur Quarry in India indicates both the method and the goal were socially transmitted—as in true imitation (Shipton et al. 2009; Shipton 2010). The transmission of some of the more complex Acheulean knapping sequences, including those used to make cleavers at Isampur Quarry, may, due to preparatory steps that were causally opaque, have required overimitation (Shipton and Nielsen 2015)—that is, the tendency to copy others with such high fidelity that they will incorporate visible but causally irrelevant actions (Nielsen et al. 2014; Nielsen and Tomaselli 2010). The existence of regional clusters of sites using complex techniques for biface production, such as Victoria West in southern Africa or Tabelbala–Tachenghit in north-western Africa (Sharon 2007), indicates that such techniques were not independently invented but socially transmitted across these regions.

We invoke Whiten's (2015) model of a 'helical curriculum' to explain the acquisition of knapping skill both among modern knappers and in the Acheulean: observation of a particular knapping sequence, followed by practice, recurring in cycles whereby new insights are made with each observation and skills are assimilated through practice until the learner achieves the skill level of whoever they are observing. It has been suggested that the abundance of bifaces at many Acheulean sites might be evidence of deliberate practice (Rossano 2003; Suddendorf et al. 2016). Practice has been shown experimentally to be important for mastering the skill of platform preparation required to produce the refined bifaces characteristic of the Late Acheulean (Stout et al. 2014).

The Acheulean is perhaps the most mysterious of all archaeological periods because it is unlike anything from recent prehistory and has no modern analogues, either from ethnography or primatology. Its most distinctive feature is the vast temporal and geographical scale over which it was produced. We suggest that the hominins who produced the Acheulean shared our propensity for imitation and overimitation, but did not have our capacity for innovation through generativity, recursion and hierarchical organisation (Shipton et al. 2013); and it was this combination of traits that gave rise to the unparalleled longevity and ubiquity of the Acheulean.

Acknowledgements The inspiration for this chapter was attending a knapping workshop with John Lord and hearing him describe how he learned to knap. This chapter benefitted from discussions with Andy Whiten during its genesis and from supportive comments from two reviewers. We thank Chris Clarkson and Bruce Bradley for providing testimony on how they learned to knap. Sheila Mishra and Sushma Deo are gratefully acknowledged for permission to study the Chirki biface collection.

References

Ashton N (2016) The human occupation of Britain during the Hoxnian interglacial. Quat Int 409:41–53

Beyene Y, Katoh S, WoldeGabriel G, Hart WK, Uto K, Sudo M, Kondo M, Hyodo M, Renne PR, Suwa G (2013) The characteristics and chronology of the earliest Acheulean at Konso, Ethiopia. Proc Natl Acad Sci USA 110:1584–1591

Bridgland DR, White MJ (2014) Fluvial archives as a framework for the Lower and Middle Palaeolithic: patterns of British artefact distribution and potential chronological implications. Boreas 43:543–555

Brumm A, Moore MW (2012) Biface distributions and the Movius line: a Southeast Asian perspective. Aust Archaeol 74:32–46

Bruner E (2010) Morphological differences in the parietal lobes within the human genus. Curr Anthropol 51:S77–S88

Bruner E, Holloway RL (2010) A bivariate approach to the widening of the frontal lobes in the genus *Homo*. J Hum Evol 58:138–146

Corbey R, Jagich A, Vaesen K, Collard M (2016) The Acheulean handaxe: more like a bird's song than a beatles' tune? Evol Anthropol 25:6–19

de la Torre I (2004) The omo revisited: evaluating the technological skills of Pliocene hominids. Curr Anthropol 45:439–465

de la Torre I, Martínez-Moreno J, Mora R (2013) Change and stasis in the Iberian Middle Paleolithic. Curr Anthropol 54:S320–S336

de la Torre I, Mora R, Arroyo A, Benito-Calvo A (2014) Acheulean technological behaviour in the Middle Pleistocene landscape of Mieso (east-central Ethiopia). J Hum Evol 76:1–25

Delagnes A, Roche H (2005) Late Pliocene hominid knapping skills: the case of Lokalalei 2C, West Turkana, Kenya. J Hum Evol 48:435–472

Eren MI, Roos CI, Story BA, von Cramon-Taubadel N, Lycett SJ (2014) The role of raw material differences in stone tool shape variation: an experimental assessment. J Archaeol Sci 49:472–487

Frith CB, Frith DW (2004) The bowerbirds. Oxford University Press, Oxford

Geribàs N, Mosquera M, Vergès JM (2010) What novice knappers have to learn to become expert stone toolmakers. J Archaeol Sci 37:2857–2870

Goren-Inbar N, Lister A, Werker E, Chech M (1994) A butchered elephant skull and associated artifacts from the Acheulian site of Gesher Benot Ya'aqov, Israel. Paléorient 20:99–112

Gowlett J (1986) Culture and conceptualisation: the Oldowan-Acheulian gradient. In: Bailey GN, Callow P (eds) Stone age prehistory: studies in memory of Charles McBurney. Cambridge University Press, Cambridge, pp 243–260

Gowlett J (2006) The elements of design form in Acheulean bifaces: modes, modalities, rules, and language. In: Goren-Inbar N, Sharon G (eds) Axe age: Acheulian toolmaking from quarry to discard. Equinox, London, pp 203–221

Gowlett J, Crompton R (1994) Kariandusi: Acheulean morphology and the question of allometry. Afr Archaeol Rev 12:3–42

Harmand S, Lewis JE, Feibel CS, Lepre CJ, Prat S, Lenoble A, Boës X, Quinn RL, Brenet M, Arroyo A (2015) 3.3-million-year-old stone tools from Lomekwi 3, West Turkana, Kenya. Nature 521:310–315

Haslam M, Roberts RG, Shipton C, Pal J, Fenwick JL, Ditchfield P, Boivin N, Dubey A, Gupta M, Petraglia M (2011) Late Acheulean hominins at the Marine Isotope Stage 6/5e transition in north-central India. Quat Res 75:670–682
Kempe M, Lycett S, Mesoudi A (2012) An experimental test of the accumulated copying error model of cultural mutation for Acheulean handaxe size. PLoS One 7:e48333
Lycett SJ (2007) Why is there a lack of Mode 3 Levallois technologies in East Asia? A phylogenetic test of the Movius–Schick hypothesis. J Anthropol Archaeol 26:541–575
Lycett SJ (2008) Acheulean variation and selection: does handaxe symmetry fit neutral expectations? J Archaeol Sci 35:2640–2648
Marshall-Pescini S, Whiten A (2008) Social learning of nut-cracking behavior in East African sanctuary-living chimpanzees (*Pan troglodytes* schweinfurthii). J Comp Psychol 122:186
McGrew WC (1992) Chimpanzee material culture: implications for human evolution. Cambridge University Press, Cambridge
Mitchell J (1994) Studying biface butchery at Boxgrove: roe deer butchery with replica handaxes. Lithics 16:64
Moncel M-H, Moigne A-M, Sam Y, Combier J (2011) The emergence of Neanderthal technical behavior: new evidence from Orgnac 3 (Level 1, MIS 8), Southeastern France. Curr Anthropol 52:37–75
Moore MW (2003) Australian Aboriginal biface reduction techniques on the Georgina River, Camooweal, Queensland. Aust Archaeol 56:22–34
Moore MW, Perston Y (2016) Experimental insights into the cognitive significance of early stone tools. PLoS One 11:e0158803
Morgan T, Uomini N, Rendell L, Chouinard-Thuly L, Street S, Lewis H, Cross C, Evans C, Kearney R, de la Torre I, Whiten A (2015) Experimental evidence for the co-evolution of hominin tool-making teaching and language. Nat Commun 6:6029
Movius HL (1948) The Lower Palaeolithic cultures of Southern and Eastern Asia. Trans Am Philos Soc 38:329–420
Nielsen M (2012) Imitation, pretend play, and childhood: essential elements in the evolution of human culture? J Comp Psychol 126:170
Nielsen M, Tomaselli K (2010) Overimitation in Kalahari Bushman children and the origins of human cultural cognition. Psychol Sci 21(5):729–736
Nielsen M, Mushin I, Tomaselli K, Whiten A (2014) Where culture takes hold: "overimitation" and its flexible deployment in Western, Aboriginal, and Bushmen children. Child Dev 85:2169–2184
Panger MA, Brooks AS, Richmond BG, Wood B (2002) Older than the Oldowan? Rethinking the emergence of hominin tool use. Evol Anthropol 11:235–245
Petraglia MD, Shipton C, Paddayya K (2005) Life and mind in the Acheulean. In: Gamble C, Porr M (eds) The hominid individual in context: archaeological investigations of Lower and Middle Palaeolithic landscapes, locales, and artefacts. Routledge, London, p 197
Piperno M, Tagliacozzo A (2001) The elephant butchery area at the Middle Pleistocene site of Notarchirico (Venosa, Basilicata, Italy), La Terra Degli Elefanti. Consiglio Nazionale delle Ricerche, Rome, pp 230–236
Putt SS, Woods AD, Franciscus RG (2014) The role of verbal interaction during experimental bifacial stone tool manufacture. Lithic Technol 39:96–112
Roberts MB, Parfitt SA (1999) Boxgrove: a Middle Pleistocene hominid site at Eartham Quarry, Boxgrove, West Sussex. English Heritage, London
Rossano MJ (2003) Expertise and the evolution of consciousness. Cognition 89:207–236
Schillinger K, Mesoudi A, Lycett SJ (2015) The impact of imitative versus emulative learning mechanisms on artifactual variation: implications for the evolution of material culture. Evol Hum Behav 36:446–455
Sharon G (2007) Acheulian large flake industries: technology, chronology, and significance. Archaeopress, Oxford

Shipton C (2010) Imitation and shared intentionality in the Acheulean. Camb Archaeol J 20:197–210
Shipton C (2016) Hierarchical organization in the Acheulean to Middle Palaeolithic transition at Bhimbetka, India. Camb Archaeol J 26:601–618
Shipton C, Clarkson C (2015) Handaxe reduction and its influence on shape: an experimental test and archaeological case study. J Archaeol Sci Rep 3:408–419
Shipton C, Nielsen M (2015) Before cumulative culture. Hum Nat 26:331–345
Shipton C, Petraglia MD (2010) Inter-continental variation in Acheulean bifaces. In: Norton C, Braun D (eds) Asian paleoanthropology. Springer, Dordrecht, pp 49–55
Shipton C, Petraglia M, Paddayya K (2009) Stone tool experiments and reduction methods at the Acheulean site of Isampur Quarry, India. Antiquity 83:769–785
Shipton C, Clarkson C, Pal JN, Jones SC, Roberts RG, Harris C, Gupta M, Ditchfield PW, Petraglia MD (2013) Generativity, hierarchical action and recursion in the technology of the Acheulean to Middle Palaeolithic transition: a perspective from Patpara, the Son Valley, India. J Hum Evol 65:93–108
Stout D (2002) Skill and cognition in stone tool production: an ethnographic case study from Irian Jaya. Curr Anthropol 43:693–722
Stout D, Semaw S (2006) Knapping skill of the earliest stone toolmakers: insights from the study of modern human novices. In: Toth N, Schick K (eds) The Oldowan: case studies into the earliest stone age. Stone Age Institute Press, Gosport, IN, pp 307–320
Stout D, Semaw S, Rogers MJ, Cauche D (2010) Technological variation in the earliest Oldowan from Gona, Afar, Ethiopia. J Hum Evol 58:474–491
Stout D, Apel J, Commander J, Roberts M (2014) Late Acheulean technology and cognition at Boxgrove, UK. J Archaeol Sci 41:576–590
Suddendorf T, Brinums M, Imuta K (2016) Shaping one's future self: the development of deliberate practice. In: Klein B (ed) Seeing the future. OUP, Oxford, pp 343–366
Tennie C, Braun DR, Premo L, McPherron SP (2016) The Island test for cumulative culture in the Paleolithic. In: Haidle MN, Conard NJ, Bolus M (eds) The nature of culture. Springer, Dordrecht, pp 121–133
Thiébaut C, Mourre V, Chalard P, Colonge D, Coudenneau A, Deschamps M, Sacco-Sonador A (2012) Lithic technology of the final Mousterian on both sides of the Pyrenees. Quat Int 247:182–198
Tixier J, Inizian M-L (1983) Préhistoire de la pierre taillée. 1. Terminologie et technologie, Valbonne, France
Utrilla P, Mazo C, Domingo R (2015) Fifty thousand years of prehistory at the cave of Abauntz (Arraitz, Navarre): a nexus point between the Ebro Valley, Aquitaine and the Cantabrian Corridor. Quat Int 364:294–305
Wenban-Smith F (2004) Handaxe typology and Lower Palaeolithic cultural development: ficrons, cleavers and two giant handaxes from Cuxton. Lithics 25:11–21
White MJ (2000) The Clactonian question: on the interpretation of core-and-flake assemblages in the British Lower Paleolithic. J World Prehist 14:1–63
Whiten A (2015) Experimental studies illuminate the cultural transmission of percussive technologies in *Homo* and *Pan*. Philos Trans R Soc Lond B Biol Sci 370:20140359
Whittaker JC (2004) American flintknappers: stone age art in the age of computers. University of Texas Press, Austin, TX
Wynn T, Tierson F (1990) Regional comparison of the shapes of later Acheulean handaxes. Am Anthropol 92:73–84

Visuospatial Integration: Paleoanthropological and Archaeological Perspectives

Emiliano Bruner, Enza Spinapolice, Ariane Burke, and Karenleigh A. Overmann

Abstract The visuospatial system integrates inner and outer functional processes, organizing spatial, temporal, and social interactions between the brain, body, and environment. These processes involve sensorimotor networks like the eye–hand circuit, which is especially important to primates, given their reliance on vision and touch as primary sensory modalities and the use of the hands in social and environmental interactions. At the same time, visuospatial cognition is intimately connected with memory, self-awareness, and simulation capacity. In the present article, we review issues associated with investigating visuospatial integration in extinct human groups through the use of anatomical and behavioral data gleaned from the paleontological and archaeological records. In modern humans, paleoneurological analyses have demonstrated noticeable and unique morphological changes in the parietal cortex, a region crucial to visuospatial management. Archaeological data provides information on hand–tool interaction, the spatial behavior of past populations, and their interaction with the environment (e.g., in domains like landscape use and navigation, the spatial relations implicit in social networks, etc.). Visuospatial integration may represent a critical bridge between extended cognition, self-awareness, and social perception. As such, visuospatial functions are relevant to the hypothesis that human evolution is characterized by changes in brain–body–environment interactions and relations, which enhance integration between internal and external cognitive components through neural plasticity and the development of a specialized embodiment capacity. We therefore advocate the investigation of

E. Bruner (✉)
Centro Nacional de Investigación sobre la Evolución Humana, Burgos, Spain
e-mail: emiliano.bruner@cenieh.es

E. Spinapolice
La Sapienza Università di Roma, Roma, Italy

A. Burke
Université de Montreal, Montreal, QC, Canada

K. A. Overmann
School of Archaeology, University of Oxford, Oxford, UK

Center for Cognitive Archaeology, University of Colorado, Colorado Springs, CO, USA

L. D. Di Paolo et al. (eds.), *Evolution of Primate Social Cognition*, Interdisciplinary Evolution Research 5, https://doi.org/10.1007/978-3-319-93776-2_19

visuospatial functions in past populations through the paleoneurological study of anatomical elements and archaeological analysis of visuospatial behaviors.

Keywords Cognitive archaeology · Embodiment · Extended cognition · Land use · Paleoneurology · Parietal lobes · Spatial cognition · Toolmaking

1 Visuospatial Evolution

Visuospatial functions have been long considered to be strictly mechanical phenomena. Activities like "handling" and "walking," for example, have been viewed as automatic processes involving sensory feedbacks and muscle adjustments (i.e., as processes that do not involve high cognitive loads and which have only limited associations with more complex cognitive functions like memory, executive functions, and consciousness). This view of visuospatial functions has persisted, despite the cybernetic paradox that we can build computers capable of performing calculations that far exceed human computational powers, but have still not been able to design a machine able to walk with fluidity. Interest in visuospatial processes has been renewed in the last decade because of an increasing attention toward parietal lobes, which include crucial areas for visuospatial management. Converging evidence from evolutionary neuroanatomy, functional neuroimaging, brain modeling, and cognitive sciences suggests that visuospatial integration may be profoundly intermingled with our most complex cognitive processes (Bruner and Iriki 2016).

The term "visuospatial integration" can be rather vague, since it embraces a variety of concepts and encompasses multiple functional aspects, and it can mean different things in neurobiological, psychological, and philosophical contexts. These subtle but relevant differences in definitions and concepts can generate misunderstandings or cultural barriers between disciplines, and this can be particularly detrimental for a topic which is clearly multidisciplinary. The word "visuospatial" conjoins concepts associated with *visual perception*, including the physiological mechanisms of vision and the associated neural codes and networks, with *spatial perception*, a basic perceptual relation between an organism and its environment. Both vision and spatiality have a physical component, as well as a "virtual" one involved in mental imagery. For its part, the word "integration" denotes functionality in combining multisensory inputs into a holistic perceptual experience.

Beyond any definition, visuospatial integration is generally something difficult to test in modern populations because of its involvement in very distinct tasks, ranging from basic spatial management to social perception. Visuospatial cognition in extinct human species is even more difficult to evaluate, because it must be detected and quantified indirectly through the partial and fragmentary information provided by the paleontological and archaeological records. Paleontology is concerned with the bones of past populations, archaeology with their material relics. Accordingly, visuospatial functions in human evolution must be inferred from the *visuospatial elements* and *behaviors* revealed by bone anatomy and cultural cues. Interpreting this indirect evidence can be challenging, as most biological and cultural processes leave no traces in the paleontological and archaeological records. Nonetheless, a proper evaluation of these biological and cultural remnants represents the only chance to test human cognitive evolution directly on the species involved in the phylogenetic process. As for other cognitive capacities, some visuospatial behavioral changes will remain silent to the paleontological and archaeological analysis, but others may have left some physical indications. Certainly, any cognitive evaluation on extinct species or populations must be framed within a broader context, in which neurobiological and behavioral data about extant groups support a general comparative and functional perspective.

Cognitive archaeology has directly addressed the evolution of spatial cognition since the late 1980s, through the pioneering work of Thomas Wynn (1989). Wynn suggested that bilateral symmetry in early Acheulean bifaces implied that the hominids who produced them about 1.8 million years ago (Mya) had developed an ability to coordinate shape and spatial information, "arguably the initial step in the evolutionary sequence of modern visuo-spatial integration" (Wynn 2014: 291; also see Wynn 2002). Similarly, by roughly 500 thousand years ago (Kya), the use of complex prepared-core techniques and the imposition of 3D symmetry in late Acheulean bifaces suggested that there were further evolutionary developments in visuospatial integration, including the capacities for "inner virtual space" (Bruner and Lozano 2014: 274) and landscape mental mapping (Wynn 2000). By 50 Kya, behavioral and morphometric change in *Homo sapiens* included greater use of projectile weapons (Lombard and Haidle 2012) and possibly a more specialized eye–hand visuospatial interface (Bruner and Lozano 2014). More recently, following the idea that cognition is embodied, embedded, extended, and enacted (see below), cognitive archaeology has begun evaluating whether such behavioral and morphological changes might signal the emergence of additional developments in visuospatial capacity, enabling in our species "a more pervasive prosthetic adaptation" (Overmann 2015: 165).

2 Beyond Neurons

Ideas about how the physical body and material environment might influence, aid, and perhaps even constitute human cognition have been introduced in recent decades. Cognition has not remained localized in brain form and function (the neurocentric view of cognition) or viewed as a computational process involving perceptual inputs, computer-like manipulation, and motor outputs (the computational model of cognition). Rather, it has become *embodied* (the body has a role in cognition—Lakoff and Johnson 2008; Prinz 2009), *situated* (cognition is influenced or perhaps even constituted by things external to the brain—Robbins and Aydede 2009), *embedded* in (cognition is influenced by aspects of the environment—Smith 1999), and *distributed* across (cognition involves resources and processes in networks of interacting agents and objects—Hutchins 1995) environments containing natural, social, and culturally modified material resources. Beyond distribution, cognition may be *extended*, the notion that physical, social, and material resources may be at least partially constitutive of cognition, perhaps even on par with psychological processing and neural activity, though the ways in which brain-external resources contribute undoubtedly differ from brain-internal ones (Clark and Chalmers 1998; Clark 2008).

The idea that physical and material elements may be included within the boundaries of cognition is especially relevant to human cognitive evolution, given the unparalleled degree and extent to which our species uses material culture. Historical ideas about brain functionality have also been updated with insights gained from research in artificial intelligence (AI). For example, traditional cognitive models have proposed that brains recreate the external world in the form of internal representations. However, breakthroughs in AI research have suggested that highly representational cognition would be inefficient and likely evolutionarily unsustainable; internal representations would also be redundant to the information-bearing capacity of the world itself, information that an organism obtains through interactivity with the world (Clark 1997).

The turn to anti-representationalism helped inspire the idea that cognition is *enactive*, constituted by the dynamic interactivity between brains, bodies, and material environments (Hutto and Myin 2013; Malafouris 2013). Thinking and doing are not separable. Performative skill, such as that involved in complex flintknapping techniques (e.g., Levallois), can therefore be viewed as mastering the enactivity of variable domains (i.e., psychological processes, motor movements, and material properties) to produce invariant results (Bril et al. 2010; Nonaka et al. 2010; Nonaka and Bril 2014; Nonaka 2011, 2013). This construct in particular highlights the importance of coordinating mechanisms like manuovisual and visuospatial integration.

As is common among primates, of all their senses, humans rely the most on *haptic perception*, active touch through the hand, and *vision* for their environmental interactions (Fig. 1). As "one of the main perturbatory channels through which the world touches us, [the hand] has a great deal to do with how this world is perceived and classified" (Malafouris 2013: 60). Haptic perception integrates sensory cues from what the fingers touch with internal sensory and positional cues about what the fingers feel and do (Hatwell et al. 2003). Its sensory dimensions are simultaneously

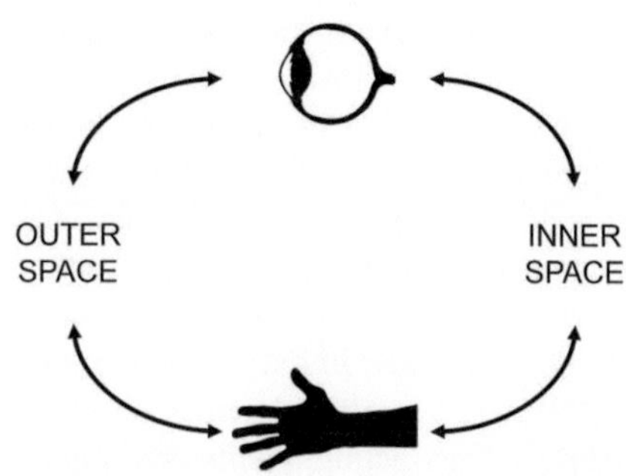

Fig. 1 Although the entire body operates as an active interface between the brain and the environment, the eye and the hand in primates represent the two main *ports* of information flow and management

internal and external, *bipolarity* that humans exploit (apparently, uniquely among animal species) in exploring their environments and making and using material culture (Mattens 2013). While sensing what objects are like by touching them undoubtedly supports the ability to explore and manipulate materiality, feeling what touches the body may be similarly important in helping broker interactions between human psychological abilities and the objects of material environments and cultures (Mattens 2013).

Haptic perception provides a relatively limited perceptual field—only what the fingers can touch—that demands the fingers explore to gain sufficient information; the need to explore, coupled with high finger mobility and control, creates in touch a degree of alterability and intentionality that other senses lack (Hatwell 2003; Helbig and Ernst 2008). Vision, in contrast, provides a wide perceptual field easily explored through saccades and attentional shifts, enabling visual acquisition of more information in less time than is possible through touch, especially for informational domains where touch is particularly limited (e.g., spatiality, shape, etc.). Proficiency in interacting with an environment thus depends upon the abilities to gain pertinent sensory information, particularly about the locations and properties of manipulated objects; coordinate the inputs of multiple sensory modalities, most importantly touch and vision; and exercise fine control over and coordination of motor movements, especially for the hands and eyes (Gallagher 2013; Tallis 2003, 2011).

In the enactivist paradigm, change in interacting neural, behavioral, and material resources cannot easily be isolated to the individual cognitive components. That is, brain function and form do not change on their own, but rather, through and because of interactivity with behavioral–material stimuli and vice versa. Further, the constant state of change in human becoming (i.e., the ongoing and unfinished evolutionary process of becoming human) relates to both the capacity for change (plasticity) and the interactivity among the components of cognition. This is *metaplasticity*, "the enactive constitutive intertwining between neural and cultural plasticity" (Malafouris 2010a). The term brain–artifact interface (BAI) has been used to recognize that "*the functional structure and anatomy of the human brain is a dynamic construct remodeled in detail by behaviourally important experiences which are mediated, and often constituted, by the use of material objects and cultural artefacts which for that reason should be seen as continuous integral parts of the human cognitive architecture*" (Malafouris 2008: 404, italics original).

BAIs are "technological mediations (material structures, processes, objects or other socio-material apparatuses and practices) that enable the configuration of a dynamic alignment or tuning between neural and cultural plasticity" (Malafouris 2010b: 265). Crucially, BAIs in the form of material structures provide "temporal anchoring and binding" that helps bridge the vastly different timescales on which neural and cultural resources operate (Malafouris 2010b: 266). While psychological responses and ontogenetic development take milliseconds to decades, materiality spans centuries to millennia or longer.

The idea that materiality not only reflects but also influences and changes brain function and form, especially over cultural and evolutionary spans of time, is perhaps not intuitive. Yet many species use or modify their behaviors and aspects of their environments in ways that can be considered tool use and culture (Laland and Janik 2006), and it is well accepted that physical anatomy and material environments influence behavior and psychological processing in most (if not all) species. Certainly, embodiment with pentadactyl limbs has strongly canalized the organization of number systems in cycles of 10 (decimal, the world's most prevalent numerical base), 5 (quinary), and 20 (vigesimal) (Comrie 2013; Menninger 1992), and manuovisual adeptness has been a primary enabler in allowing humans to use materiality to decompose tasks, collaboratively problem-solve, and—perhaps the most unique—accumulate and distribute cognitive effort over space and time (Beer 2003; Hutchins 1995).

Human manipulation of materiality into novel forms can also create new behavioral and psychological stimuli with the potential to change these cognitive components. Such interactivity opens up multiple cascading possibilities for further changes to the cognitive system and over time may yield the kind of anatomical and behavioral differences that can be detected in the archaeological and fossil records. Examples include the change in spatial competence inferred from increasing symmetry in the shape of stone tools (Wynn 1989), as well as the intensification of the ability to incorporate materiality as a component of cognition inferred by comparing morphological change in brains and bodies in early *H. sapiens* and Neanderthals (Overmann 2015).

Given the importance of visuospatial integration to human becoming, a multidisciplinary approach is required to investigate the development of complexity in systemic behaviors arising from the interaction of brains (paleoneurology), bodies (paleontology), and materiality (archaeology). In this paper, we introduce three main sources of evidence for making inferences about visuospatial functions in extinct populations: neuroanatomy, technology, and ecology.

3 Evolving Parietal Lobes

The first source of evidence for investigating visuospatial functions in extinct species comes from their cranial remains: morphological changes in the neuroanatomical regions involved in visuospatial integration, particularly the parietal lobes. This region of the brain has been a particular research focus in the last decades (Mountcastle 1995; Zilles and Palomero-Gallagher 2001). Initial interest examined general parietal functions like visuospatial integration and attention (Andersen and Buneo 2002; Andersen et al. 1997; Rushworth et al. 2001; Wardak et al. 2005), while more recently the analysis of the parietal cortex has been expanded to include more specialized processes like numbering (Ansari 2008; Cantlon et al. 2006). Studying the functions of the parietal lobes is challenging, for several reasons. First, as previously mentioned, relatively little attention has been given to them until fairly recently because of the perception that visuospatial functions were only associated with simple, mechanical behaviors. Second, these regions are involved in many different functions, with consequent operational difficulties in the design of reliable and comprehensive experimental paradigms. A third limit comes from the anatomy of these areas. The parietal elements primarily involved in visuospatial management are the intraparietal sulcus and precuneus, both of which lie hidden in the depths of the cerebral volume (Ebeling and Steinmetz 1995). Such internal folds are more difficult to analyze in extant brains because of their secluded position within the brain mass. Dealing with fossil species, paleoneurology largely concerns the morphological study of the outer surface, and changes of the inner cortical areas can be only partially evaluated. Their position also restricted those traditional approaches associating brain anatomy and behavior after physical or vascular lesions. In fact, individuals experiencing injuries or pathologies in deep cortical regions are less likely to survive, relative to individuals injured in more superficial brain districts. As a result, there are less data available to correlate regional damage with functional impairment. Finally, these regions are phylogenetically diverse. Differences among primate species are often so apparent that direct comparisons between corresponding areas are not easily established. As a result, parietal homologies between humans and nonhuman primates have yet to be reliably established.

In humans, the intraparietal sulcus is larger and more complex than in other primates (Grefkes and Fink 2005; Orban et al. 2006; Vanduffel et al. 2002). Its areas are critical for eye–hand coordination (Battaglia-Mayer et al. 2006; Orban and Caruana 2014; Sakata et al. 1997). As might be expected, the intraparietal sulcus has been directly implicated in toolmaking (Stout and Chaminade 2007; Orban and Caruana 2014; Stout et al. 2015). The precuneus is a major functional and structural brain node, and it is involved in a wide array of behavioral processes. The precuneus appears to support two primary functions: it is a crucial node for visuospatial integration and a main connectivity hub for brain networks. As visuospatial node, it coordinates external and internal information to bridge the somatosensory cortex with vision and egocentric memory (Cavanna and Trimble 2006; Margulies et al. 2009; Zhang and Li 2012). The precuneus has been implicated in generating self-centered mental representations, as well as coordinating spatial, chronological, and social information (Freton et al. 2014; Land 2014; Peer et al. 2015). It is particularly connected with the frontal cortex, as a key element of a highly integrated frontoparietal system (Caminiti et al. 2015; Jung and Haier 2007). It is part of a network found only in humans thus far that joins the frontal and upper parietal areas and which is thought to be essential to the shift from *emulation* (reproducing results, a primate ability) to *imitation* (reproducing processes, a human ability) (Hecht et al. 2013). Its upper areas are strongly involved in visuospatial managements, through direct connections with the intraparietal sulcus and with the supramarginal gyrus, while its lower areas are more integrated with a central hub of the default mode network (Bzdok et al. 2015; Utevsky et al. 2014), a system that coordinates resting (or default) neural activity in the absence of specific cognitive tasks (Buckner et al. 2008; Hagmann et al. 2008; Meunier et al. 2010). Interestingly, the precuneus is also particularly active in metabolic terms (Cavanna and Trimble 2006; Sotero and Iturria-Medina 2011), suggesting that its evolutionary changes may have been associated with thermal constraints due to the brain's energy consumption and heat release (Bruner et al. 2014a). Interestingly, Alzheimer's disease, a pathology

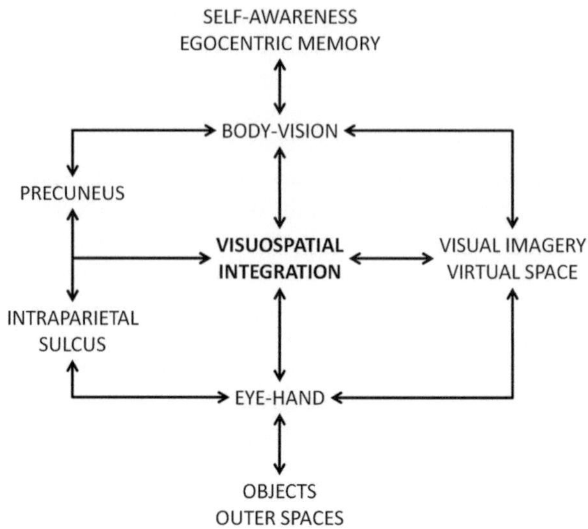

Fig. 2 The dorsal and medial parietal areas (precuneus and intraparietal sulcus) are crucial for processes of visuospatial integration, generating a bridge between embodied experience and cognition

characterized by a combination of features found only in our species, is associated with early metabolic impairments in these deep parietal regions (Zhang and Li 2012; Bruner and Jacobs 2013).

The impression that visuospatial integration supports simple, mechanical behaviors has been significantly challenged by recent experimental evidence of neural change associated with interactions between body and tools (Iriki and Taoka 2012; Maravita and Iriki 2004). Tools induce different cognitive and neural responses when positioned outside the range of the body (extra-personal space) or when positioned within the range of physical interaction (peri-personal space). Most importantly, tools are incorporated into the body schemata when they are touched or handled: the brain interprets an object that is handled as an extension of the body. The classic example of this effect is the blind man's stick, whose neural incorporation extends tactile perception to the tip (Malafouris 2008). In this sense, objects function as "extra-neural" elements of the cognitive system, with the deep parietal areas serving to integrate internal neural processes with external environment features and managing the body's sensorimotor capacities as the interface between the two (Fig. 2).

Although the whole body may act as an active interface, in primates the eye and the hand represent the main "ports" through which information is directed inward and outward, through an active integration of sensorimotor signals. In this process, the hand is not merely a biomechanical structure but also functions as a sensory device associated with neural feedback mechanisms that enables it to perform as an active biological interface, and not a mere passive anatomical component (Ingber 2008; Turvey and Carello 2011). It is interesting to note that handling processes (e.g., grasping objects, manipulating tools, and recognizing characteristics like size, texture, and quantity) are functionally subserved by distinct neural and cognitive mechanisms (Goodale et al. 1994). Therefore, "producing a tool" may rely on different processes than "using a tool" does, which in turn may only partially correspond to "feeling a tool" in terms of perception. Evolutionary changes in these processes have likely been complementary, but not necessarily identical or always integrated.

Coordinating the outer and inner worlds is a main target in terms of ecology and adaptations, and it largely relies on visuospatial issues. The *outer world* is the environment, and, for humans, environment means also culture, including technology. Tools enhance and influence our sensory and computational capacities, constrain and channel reasoning and decision-making processes, store information, and activate, catalyze, or induce responses in specific neural circuits. It is reasonable to believe that, among human populations, the less technological complexity the more the natural environment may be relevant in cultural terms. Conversely, the natural environment may lose importance as technology becomes a dominant mode of outer extension.

The *inner world* is the neural system, its organic extensions (somatosensory elements), and its intrinsic codes and representations. As previously mentioned, brains have historically been thought to recreate the external world internally, with the word *representation* used to describe the resultant stable and environment-independent simulations. However, there is no reason to think that representations are necessarily the result of disembodied processes (Prinz and Barsalou 2000). The current debate on this issue may be largely a matter of terminology, probably because the exact nature of such "representation" is not clear at all in biological or neurobiological terms. Physiologically, a representation may simply be a neural map that encodes sensory information (e.g., gained through visual means). Such mapping is not necessarily complete and self-sufficient, as it may rely on external components engaged through interactivity. And it is not necessarily stable, since it may depend on constant inputs and feedbacks from the body and the environment. As a matter of fact, humans generally think in images and language. Images need a virtual space in which we can perform simulations and mental experiments. Such "imaged space," whatever its actual biological background or neural substrate, is visuospatial in nature, largely self-centered, and based on spatial and chronological managements crucially coordinated by the precuneus (*visual imagery*; see Fletcher et al. 1995), which represent a functional bridge between the body and vision. Interestingly, also regarding language, hypotheses concerning its evolution have associated its mechanisms with praxis, gesture, and movement (Binkofski and Buccino 2004; Buccino et al. 2005; Jirak et al. 2010; Marino et al. 2012), highlighting an additional importance of visuospatial managements.

Parietal growth and development have been associated with specific genetic factors (Chen et al. 2012), but they are also sensitive to training and behavioral effects (Hihara et al. 2006; Iriki and Taoka 2012; Quallo et al. 2009). Therefore, observed anatomical, physiological, or behavioral differences between individuals or groups (species or populations) can be primarily due to specific genetic characters or else be the secondary effects of cultural training. The integration between biological and cultural factors may also promote hybrid evolutionary processes. For example, at the end of the nineteenth century, James Baldwin hypothesized that learned behaviors can change the selective environment, channeling the following genetic selection toward culture-based directions (Baldwin 1896; Crispo 2007; Sznajder et al. 2012). In other words, the "Baldwin effect" holds that "under some conditions, learned behaviors can affect the direction and rate of evolutionary change by natural selection" (Depew 2003: 3). Taking into account all these possibilities, evolutionary changes of the parietal functions in general and visuospatial capacities in particular can have been influenced by genetic, epigenetic, environmental, and cultural factors, and the extent to which each of these may have contributed to observed neural and behavioral differences between humans and nonhuman primates is currently unclear.

Paleoneurological analyses have provided compelling evidence of evolutionary change in parietal morphology. Though paleoanthropologists have long observed parietal anatomical differences and variations in early hominids and between extant and archaic humans (Dart 1925; Holloway 1981; Weidenreich 1941), quantitative data on evolutionary changes in parietal shape, size, and volume were not available until the last decade (Fig. 3). Shape analyses suggest that the globularity of the modern human head is primarily related to an enlargement of the parietal bones (Bruner et al. 2004, 2011) that is in turn associated with an expansion of parietal lobes (Bruner 2004; Bruner et al. 2003). During development, this globularization takes place in an early stage present only in modern humans and absent in Neanderthals and extant apes (Gunz et al. 2010; Neubauer et al. 2009, 2010; Scott et al. 2014). Most studies suggest that this stage occurs after birth, although there is the possibility that it takes place during gestation (Ponce de León et al. 2016).

The pattern of parietal expansion observed in modern humans is comparable to the pattern of enlargement of the precuneus observed in extant individuals, which is a major source of intraspecific variation in human brains (Bruner et al. 2014a, b, 2015). This same pattern also represents a key difference between human and chimpanzee midsagittal brain morphology, because of a noticeable enlargement of the precuneus in our species (Bruner et al. 2017). Interestingly, early modern humans may have lacked the parietal bulging that characterizes *H. sapiens* (Bruner and Pearson 2013), with some later specimens displaying an intermediate phenotype (Bruner et al. 2018). This evidence is still based on very few individuals, but, if confirmed, it separates the origin of the modern human lineage from the origin of the modern human brain form. This raises once more the question of whether such changes were gradual or discrete and whether associated with genetic selection or environmental feedback (Bruner and Iriki 2016).

Apart from these gross neuroanatomical differences dealing with size and proportions, comparative analyses to date have failed to localize any specific structural or functional difference of the parietal cortex between humans and other primates (Caminiti et al. 2015; Caspers et al. 2011; Mars et al. 2011; Scheperjans et al. 2008). Also, its role in the default mode network seems comparable in different species (Barks et al. 2015; Rilling et al. 2007). All these results suggest that the general organization, connectivity, and processes of the parietal lobes are generally conserved. Therefore, we must evaluate whether differences in human parietal anatomy and functions have been a matter of new elements and mechanisms (i.e., derived

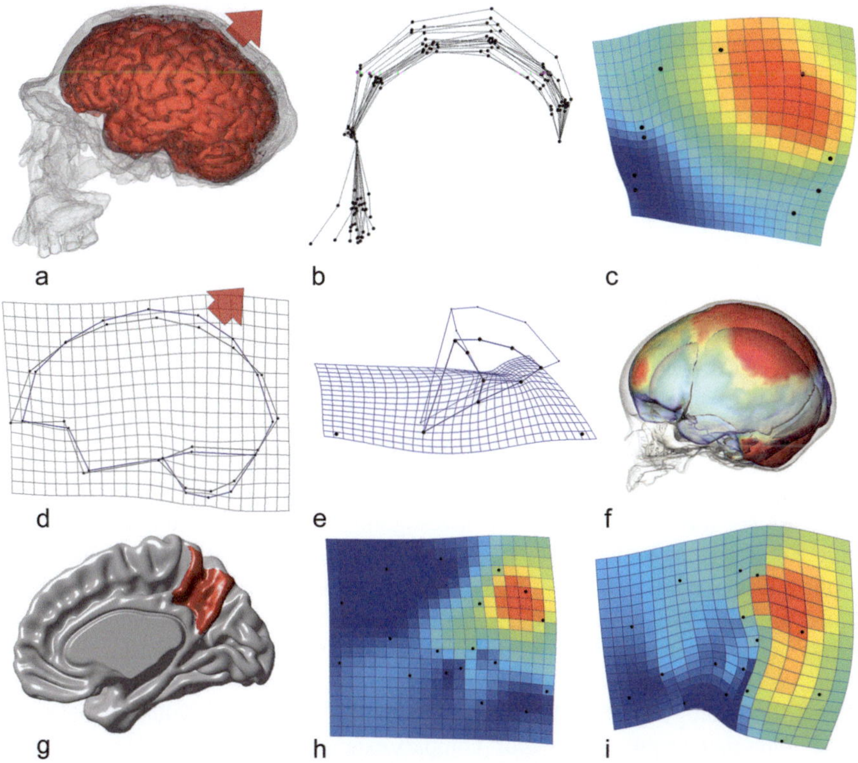

Fig. 3 A general enlargement of the parietal areas represents the main morphological difference between modern humans and other hominids. (**a**) Parietal bone enlargement is the main cranial feature differentiating modern and non-modern skulls (**b**, **c**) (red, expansion; blue, reduction) and endocasts. (**d**) Such geometric dilation is associated with the dorsal parietal surface (**e**) and an early ontogenetic stage that has been described only in *Homo sapiens*. (**f**) In modern humans, a major source of variability is the size of the precuneus (**g**, **h**), which is also the main midsagittal brain difference between humans and chimpanzees. (**i**) Images after Bruner (2004, 2010), Bruner et al. (2004, 2014a, 2015, 2017), Gunz et al. (2010)

characters), the reuse of primitive traits (exaptation), or simply a matter of the degree of expression of shared abilities (a plesiomorphic scheme enhanced in terms of capacity, possibly with threshold effects).

Although the parietal surface displays the most noticeable morphological change between modern and non-modern humans, other brain regions represent important nodes for functions associated with spatial behavior. For example, spatial mapping relies on medial temporal elements, and there is direct quantitative evidence of larger temporal lobes in our species compared to extant apes (Rilling 2006; Rilling and Seligman 2002). Actually, the temporal poles in extant humans are also more projected relative to those of other human species, at least when the morphology and position of the middle endocranial fossa are considered (Bastir et al. 2008).

In sum, we should consider three distinct issues: (1) *H. sapiens* is characterized by a unique morphological expansion of the parietal bones and lobes; (2) the intraparietal sulcus and the precuneus, implicated in a number of critical functions related to visuospatial behaviors, are localized in those same areas and represent a relevant source of variability both within and between extant and extinct species; and (3) the possibility that these two facts are somehow related. It remains to be established to what extent these morphological changes are the result of genetic, selection, and environmental influences. In all cases, taking into consideration the conspicuous parietal changes and the importance of culture and tooling in our species, the exploration and analysis of visuospatial functions merit attention in evolutionary anthropology and behavioral neuroanatomy.

4 Tools and Tooling: Changing Perspectives

The second line of evidence for investigating visuospatial functions in past populations comes from their tools and the elements and processes associated with tooling. Extant humans evolved from ancestors who had the ability to make and use tools, and material culture plays a major role in current evolutionary theories: behaviors with materiality are an essential part of hominid evolution, bridging adaptive responses, natural selection, and cognitive capacities (Foley and Lahr 2003). Technology and tooling are thus hallmarks of human evolution, and there is no doubt that we provide the most complex example of technology-based species. Because stone was an essential resource for Paleolithic groups, and because it has the property to persist sometimes over million years, we largely rely on it to make inferences about the behavior of extinct prehistoric populations.

The interaction between the brain and body is a focal point for thinking about the change in tool form and use over time and about the interaction between hominids and artifacts. Because they are made of a material whose durability makes them more likely to preserve, lithic artifacts are the most common archaeological objects. Lithics are often the only vestige of past behavior for an extinct population or species. The stone tool archaeological record covers an enormous span of time: from over 3 Mya to the present (Fig. 4). Thus the interaction of brains and tools through the interface of hands and eyes fundamentally underpins and pervades the evolutionary history of our species. As Robert Foley (1987: 381) noted, "covariation between hominid fossil morphology and artefact variability is an essential starting-point for understanding the evolution of human behavior," with visuospatial integration offering a behavioral bridge between evolutionary neuroscience and prehistoric archaeology.

An attempt to account for the evolution of tooling is the "mode" model (Clark 1969), based on the relation between the core and the product. Mode 1 represents the simplest flaking technology, linked to the Oldowan industry. Mode 2 corresponds to the Acheulean (shaping and producing larger flakes) and mode 3 to the prepared-core technologies of the Middle Paleolithic sensu lato. Finally, modes 4 and 5 are

Fig. 4 Approximate chronology of major lithic groups (key: My, million years; Ky, thousand years). Drawings show (from left to right) examples of the Oldowan, Acheulean, Mousterian, and Aurignacian industries (images of the tools redrawn after Didier Descouens, Creative Commons License)

associated with the production of blades and microliths, respectively. The sympatry of different hominid species sharing the same technocomplex challenges the assumption of a one-to-one relationship between a particular species and cultural innovations represented by specific lithic technologies. Nonetheless, the mode system enables an initial tentative identification of distinct cognitive schemes with different evolutionary and adaptive implications. Most importantly, the model enables us to make plausible associations between particular cognitive processes and the adaptive behaviors producing specific technocomplexes (Foley and Lahr 2003).

Considered in total, the rate of change in lithic technologies is relatively slow, with very long periods without significant innovations followed by apparent changes associated with novel patterns. These innovations often seem to be the combination of pre-existing technical knowledge with new independent ideas (Beaune et al. 2009), and in this sense they could be considered a form of exaptation (sensu Gould and Vrba 1982).

The shift from *tool-assisted* to *tool-dependent* foraging may signal an important cognitive change in hominids (Plummer 2004), one that would have introduced culture ipso facto into their cognitive and neural niches (Iriki and Taoka 2012). During handling, tools are included in the "body schemata" of the brain and thus are integrated within the neurological mechanisms that affect the way we move and we perceive the space around us (Maravita and Iriki 2004). The cognitive foundations connected with toolmaking are controversial, involving many distinct areas and processes (Stout and Chaminade 2007), and lithic studies have often been used to provide a different order of understanding about hominid cognitive capacities, from mental imagery to planning specific knapping goals. However, exactly what that cognitive change may have consisted of has been difficult to define. Lithic production has been decomposed into sequences of steps and decisions (the *chaîne opératoire* approach) in attempts to understand, among other aspects, its cognitive underpinnings. For example, the concept of "method" in lithic studies implies the repeated organization of the knapping operation, traditionally linked to a "mental preparation" before the knapping itself (Tixier 1967). To illustrate this point, Pelegrin (2009) compared the knapping process to a game of chess, in which many preparatory steps are aimed at producing the conditions needed for the desired outcome. Such analyses have been used to argue for different understandings of hominid cognitive capacities, from mental imagery to planning specific knapping goals, and indeed, the full extent of the cognitive foundations of toolmaking remains unsettled, as they involve multiple, distinct brain regions and processes (Stout and Chaminade 2007). Further, cognitive models of lithic production have not yet fully incorporated insights like the neurological incorporation of tools, when they are handled, into mechanisms affecting movement and spatial perception (Maravita and Iriki 2004).

However, if we interpret the mind as an integrated process between the brain, body, and environment, the analogy of a computer-like process performed by activity in the brain becomes too reductive. If the body and the material culture are indeed implicit components of cognition, both hands and tools must be the target of a more comprehensive behavioral and cognitive analysis. Certainly, the hand/tool relationship is not unidimensional: the hand not only produces but also uses the tool, actions that may be managed by complementary but distinct cognitive processes. Hand and tool variation can be the result of reciprocal influence channeling their properties through information that is external to the brain and central nervous system but which is necessary to and implicit in the cognitive responses.

Managing relationships between the brain, body, and tools relies on visuospatial processes, and for this reason, changes associated with the body and material culture can reveal underlying differences in visuospatial behavior, allowing inferences about the change in visuospatial functions. Evidence of visuospatial behaviors can be direct or indirect.

Direct information providing evidence of visuospatial behaviors can be obtained from tool manufacture and use and from hand morphology, as well as the integration of the two. Behaviors associated with tools and tooling can be investigated through experimental replication of prehistoric tools, observation of stone making in contemporary human groups, and observation of the tool use in apes (Marzke and Shackley 1986; Toth et al. 1993; Schick et al. 1999).

Following Napier (1956), hand morphology and tool evolution have been major topics in anthropology and functional anatomy (Marzke 2013; Marzke and Marzke 2000). Bipedalism, which freed the hands and upper limbs from locomotion, set the basic conditions for the evolutionary emergence of tooling, with locomotor and manipulation abilities evolving in mosaic fashion rather than following a gradual trend (Kivell et al. 2011). Morphological advantages of the human hand with respect to apes include precision grip (Napier 1956) and "squeeze" grip (Marzke 1997), facilitating the ability to handle a stone hammer. Precision movements of the wrist are also considered significant. However, the presence of pad-to-pad precision grasping in the genus *Orrorin* suggests that some of the basic features of the human-like hand may be plesiomorphic for hominids (Almécija et al. 2015).

Grasping patterns can be classified in terms of hand–tool interaction (Feix et al. 2015), and the hand's functional morphology can be analyzed within a specific archaeological context (Williams et al. 2012). This physical interaction requires a proper integration between the motor system, the properties of the object, and the action itself (Vankov and Kokinov 2013). These kinds of studies can be modeled by using different quantitative (Liu and Zhan 2013) and comparative (Pouydebat et al. 2009; Rolian et al. 2011) approaches, and their applications range from evolution to robotics (Kappassov et al. 2015). Interestingly, Neanderthals and coeval modern humans in the Levant may have differed in their hand anatomy while using similar technology (Churchill 2001; Niewoehner 2001).

Although precision and hand dexterity are crucial for tooling, they are not a necessary requisite for cognitive extension. A machine (e.g., a mechanic arm) can be far more precise than a human hand, but its close program prevents any cognitive mechanisms. Therefore, we should consider whether precision is a cause or an evolutionary consequence of prosthetic capacity. As previously mentioned, beyond purely mechanical issues, the body responds to tactile stimulations according to a system of tensions and stresses in which the process of converting mechanical signals into biochemical responses is synchronized with perceptive feedback and hence directly with behavioral and cognitive experience (Ingber 2008; Turvey and Carello 2011). How such relationships might influence tool manufacture and use, however, remains to be investigated.

Indirect information of visuospatial behaviors can be obtained from remains that imply specific activities. For example, Neanderthals and their common ancestor with *H. sapiens* displayed frequent and multiple scratches on their anterior teeth, suggesting they used their mouths as a "third hand" to a greater extent than contemporary *H. sapiens* or any extant human population (Lozano et al. 2008). Because Neanderthals relied more extensively on the mouth for handling and also lacked the dorsal parietal expansion characteristic of contemporary *H. sapiens*, Bruner and Lozano (2014, 2015) have hypothesized that Neanderthals may have lacked the specialized visuospatial system and embodying capacity possessed by modern humans. The hypothesis can be evaluated through expanded analyses of the paleontological and archaeological information, investigation of mouth–hand exploratory behaviors during ontogenetic development, examination of ethnographic parallels in living populations of hunter–gatherers, and functional analyses of mouth–hand integrative relationships.

Since making stone tools is a social behavior (i.e., learned and reproduced through social processes of observation, reproduction, practice, and perhaps language), it depends on cognitive capacities, processes, and skills that exceed those of the individual (e.g., Lycett 2011; Tostevin 2013). Even *expedient* tooling implies accumulated knowledge, social transmission and learning, and an understanding of the mineral environment, factors that become more critical as tool complexity increases. These features suggest that tooling as a cognitive issue is intimately connected with the social environment. Since tools and social systems can exert reciprocal influences, the archaeological record may be able to detect and relate change in both components.

About tooling and social structure, handling capacities have often been associated with language (Binkofski and Buccino 2004; Leroi-Gourhan 1964). In extant populations, language appears to play an active role in learning and teaching manufacturing techniques (Uomini and Meyer 2013). As mentioned, it has been recently hypothesized that language comprehension may be strongly embodied, requiring a response of the motor system to code words and actions (Corballis 1999; Gentilucci and Corballis 2006; also see Buccino et al. 2005; Marino et al. 2012). This possibility suggests that we should consider a scenario involving the hand–tool circuit, the language, and the social system, with visuospatial capacity integrating and managing the relations between the brain, body, and environment.

5 Evolving Spaces

A third way to investigate visuospatial functions is to examine the spatial behavior of past human populations. Complex patterns of landscape use can be reconstructed through the distribution of archaeological sites, archaeozoological evidence of seasonal patterns of occupation, and evidence for the circulation of raw materials and artifacts which act as proxies for human mobility. Patterns of mobility and the maintenance of social networks within settlement systems are activities that rely upon spatial cognition and wayfinding ability.

Wayfinding, the ability to plan and carry out navigation tasks, implies an ability to create abstract mental representations of space. Two distinct representations of the world are linked by a process of translation: first, an *egocentric* representation that involves "transient action-oriented egocentric self-object associations" and, second, an *allocentric* representation with a "more enduring … object-object or environment-object association" (Burgess 2006: 555). The allocentric representation implies the adoption of a "survey" perspective, also known as a "bird's-eye view." The translation process, known as *spatial updating*, allows agents to translate their current position onto an enduring, allocentric representation of space which functions as a "cognitive map," an internal representation of the world which acts as a framework for planning further action and is used metaphorically during memory recall (Golledge 2003: 30).

The hippocampal formation plays an important role in generating spatial representations (Roche et al. 2005; Nadel and Hardt 2004; Spiers and Maguire 2006; Maguire et al. 1999, 2000, 2003; Etienne and Jeffery 2004). The parahippocampal and perirhinal cortices are the main links between the hippocampal formation and the sensory systems (Roche et al. 2005), thus playing a critical role in path integration (Whitlock et al. 2008). The posterior parietal cortex integrates the sensory and motor systems and plays a key role in planning specific actions and enabling goal-directed movement (Spiers and Maguire 2006; Whitlock et al. 2008). Together, these different regions of the brain contribute to the creation of the spatial representations used during wayfinding.

Spatial representations imply the motivated and deliberate encoding of environmental information; they are built up from direct experience of the world (or from external aids, such as maps) and constantly updated as an agent moves through space. There are two distinct strategies used during this process: path integration and landmark recognition (Foo et al. 2005). *Path integration* (aka "dead reckoning") is the ability to update one's position on the allocentric representation using inertial signals processed by the vestibular system (Etienne and Jeffery 2004) and Euclidian geometrics (bearings and path lengths or distance and direction). *Landmark recognition* (the ability to select relevant features of the landscape and remember their position) is used when adopting an egocentric perspective and also comes into play during the process of translating between spatial perspectives to create an integrated cognitive map. Path integration and landmark recognition are not mutually exclusive processes since external, location-based cues can be combined with inertial signals during wayfinding; humans (and other animals) use these wayfinding strategies contextually, and individual preferences are also evident.

Individual differences in spatial ability may reflect differences in cognitive style or underlying differences in cognitive ability (Bosco et al. 2004; Burke et al. 2012; Saucier et al. 2002). A link between spatial ability and navigation outcomes is apparent, and tests of mental rotation, visualization, and memory are good predictors of successful wayfinding (Allen 1999). A training effect has also been demonstrated (Feng et al. 2007), with structural changes to the brain observable in highly trained individuals (e.g., the well-known case study involving London cab drivers; see Maguire et al. 2000). In other words, the brain is capable of responding physiologically to the demands of regularly performed spatial tasks.

The archaeological record suggests that modern humans developed complex, spatially extensive social networks during the course of the Middle Stone Age (280–50 Kya) and developed the tools and social conventions necessary to maintain them (d'Errico et al. 2003, 2005; Gamble 1998; Wadley 2001). The symbolic use of material culture to define social relationships is considered typical of modern human behavior and is thought to emerge during the Middle Stone Age (Henshilwood and Marean 2003; Wadley 2001). The ethnographic record shows that the reciprocal exchange of artifacts functions to maintain social networks (Weissner 1984) and the manufacture of items of personal adornment (e.g., beads, pendants) during the Paleolithic could signal the symbolic use of material culture as a marker of social identity, used to cement social relationships and maintain social networks. Supporting evidence for the development of more spatially extensive networks during the Upper Paleolithic of Europe includes the long-distance circulation of "exotic" raw materials (Feblot-Augustin 1993, 1997).

Hunter–gatherers use social networks to collect information about resource availability (Whallon 2006; Whallon et al. 2011). They also act as a form of "social safety net" since the social obligations engendered through reciprocal exchanges between members of a network can be exploited in times of resource stress (Kelly

2013; Cashdan 1985; Weissner 1982; Smith 1988). Keeping track of dynamic, spatially extensive social networks would have placed additional cognitive demands on the brain and may have enhanced spatial ability, as well as contributing to the dispersibility of modern human populations (Burke 2012).

While the development of complex, spatially extensive social networks might have selected for enhanced spatial cognition, the reverse may also be true. That is, it is possible that structural changes in the human brain, such as the enlargement of the parietal and temporal regions described earlier in this chapter, enhanced spatial cognition and facilitated the development of dynamic, long-distance social networks. Similarly, it has been hypothesized that enhanced working memory (associated with neurological changes in the frontal lobes) is a key aspect of modern human cognition that promoted planning skills and the development of managed foraging systems, albeit at a much later date (Wynn and Coolidge 2003). Whether or not structural changes to the brain translated into more efficient wayfinding remains to be seen, but it is possible that the role of the parietal and temporal regions in coordinating goal-directed movement and encoding environmental signals was enhanced with implications for path integration and wayfinding ability. This would have permitted modern humans to travel and explore new regions with greater ease, facilitating the development of modern human social networks and enhancing the ability of human groups to disperse widely—a phenomenon observable in the archaeological record of the Late Pleistocene.

6 Spaces and Minds: Visuospatial Cognition and Social Aspects

If humans experienced a change or evolved a specialization in their ability to extend their cognitive capacities beyond the brain, then visuospatial functions may well represent a focal point for future investigation in anthropology and archaeology (Bruner and Iriki 2016). The visuospatial system can be partially investigated in paleoanthropology tracking evolutionary changes in the anatomical elements involved in visuospatial functions (Bruner et al. 2016). Visuospatial integration can be studied in physical and cultural anthropology by using a comparative approach and by testing differences in visuospatial behaviors associated with specific tools or environments. Following a traditional approach in human ethology (Eibl-Eibesfeldt 1989), universal visuospatial behaviors (i.e., those considered to be independent of culture) can reveal shared human capacities channeling the cognitive experience. In terms of neurobiology, visuospatial processes can be investigated in extant human populations through functional imaging and psychometrics. Taking into account that modern humans rely on dental manipulation less than other hominids did, it may be interesting to evaluate in greater detail the neurological mechanisms involved in the ontogenetic transition from mouth-based to hand-based exploration in infants (Haggard and De Boer 2014; Rochat 1989, 1993). Regarding cognitive performance, preliminary psychometric analyses have thus far not revealed correlations between precuneus dimensions and standard visuospatial cognitive tests (Bruner et al. 2015). Nonetheless, there is a possibility that psychometric evaluation of parietal functions should be based on batteries of tests designed to explore integrative aspects of visuospatial cognition or embodiment capacities. In archaeology, visuospatial functions can be investigated by interpreting visuospatial behaviors, following the principle of cognitive archaeology (Coolidge and Wynn 2005, 2009; Coolidge et al. 2015). Tool forms and structures, as well as activities like hunting, visual arts, and patterns of land use, can provide direct and indirect evidence of particular behaviors, which may yield variables and parameters that can quantitatively support specific cognitive hypotheses.

We are now realizing that the body and material culture may play a major role in human cognition (Byrge et al. 2014; Haggard 2005). This relationship may become even more complex when different bodies interact (i.e., when dealing with interpersonal and social relationships). In fact, the spatial and chronological perception associated with the self and others strongly influences how social structures are perceived (Hills et al. 2015; Maister et al. 2015). The individual self may think of sociality in terms of a "social space" with spatial and chronological properties. Through autonoetic perspectives based on the recognition of the self in time and space, we manage our representations of relationships between persons, environments, and objects (Peer et al. 2015). Probably not by chance, there is a special link in primates between brain dimension, social structure, and touch (Dunbar 1998, 2008): brain size appears proportional to group size, which in turn appears proportional to grooming activity. These three factors exert a reciprocal influence in the sense that they can promote or constrain each other. Whatever the actual causes and effects, brain size, social structure, and physical contact are profoundly associated. Interestingly, the relationship between touch and group behaviors is mediated by endorphins, representing an organic bridge between touch, brain, and society (Dunbar 2010; Machin and Dunbar 2011; Suvilehto et al. 2015; Nummenmaa et al. 2016).

Because of the limited information available on the mechanisms of visuospatial integration and embodiment, the first priority will be to identify experimental paradigms that will enable the exploration of these functional networks and the evaluation of specific cognitive hypotheses. Although many evolutionary processes cannot be tested and although cognition cannot be directly investigated in fossil species, an effort must be made to assess the consistency of hypotheses in cognitive archaeology through quantitative comparisons and multidisciplinary convergence. The second priority is to look for evidence of changes in the paleontological and archaeological records that could possibly relate to visuospatial behaviors. The evolutionary perspective can reveal gradual or discontinuous variations in the way we perceive and interact with the external environment, integrating and enhancing our neural processes with additional extended resources.

Acknowledgments We are grateful to the many friends, colleagues, and students who helped us develop the topics introduced in this paper: Fred Coolidge, Tom Wynn, Lambros Malafouris, Duilio Garofoli, Eiluned Pearce, Atsushi Iriki, Jim Rilling, Annapaola Fedato, María Silva Gago, Sofia

Pereira Pedro, Roberto Colom, Manuel Martin-Loeches, Ralph Holloway, Marina Lozano, Joseba Rios Garaizar, and Sileshi Semaw. EB is supported by the Spanish government (MINECO CGL2015-65387-C3-3-P). EES is supported by a SIR (RBSI142SRD) grant from the MIUR (Italian Ministry of University and Research).

References

Allen GL (1999) Spatial abilities, cognitive maps, and wayfinding: basis for individual differences in spatial cognition and behavior. In: Golledge RG (ed) Wayfinding behavior. Johns Hopkins University Press, Baltimore, MD

Almécija S, Wallace IJ, Judex S, Alba DM, Moyà-Solà S (2015) Comment on ape human-like hand use in *Australopithecus africanus*. Science 348:1101–1101

Andersen RA, Buneo CA (2002) Intentional maps in posterior parietal cortex. Annu Rev Neurosci 25:189–220

Andersen RA, Snyder LH, Bradley DC, Xing J (1997) Multimodal representation of space in the posterior parietal cortex and its use in planning movements. Annu Rev Neurosci 20:303–330

Ansari D (2008) Effects of development and enculturation on number representation in the brain. Nat Rev Neurosci 9:278–291

Baldwin MJ (1896) A new factor in evolution. Am Nat 30:441–451

Barks SK, Parr LA, Rillin JL (2015) The default mode network in chimpanzees (*Pan troglodytes*) is similar to that of humans. Cereb Cortex 25:538–544

Bastir M, Rosas A, Lieberman DE, O'Higgins P (2008) Middle cranial fossa and the origin of modern humans. Anat Rec 291:130–140

Battaglia-Mayer A, Archambault PS, Caminiti R (2006) The cortical network for eye–hand coordination and its relevance to understanding motor disorders of parietal patients. Neuropsychologia 44:2607–2620

Beer RD (2003) The dynamics of active categorical perception in an evolved model agent. Adapt Behav 11:209–243

Binkofski F, Buccino G (2004) Motor functions of the Broca's region. Brain Lang 89:362–369

Bosco A, Longoni AM, Vecchi T (2004) Gender effects in spatial orientation: cognitive profiles and mental strategies. Appl Cogn Psychol 18:519–532

Bril B, Rein R, Nonaka T, Wenban-Smith F, Dietrich G (2010) The role of expertise in tool use: skill differences in functional action adaptations to task constraints. J Exp Psychol Hum Percept Perform 36:825–839

Bruner E (2004) Geometric morphometrics and paleoneurology: brain shape evolution in the genus *Homo*. J Hum Evol 47:279–303

Bruner E (2010) Morphological differences in the parietal lobes within the human genus. Curr Anthropol 51:S77–S88

Bruner E, Iriki A (2016) Extending mind, visuospatial integration, and the evolution of the parietal lobes in the human genus. Quat Int 405:98–110

Bruner E, Jacobs HIL (2013) Alzheimer's disease: the downside of a highly evolved parietal lobe? J Alzheimers Dis 35:227–240

Bruner E, Lozano M (2014) Extended mind and visuo-spatial integration: three hands for the Neandertal lineage. J Anthropol Sci 92:273–280

Bruner E, Lozano M (2015) Three hands: one year later. J Anthropol Sci 93:191–195

Bruner E, Pearson O (2013) Neurocranial evolution in modern humans: the case of Jebel Irhoud 1. Anthropol Sci 121:31–41

Bruner E, Manzi G, Arsuaga JL (2003) Encephalization and allometric trajectories in the genus *Homo*: evidence from the Neandertal and modern lineages. Proc Natl Acad Sci U S A 100:15335–15340

Bruner E, Saracino B, Ricci F, Tafuri M, Passarello P, Manzi G (2004) Midsagittal cranial shape variation in the genus *Homo* by geometric morphometrics. Coll Anthropol 28:99–112

Bruner E, de la Cuétara JM, Holloway R (2011) A bivariate approach to the variation of the parietal curvature in the genus *Homo*. Anat Rec 294:1548–1556

Bruner E, Rangel de Lázaro G, de la Cuétara JM, Martín-Loeches M, Colóm R, Jacobs HIL (2014a) Midsagittal brain variation and MRI shape analysis of the precuneus in adult individuals. J Anat 224:367–376

Bruner E, de la Cuétara JM, Masters M, Amano H, Ogihara N (2014b) Functional craniology and brain evolution: from paleontology to biomedicine. Front Neuroanat 8:1–19

Bruner E, Román FJ, de la Cuétara JM, Martín-Loeches M, Colóm R (2015) Cortical surface area and cortical thickness in the precuneus of adult humans. Neuroscience 286:345–352

Bruner E, Preuss T, Chen X, Rilling J (2017) Evidence for expansion of the precuneus in human evolution. Brain Struct Funct 222:1053–1060

Bruner E, Amano H, Pereira-Pedro S, Ogihara N (2018) The evolution of the parietal lobes in the genus homo. In: Bruner E, Ogihara N, Tanabe H (eds) Digital endocasts. Springer, Tokyo, pp 219–237

Bruner E, Lozano M, Lorenzo C (2016) Visuospatial integration and human evolution: the fossil evidence. J Anthropol Sci 94:81–97

Buccino G, Riggio L, Melli G, Binkofski F, Gallese V, Rizzolatti G (2005) Listening to action-related sentences modulates the activity of the motor system: a combined TMS and behavioral study. Cogn Brain Res 24:355–363

Buckner RL, Andrews-Hanna JR, Schacter DL (2008) The brain's default network. Ann N Y Acad Sci 1124:1–38

Burgess N (2006) Spatial memory: how egocentric and allocentric combine. Trends Cogn Sci 10:551–557

Burke A (2012) Spatial abilities, cognition and the pattern of Neanderthal and modern human dispersals. Quat Int 247:230–235

Burke A, Kandler A, Good D (2012) Women who know their place: sex-based differences in spatial abilities and their evolutionary significance. Hum Nat 23:133–148

Byrge L, Sporns O, Smith LB (2014) Developmental process emerges from extended brain–body–behavior networks. Trends Cogn Sci 18:395–403

Bzdok D, Heeger A, Langner R, Laird AR, Fox PT, Palomero-Gallagher N, Vogt BA, Zilles K, Eickhoff SB (2015) Subspecialization in the human posterior medial cortex. NeuroImage 106:55–71

Caminiti R, Innocenti GM, Battaglia-Mayer A (2015) Organization and evolution of parieto-frontal processing streams in macaque monkeys and humans. Neurosci Biobehav Rev 56:73–96

Cantlon JF, Brannon EM, Carter EJ, Pelphrey KA (2006) Functional imaging of numerical processing in adults and 4-y-old children. PLoS Biol 4:844–854

Cashdan EA (1985) Coping with risk: reciprocity among the Basarwa of Northern Botswana. Man 20(3):454–474

Caspers S, Eickhoff SB, Rick T, von Kapri A, Kuhlen T, Huang R, Shah NJ, Zilles K (2011) Probabilistic fiber tract analysis of cytoarchitectonically defined human inferior parietal lobule areas reveals similarities to macaques. NeuroImage 58:362–380

Cavanna AE, Trimble MR (2006) The precuneus: a review of its functional anatomy and behavioural correlates. Brain 129:564–583

Chen CH, Gutierrez ED, Thompson W, Panizzon MS, Jernigan TL, Eyler LT, Fennema-Notestine-C, Jak AJ, Neale MC, Franz CE, Lyons MJ, Grant MD, Fischl B, Seidman LJ, Tsuang MT, Kremen WS, Dale AM (2012) Hierarchical genetic organization of human cortical surface area. Science 335:1634–1636

Churchill SE (2001) Hand morphology, manipulation, and tool use in Neandertals and early modern humans of the Near East. Proc Natl Acad Sci U S A 98:2953–2955

Clark JGD (1969) World prehistory: a new outline. Cambridge University Press, Cambridge

Clark A (1997) Being there: putting brain, body, and world together again. MIT Press, Cambridge

Clark A (2008) Supersizing the mind: embodiment, action, and cognitive extension. Oxford University Press, New York
Clark A, Chalmers DJ (1998) The extended mind. Analysis 58:7–19
Comrie B (2013) Numeral bases. In: Dryer MS, Haspelmath M (eds) The world atlas of language structures online. Max Planck Institute for Evolutionary Anthropology, Leipzig
Coolidge FL, Wynn T (2005) Working memory, its executive functions, and the emergence of modern thinking. Camb Archaeol J 15:5–26
Coolidge FL, Wynn T (2009) The rise of *Homo sapiens*: the evolution of modern thinking. Wiley, Oxford
Coolidge FL, Wynn T, Overmann KA, Hicks JM (2015) Cognitive archaeology and the cognitive sciences. In: Bruner E (ed) Human paleoneurology. Springer, Cham
Corballis MC (1999) The gestural origins of language: human language may have evolved from manual gestures, which survive today as a 'behavioral fossil' coupled to speech. Am Sci 87:138–145
Crispo E (2007) The Baldwin effect and genetic assimilation: revisiting two mechanisms of evolutionary change mediated by phenotypic plasticity. Evolution 61:2469–2479
d'Errico F, Henshilwood CS, Lawson G, Vanhaeren M, Tillier A-M, Soressi M, Bresson F, Maureille B, Nowell A, Lakarra J, Backwell L, Julien M (2003) Archaeological evidence for the emergence of language, symbolism, and music–an alternative multidisciplinary perspective. J World Prehist 17:1–70
d'Errico F, Henshilwood CS, Vanhaeren M, van Niekerk K (2005) *Nassarius kraussianus* shell beads from Blombos Cave: evidence for symbolic behaviour in the Middle Stone Age. J Hum Evol 48:3–24
Dart RA (1925) *Australopithecus africanus*: the man-ape of South Africa. Nature 115 (2884):195–199
de Beaune SA, Coolidge FL, Wynn T (eds) (2009) Cognitive archaeology and human evolution. Cambridge University Press, Cambridge
Depew DJ (2003) Baldwin and his many effects. In: Weber B, Depew DJ (eds) Evolution and learning: the Baldwin effect reconsidered. MIT Press, Cambridge, MA
Dunbar RIM (1998) The social brain hypothesis. Evol Anthropol 6:178–190
Dunbar RIM (2008) Mind the gap: or why humans aren't just great apes. Proc Br Acad 154:403–423
Dunbar RIM (2010) The social role of touch in humans and primates: behavioural function and neurobiological mechanisms. Neurosci Biobehav Rev 34:260–268
Ebeling U, Steinmetz H (1995) Anatomy of the parietal lobe: mapping the individual pattern. Acta Neurochir 136:8–11
Eibl-Eibesfeldt I (1989) Human ethology. Aldine de Gruyter, Hawthorne, NY
Etienne AS, Jeffery KA (2004) Path integration in mammals. Hippocampus 14:180–192
Feblot-Augustin J (1993) Mobility strategies in the late middle Palaeolithic of central Europe and Western Europe: elements of stability and variability. J Anthropol Archaeol 112:211–265
Feblot-Augustin J (1997) La Circulation des Matières Premières au Paléolithique. CNRS, Liège
Feix T, Romero J, Schmiedmayer HB, Dollar AM, Kragic D (2015) The GRASP taxonomy of human grasp types. IEEE Trans Hum Mach Syst 46:66–77
Feng J, Spence I, Pratt J (2007) Playing an action video game reduces gender differences in spatial cognition. Psychol Sci 18:850–855
Fletcher PC, Frith CD, Baker SC, Shallice T, Frackowiak RSJ, Dolan RJ (1995) The mind's eye—precuneus activation in memory-related imagery. NeuroImage 2:195–200
Foley R (1987) Hominid species and stone-tool assemblages: how are they related? Antiquity 61:380–392
Foley R, Lahr MM (2003) On stony ground: lithic technology, human evolution, and the emergence of culture. Evol Anthropol 12:109–122
Foo P, Warren WH, Duchon A, Tarr MJ (2005) Do humans integrate routes into a cognitive map? Map- versus landmark-based navigation of novel shortcuts. J Exp Psychol 31:195–215

Freton M, Lemogne C, Bergouignan L, Delaveau P, Lehéricy S, Fossati P (2014) The eye of the self: precuneus volume and visual perspective during autobiographical memory retrieval. Brain Struct Funct 219:959–968
Gallagher S (2013) The enactive hand. In: Radman Z (ed) The hand, an organ of the mind: what the manual tells the mental. MIT Press, Cambridge
Gamble C (1998) Palaeolithic society and the release from proximity: a network approach to intimate relations (human social evolution). World Archaeol 29:426–449
Gentilucci M, Corballis MC (2006) From manual gesture to speech: a gradual transition. Neurosci Biobehav Rev 30:949–960
Golledge RG (2003) Human wayfinding and cognitive maps. In: Rockman M, Steele J (eds) Colonization of unfamiliar landscapes: the archaeology of adaptation. Routledge, London
Goodale MA, Meenan JP, Bülthoff HH, Nicolle DA, Murphy KJ, Racicot CI (1994) Separate neural pathways for the visual analysis of object shape in perception and prehension. Curr Biol 4:604–610
Gould SJ, Vrba ES (1982) Exaptation: a missing term in the science of form. Paleobiology 8:4–15
Grefkes C, Fink GR (2005) The functional organization of the intraparietal sulcus in humans and monkeys. J Anat 207:3–17
Gunz P, Neubauer S, Maureille B, Hublin J-J (2010) Brain development after birth differs between Neanderthals and modern humans. Curr Biol 20:R921–R922
Haggard P (2005) Conscious intention and motor cognition. Trends Cogn Sci 9:290–295
Haggard P, de Boer L (2014) Oral somatosensory awareness. Neurosci Biobehav Rev 47:469–484
Hagmann P, Cammoun L, Gigandet X, Meuli R, Honey CJ, Wedeen VJ, Sporns O (2008) Mapping the structural core of human cerebral cortex. PLoS Biol 6:1–15
Hatwell Y (2003) Introduction: touch and cognition. In: Hatwell Y, Streri A, Genatz E (eds) Touching for knowing: cognitive psychology of haptic manual perception. John Benjamins, Amsterdam
Hatwell Y, Streri A, Gentaz E (2003) Touching for knowing: cognitive psychology of haptic manual perception. John Benjamins, Amsterdam
Hecht EE, Gutman DA, Preuss TM, Sanchez MM, Parr LA, Rilling JK (2013) Process versus product in social learning: comparative diffusion tensor imaging of neural systems for action execution–observation matching in macaques, chimpanzees, and humans. Cereb Cortex 23:1014–1024
Helbig HB, Ernst MO (2008) Haptic perception in interaction with other senses. In: Grunwald M (ed) Human haptic perception: basics and applications. Birkhäuser, Basel
Henshilwood CS, Marean C (2003) The origin of modern human behavior: critique of the models and their test implications. Curr Anthropol 44:627–651
Hihara S, Notoya T, Tanaka M, Ichinose S, Ojima H, Obayashi S, Fujii N, Iriki A (2006) Extension of corticocortical afferents into the anterior bank of the intraparietal sulcus by tool-use training in adult monkeys. Neuropsychologia 44:2636–2646
Hills TT, Todd PM, Lazer D, Redish AD, Couzin ID (2015) Exploration versus exploitation in space, mind, and society. Trends Cogn Sci 19:46–54
Holloway RL (1981) Exploring the dorsal surface of hominoid brain endocasts by stereoplotter and discriminant analysis. Philos Trans R Soc Lond Ser B Biol Sci 292:155–166
Hutchins E (1995) Cognition in the wild. MIT Press, Cambridge, MA
Hutto DD, Myin E (2013) Radicalizing enactivism: basic minds without content. MIT Press, Cambridge, MA
Ingber D (2008) Tensegrity-based mechanosensing from macro to micro. Prog Biophys Mol Biol 97:163–179
Iriki A, Taoka M (2012) Triadic (ecological, neural, cognitive) niche construction: a scenario of human brain evolution extrapolating tool use and language from the control of reaching actions. Philos Trans R Soc Lond Ser B Biol Sci 367:10–23
Jirak D, Menz MM, Buccino G, Borghi AM, Binkofski F (2010) Grasping language—a short story on embodiment. Conscious Cogn 19:711–720

Jung RE, Haier RJ (2007) The parieto-frontal integration theory (P-FIT) of intelligence: converging neuroimaging evidence. Behav Brain Sci 30:135–187
Kappassov Z, Corrales JA, Perdereau V (2015) Tactile sensing in dexterous robots hands. Robot Auton Syst 74:195–220
Kelly RL (2013) The lifeways of hunter-gatherers. Cambridge University Press, Cambridge
Kivell TL, Kibii JM, Churchill SE, Schmid P, Berger LR (2011) *Australopithecus sediba* hand demonstrates mosaic evolution of locomotor and manipulative abilities. Science 333:1411–1417
Lakoff G, Johnson M (2008) Metaphors we live by. University of Chicago Press, Chicago
Laland KN, Janik VM (2006) The animal cultures debate. Trends Ecol Evol 21:542–547
Land MF (2014) Do we have an internal model of the outside world. Philos Trans R Soc Lond B Biol Sci 369:1–6
Leroi-Gourhan A (1964) Le geste et la parole. Armand Colin, Paris
Liu X, Zhan Q (2013) Description of the human hand grasp using graph theory. Med Eng Phys 35:1020–1027
Lombard M, Haidle MN (2012) Thinking a bow-and-arrow set: cognitive implications of Middle Stone Age bow and stone-tipped arrow technology. Camb Archaeol J 22:237–264
Lozano M, Bermúdez de Castro JM, Carbonell E, Arsuaga JL (2008) Non-masticatory uses of anterior teeth of Sima de los Huesos individuals (Sierra de Atapuerca, Spain). J Hum Evol 55:713–728
Lycett SJ (2011) Most beautiful and most wonderful: those endless stone tool forms. J Evol Psychol 9:143–171
Machin AJ, Dunbar RIM (2011) The brain opiod theory of social attachment: a review of the evidence. Behaviour 148:985–1025
Maguire EA, Burgess N, O'Keefe J (1999) Human spatial navigation: cognitive maps, sexual dimorphism, and neural substrates. Curr Opin Neurobiol 9:171–177
Maguire EA, Gadian DG, Johnsrude IS, Good CD, Ashburner J, Frackowiak RSJ, Frith CD (2000) Navigation-related structural change in the hippocampi of taxi drivers. Proc Natl Acad Sci U S A 97:4398–4403
Maguire EA, Spiers HG, Good CD, Hartley T, Frackowiak RSJ, Burgess N (2003) Navigation expertise and the human hippocampus: a structural brain imaging analysis. Hippocampus 13:250–259
Maister L, Slater M, Sanchez-Vives MV, Tsakiris M (2015) Changing bodies changes minds: owning another body affects social cognition. Trends Cogn Sci 19:6–12
Malafouris L (2008) Beads for a plastic mind: the 'blind man's stick' (BMS) hypothesis and the active nature of material culture. Camb Archaeol J 18:401–414
Malafouris L (2010a) Metaplasticity and the human becoming: principles of neuroarchaeology. J Anthropol Sci 88:49–72
Malafouris L (2010b) The brain-artefact interface (BAI): a challenge for archaeology and cultural neuroscience. Soc Cogn Affect Neurosci 5:264–273
Malafouris L (2013) How things shape the mind: a theory of material engagement. MIT Press, Cambridge, MA
Maravita A, Iriki A (2004) Tools for the body (schema). Trends Cogn Sci 8:79–86
Margulies DS, Vincent JL, Kelly C, Lohmann G, Uddin LQ, Biswal BB, Villringer A, Castellanos FX, Milham MP, Petrides M (2009) Precuneus shares intrinsic functional architecture in humans and monkeys. Proc Natl Acad Sci U S A 106:20069–20074
Marino BFM, Gallese V, Buccino G, Riggio L (2012) Language sensorimotor specificity modulates the motor system. Cortex 48:849–856
Mars R, Jbabdi S, Sallet J, O'Reilly JX, Croxson PL, Olivier E, Noonan MAP, Bergmann C, Mitchell AS, Baxter MG, Behrens TEJ, Johansen-Berg H, Tomassini V, Miller KL, Rushworth MFS (2011) Diffusion-weighted imaging tractography-based parcellation of the human parietal cortex and comparison with human and macaque resting-state functional connectivity. J Neurosci 31:4087–4100

Marzke MW (1997) Precision grips, hand morphology, and tools. Am J Phys Anthropol 102:91–110
Marzke MW (2013) Tool making, hand morphology and fossil hominins. Philos Trans R Soc B 368:1–8
Marzke MW, Marzke RF (2000) Evolution of the human hand: approaches to acquiring, analysing and interpreting the anatomical evidence. J Anat 197:121–140
Marzke MW, Shackley MS (1986) Hominid hand use in the Pliocene and Pleistocene: evidence from experimental archaeology and comparative morphology. J Hum Evol 15:439–460
Mattens F (2013) Perception and representation: mind the hand. In: Radman Z (ed) The hand, an organ of the mind: what the manual tells the mental. MIT Press, Cambridge, MA
Menninger K (1992) Number words and number symbols: a cultural history of numbers. Dover, New York
Meunier D, Lambiotte R, Bullmore ET (2010) Modular and hierarchically modular organization of brain networks. Front Neurosci 4:1–11
Mountcastle VB (1995) The parietal system and some higher brain functions. Cereb Cortex 5:377–390
Nadel L, Hardt O (2004) The spatial brain. Neuropsychology 18:473–476
Napier JR (1956) The prehensile movements of the human hand. J Bone Joint Surg 38B:902–913
Neubauer S, Gunz P, Hublin JJ (2009) The pattern of endocranial ontogenetic shape changes in humans. J Anat 215:240–255
Neubauer S, Gunz P, Hublin J-J (2010) Endocranial shape changes during growth in chimpanzees and humans: a morphometric analysis of unique and shared aspects. J Hum Evol 59:555–566
Niewoehner WA (2001) Behavioral inferences from the Skhul/Qafzeh early modern human hand remains. Proc Natl Acad Sci U S A 98:2979–2984
Nonaka T (2011) What is the behavior of a C4 quadriplegic mouth calligrapher constant function of? BIO Web Conf 1:1–4
Nonaka T (2013) Motor variability but functional specificity: the case of a C4 tetraplegic mouth calligrapher. Ecol Psychol 25:131–154
Nonaka T, Bril B (2014) Fractal dynamics in dexterous tool use: the case of hammering behavior of bead craftsmen. J Exp Psychol 40:218–231
Nonaka T, Bril B, Rein R (2010) How do stone knappers predict and control the outcome of flaking? implications for understanding early stone tool technology. J Hum Evol 59:155–167
Nummenmaa L, Tuominen L, Dunbar R, Hirvonen J, Manninen S, Arponen E, Machin A, Hari R, Jääskeläinen IP, Sams M (2016) Social touch modulates endogenous μ-opioid system activity in humans. NeuroImage 138:242–247
Orban GA, Caruana F (2014) The neural basis of human tool use. Front Psychol 5:1–12
Orban GA, Claeys K, Nelissen K, Smans R, Sunaert S, Todd JT, Wardak C, Durand JB, Vanduffel W (2006) Mapping the parietal cortex of human and non-human primates. Neuropsychologia 44:2647–2667
Overmann KA (2015) Teeth, tools and human becoming. J Anthropol Sci 93:163–167
Peer M, Salomon R, Goldberg I, Blanke O, Arzy S (2015) Brain system for mental orientation in space, time, and person. Proc Natl Acad Sci U S A 112:11072–11077
Pelegrin J (2009) Cognition and the emergence of language: a contribution from lithic technology. In: de Beaune SA, Coolidge FL, Wynn T (eds) Cognitive archaeology and human evolution. Cambridge University Press, Cambridge
Plummer T (2004) Flaked stones and old bones: biological and cultural evolution at the dawn of technology. Yearb Phys Anthropol 47:118–164
Ponce de León MS, Bienvenu T, Akazawa T, Zollikofer CPE (2016) Brain development is similar in Neanderthals and modern humans. Curr Biol 26:R665–R666
Pouydebat E, Gorce P, Coppens Y, Bels V (2009) Biomechanical study of grasping according to the volume of the object: human versus non-human primates. J Biomech 42:266–272
Prinz J (2009) Is consciousness embodied? In: Robbins P, Aydede M (eds) The Cambridge handbook of situated cognition. Cambridge University Press, Cambridge

Prinz JJ, Barsalou LW (2000) Steering a course for embodied representation. In: Dietrich E, Markman A (eds) Cognitive dynamics: conceptual change in humans and machines. MIT Press, Cambridge

Quallo MM, Price CJ, Ueno K, Asamizuya T, Cheng K, Lemon RN, Iriki A (2009) Gray and white matter changes associated with tool-use learning in macaque monkeys. Proc Natl Acad Sci U S A 106:18379–18384

Rilling JK (2006) Human and non-human primate brains: are they allometrically scaled versions of the same design? Evol Anthropol 15:65–67

Rilling JK, Seligman RA (2002) A quantitative morphometrics comparative analysis of the primate temporal lobe. J Hum Evol 42:505–533

Rilling JK, Barks SK, Parr LA, Preuss TM, Faber TL, Pagnoni G, Bremner JD, Votaw JR (2007) A comparison of resting-state brain activity in humans and chimpanzees. Proc Natl Acad Sci U S A 104:17146–17151

Robbins P, Aydede M (eds) (2009) The Cambridge handbook of situated cognition. Cambridge University Press, Cambridge

Rochat P (1989) Object manipulation and exploration in 2- to 5-month-old infants. Dev Psychobiol 25:871–884

Rochat P (1993) Hand-mouth coordination in the newborn: morphology, determinants, and early development of a basic act. In: Savelsbergh GJP (ed) The development of coordination in infancy. Elsevier, Amsterdam

Roche RAP, Mangaoang MA, Commins S, O'Mara SM (2005) Hippocampal contributions to neurocognitive mapping in humans: a new model. Hippocampus 15:622–641

Rolian C, Lieberman DE, Zermeno JP (2011) Hand biomechanics during simulated stone tool use. J Hum Evol 61:26–41

Rushworth MFS, Paus T, Sipila PK (2001) Attention systems and the organization of the human parietal cortex. J Neurosci 21:5262–5271

Sakata H, Taira M, Kusunoki M, Murata A, Tanaka Y (1997) The parietal association cortex in depth perception and visual control of hand action. Trends Neurosci 20:350–357

Saucier DM, Green SM, Leason J, MacFadden A, Bell S, Elias LJ (2002) Are sex differences in navigation caused by sexually dimorphic strategies or by differences in the ability to use the strategies? Behav Neurosci 116:403–410

Scheperjans F, Eickhoff SB, Hömke L, Mohlberg H, Hermann K, Amunts K, Zilles K (2008) Probabilistic maps, morphometry, and variability of cytoarchitectonic areas in the human superior parietal cortex. Cereb Cortex 18:2141–2157

Schick K, Toth N, Garufi G, Savage-Rumbaugh ES, Rumbaugh D, Sevcik R (1999) Continuing investigations into stone tool-making and tool-using capabilities of a bonobo (*Pan paniscus*). J Archaeol Sci 26:821–832

Scott N, Neubauer S, Hublin JJ, Gunz P (2014) A shared pattern of postnatal endocranial development in extant hominoids. Evol Biol 41:572–594

Smith EA (1988) Risk and uncertainty in the "original affluent society": evolutionary ecology of resource-sharing and land tenure. In: Ingold T, Riches D, Woodburn J (eds) Hunters and gatherers: history, evolution, and social change, vol 1. Berg, Oxford, pp 222–252

Smith BC (1999) Situatedness/embeddedness. In: Wilson RA, Keil FC (eds) The MIT encyclopedia of the cognitive sciences. MIT Press, Cambridge

Sotero RC, Iturria-Medina Y (2011) From blood oxygenation level dependent (BOLD) signals to brain temperature maps. Bull Math Biol 73:2731–2747

Spiers HJ, Maguire EA (2006) Thoughts, behaviour, and brain dynamics during navigation in the real world. NeuroImage 31:1826–1840

Stout D, Chaminade T (2007) The evolutionary neuroscience of tool making. Neuropsychologia 45:1091–1100

Stout D, Hecht E, Khreisheh N, Bradley B, Chaminade T (2015) Cognitive demands of lower Paleolithic toolmaking. PLoS One 10:1–18

Suvilehto JT, Glerean E, Dunbar RIM, Hari R, Nummenmaa L (2015) Topography of social touching depends on emotional bonds between humans. Proc Natl Acad Sci U S A 112:13811–13816

Sznajder B, Sabelis MW, Egas M (2012) How adaptive learning affects evolution: reviewing theory on the Baldwin effect. Evol Biol 39:301–310

Tallis R (2003) The hand: a philosophical inquiry into human being. Edinburgh University Press, Edinburgh

Tallis R (2011) Aping mankind: neuromania, darwinitis and the misrepresentation of humanity. Acumen, Durham

Tixier J (1967) Procédés d'analyse et questions de terminologie dans l'étude des ensembles industriels du Paléolithique récent et de l'Épipaléolithique en Afrique du Nord-Ouest. In: Bishop WW, Clark JD (eds) Background to evolution in Africa. University of Chicago Press, Chicago

Tostevin GB (2013) Seeing lithics: a middle-range theory for testing for cultural transmission in the Pleistocene. Oxbow Books, Oakville

Toth N, Schick KD, Savage-Rumbaugh ES, Sevcik RA, Rumbaugh DM (1993) *Pan* the tool-maker: investigations into the stone tool-making and tool-using capabilities of a bonobo (*Pan paniscus*). J Archaeol Sci 20:81–91

Turvey MT, Carello C (2011) Obtaining information by dynamic (effortful) touching. Philos Trans R Soc B 366:3123–3132

Uomini N, Meyer GF (2013) Shared brain lateralization patterns in language and Acheulean stone tool production: a functional transcranial Doppler ultrasound study. PLoS One 8:1–8

Utevsky AV, Smith DV, Huettel SA (2014) Precuneus is a functional core of the default-mode network. J Neurosci 34:932–940

Vanduffel W, Fize D, Peuskens H, Denys K, Sunaert S, Todd JT, Orban GA (2002) Extracting 3D from motion: differences in human and monkey intraparietal cortex. Science 298:413–415

Vankov I, Kokinov B (2013) The role of the motor system in conceptual processing: effects of object affordances beyond response interference. Acta Physiol 143:52–57

Wadley L (2001) What is cultural modernity? A general view and a South African perspective from Rose Cottage Cave. Camb Archaeol J 11:201–221

Wardak C, Hamed SB, Duhamel JR (2005) Parietal mechanism of selective attention in monkeys and humans. In: Dehaene S, Duhamel JR, Hauser MD, Rizzolatti G (eds) From monkey brain to human brain. MIT Press, Cambridge, MA

Weidenreich F (1941) The brain and its role in the phylogenetic transformation of the human skull. Trans Am Philos Soc XXXI:321–442

Weissner P (1982) Risk, reciprocity and social influences on !Kung San economics. In: ELR L (ed) Politics and history in band societies. Cambridge University Press, Cambridge

Weissner P (1984) Reconsidering the behavioral basis for style: a case study among the Kalahari San. J Anthropol Archaeol 3:190–234

Whallon R (2006) Social networks and information: non-utilitarian mobility among hunter-gatherers. J Anthropol Archaeol 25:259–270

Whallon R, Lovis WA, Hitchcock RK (2011) Information and its role in hunter-gatherer bands, Ideas, debates and perspectives, vol 5. UCLA/Cotsen Institute of Archaeology Press, Los Angeles, CA

Whitlock JR, Sutherland RJ, Witter MP, Moser M-B, Moser EI (2008) Navigating from hippocampus to parietal cortex. Proc Natl Acad Sci U S A 105:14755–14762

Williams EM, Gordon AD, Richmond BG (2012) Hand pressure distribution during Oldowan stone tool production. J Hum Evol 62:520–532

Wynn T (1989) The evolution of spatial competence. University of Illinois Press, Chicago

Wynn T (2000) Symmetry and the evolution of the modular linguistic mind. In: Carruthers P, Chamberlain A (eds) Evolution and the human mind: modularity, language and meta-cognition. Cambridge University Press, Cambridge

Wynn T (2002) Archaeology and cognitive evolution. Behav Brain Sci 25:389–402

Wynn T (2014) Commentary on 'extended mind and visuo-spatial integration: three hands for the Neandertal lineage'. J Anthropol Sci 92:291–293
Wynn T, Coolidge FL (2003) The role of working memory in the evolution of managed foraging. Bef Farm 2:1–16
Zhang S, Li CR (2012) Functional connectivity mapping of the human precuneus by resting state fMRI. NeuroImage 59:3548–3562
Zilles K, Palomero-Gallagher N (2001) Cyto-, myelo-, and receptor architectonics of the human parietal cortex. NeuroImage 14:S8–S20

MIX
Papier aus verantwortungsvollen Quellen
Paper from responsible sources
FSC® C105338

If you have any concerns about our products,
you can contact us on
ProductSafety@springernature.com

In case Publisher is established outside the EU,
the EU authorized representative is:
Springer Nature Customer Service Center GmbH
Europaplatz 3, 69115 Heidelberg, Germany

Printed by Libri Plureos GmbH
in Hamburg, Germany